Inside *MicroStation*®

Updated for MicroStation/J and ProjectBank DGN

And Introducing V8

Fifth Edition

Frank Conforti

ONWORD PRESS

THOMSON LEARNING™

Australia, Canada, Mexico, Singapore, Spain, United Kingdom, United States

ONWORD PRESS

THOMSON LEARNING

INSIDE MicroStation, Fifth Edition
by Frank Conforti

Publisher:
Alar Elken

Development Editor:
Daril Bentley

Production Coordinator:
Larry Main

Executive Editor:
Sandy Clark

Editorial Assistant:
Jaimie Wetzel

Manufacturing Coordinator:
Betsy Hough

Acquisitions Editor:
James Gish

Executive Marketing Manager:
Maura Theriault

Technology Project Manager:
David Porush

Managing Editor:
Carol Leyba

Executive Production Manager:
Mary Ellen Black

Cover Design:
Cammi Noah

Permissions
Grateful acknowledgment is made to the Frank Lloyd Wright Foundation for permission to reproduce the image of the Martin house Tree of Life window in Chapter 2 (see the note in Chapter 2 concerning the use of). The drawings and designs of Frank Lloyd Wright are copyright © The Frank Lloyd Wright Foundation, Scottsdale, AZ. All rights reserved. Used with permission.

For more information, contact
OnWord Press
An imprint of Thomson Learning
Box 15-015
Albany, New York USA 12212-15015

Library of Congress Cataloging-in-Publication Data
Conforti, Frank
    Inside MicroStation : updated for MS/J and Project Bank DGN / Frank Conforti.—5th ed.
        p. cm.
    Rev. ed. of: Inside MicroStation, 4th ed. / by Ranjit Sahai and the OnWord Press development team. 1996.
    ISBN 1-56690-161-8
    1. MicroStation. 2. Computer-aided design. I. Sahai, Ranjit S. Inside MicroStation. II. Title.

TA174.C63 1999
620'.0042'02855369—dc21
                                    99-030655
                                    Rev.

About the Author

Frank Conforti has been involved in computer-aided design for over 20 years. He has been a designer, CAD administrator, consultant, and journalist, and an author of several CAD-related books. Frank joined Bentley Systems in 1995 to help start the MicroStation Institute (now the Bentley Institute), Bentley's premiere end-user training program. Frank is currently Product Evangelist for Bentley Core Technologies, working with both software developers and end users to better communicate their needs and ideas.

Acknowledgments

In the fast-paced world of technical book writing, it is easy to overlook the many people who contribute to a project like this. First, I would like to thank all of the individuals here at Bentley Systems, Inc., for their unwavering support of this project. I would especially like to thank Tom Anderson, Jerry King, Chris Bober, and Keith Bentley for giving me the time to update this book. Without that time, it would have been several more years before I would have finished it! I would also like to thank all of the folks in Bentley Software, including the Certification, Technical Support, and Product Release groups, as well as the folks who actually write the software. I am sure my incessant questions interrupted more than one train of thought.

Of course, there are the folks at OnWord Press, without whom this book would still just be a fanciful idea. Specifically, I would like to thank Carol Leyba for her dedication to quality and design, Daril Bentley for his attention to detail (and making me answer those tough questions) and, finally, Jim Gish, without whom the entire project would have simply ground to a halt. Thanks all; let's hope the next project goes as smoothly as this one did!

I dedicate this book to Beccie, my wife of 21 years. Without your patience and unwavering support this book would not have made it to that first edition a decade ago, much less to this fifth edition. Thank you again!

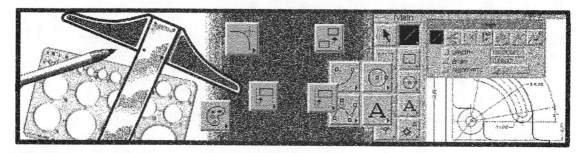

CONTENTS

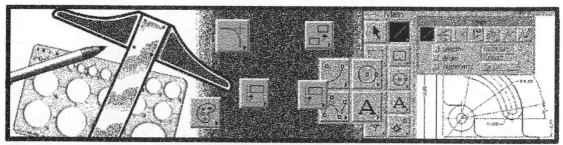

LIST OF EXERCISES

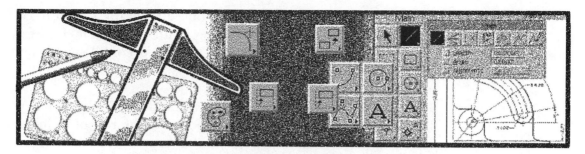

FOREWORD

MICROSTATION IS JUST A SOFTWARE PROGRAM. For many of us at Bentley Systems, though, it has been an ongoing multiyear project to continually try to make MicroStation a better, faster, easier-to-use, and more capable software program. We have spent many long hours working on and refining the various aspects of Micro-Station as it has "grown up" with the technology landscape that surrounds it. There are so many dimensions to the problems to which our users apply MicroStation that it has grown to be a fairly large software program. But it is still just a software program.

On the other hand, I am often reminded that MicroStation the software program can be a key part of the way many companies and individuals make their living. MicroStation is often such a critical part of their workflows that effective use of the program can be extremely important. Thankfully, MicroStation users have always been very eager to help one another and to share their experience.

It is very rewarding to us that there is a large community of individuals for whom MicroStation is important enough to spend their professional, and often personal, time making use of. They have literally created a worldwide MicroStation community of people who help one another use MicroStation more effectively. That commitment is not taken lightly by those of us creating the software, because we know that people like Frank Conforti can make their living doing anything they want. Through their investment in time, they have implicitly expressed confidence in our ability to make the software do what *they* want.

I would like to express my personal gratitude to Frank for taking the time to create and update this book for the rest of us in the MicroStation community. I would also like to personally thank readers of this book for your interest in and time spent in learning about MicroStation. Hopefully, our software program will live up to your expectations. However, if you ever have suggestions for us regarding the improvement of MicroStation, I also hope you will follow the lead of many MicroStation users over the years and let us know.

I hope you enjoy this book and that it helps you toward many long years of happy and profitable use of MicroStation.

Keith Bentley

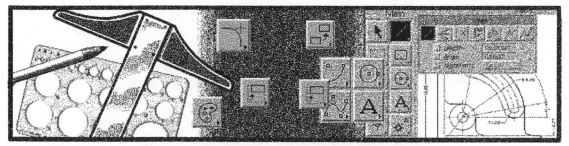

INTRODUCTION

How to Read This Book

Congratulations on purchasing one of the best-selling MicroStation books available. This is not mentioned lightly. *Inside MicroStation* is the cornerstone of the OnWord Press MicroStation line. What this means to you is that at the conclusion of this book, if you have followed the exercises, you will understand MicroStation.

As MicroStation has evolved, so has this book. This fifth edition has been updated to reflect not only the new features of MicroStation/J but some interesting information about MicroStation V8, a release that promises to revolutionize the way we all work together. One major change in this edition is the more holistic integration of AccuDraw, MicroStation's revolutionary user input system. From the outset, you are introduced to this tool and how it works within all aspects of the design process. This edition also includes a significantly updated introduction to 3D design, focusing on how you use AccuDraw to navigate within this exciting environment.

Inside MicroStation is written to guide you through the intricacies of the MicroStation CAD software package. During the course of reading this book you will learn how to use MicroStation in your design process.

As with any complex system, it is impossible to cover every aspect of the program in one book. The intention then is to expose you to as much of MicroStation as possible, while guiding you through the individual commands and their uses. Along the way, tips, tricks, and traps are pointed out as new concepts and commands are introduced.

How *Inside MicroStation* Is Organized

Inside MicroStation is organized by subject matter, starting with the basics and working through to the complex issues. *Inside MicroStation* roughly follows the design process, starting with simple drawing placement, and then moving through construction and manipulation commands. Along the way you will get hands-on experience, supported by illustrations of various commands and detailed exercises that take you through the design process one step at a time.

Illustrations, Exercises, and Typographical Conventions

In the course of reading this book you will find a number of exercises, illustrations, and notes. These are all designed to clearly illustrate concepts discussed in the text. In fact, a lot can be learned about MicroStation by simply studying the illustrations and exercises.

To make the illustrations and exercises as clear as possible, each is presented in a consistent format. The following sections describe components of this book.

Command Illustrations

To help orient you to MicroStation, a "standard" illustration of individual commands was developed. The following diagram shows a typical tool illustration from this book.

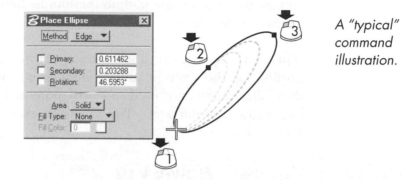

A "typical" command illustration.

Each illustration contains specific information on how the tool operates. First, you are shown its location on the toolbox. The option fields associated with the tool are always shown on the Tool Settings window. Next, you are shown an example of the tool in action. Each step of using the tool identifies the type of user input required, whether that is a left mouse button press (i.e., a data point), text entry via the keyboard, or a mouse gesture. The result of using the tool is also shown.

Exercises

Each exercise is designed to illustrate an important aspect of MicroStation. Step-by-step instructions are given, showing you how to use MicroStation to solve a problem. You can work through these exercises and create design files as you go. There is a companion data set available on the companion web site, the URL address for which is provided at the end of this introduction. This data set contains the files used with each exercise.

The file *EXERFILE.TXT*, included in the data set, provides additional information about the files in the data set. Additional instructions on the companion data set are included in the *README.TXT* file.

Each exercise contains a number of important features. Each step is individually numbered, and includes directives such as "Select the Place Line tool" as well as instructional narrative about the particular tool or operation. Furthermore, in the case of specific commands or tools, the path to the tool is shown, for example, as Settings > Design File.

Where appropriate, an illustration is provided, highlighting the portion of the exercise under discussion. Whenever possible, these illustrations are used to clarify potentially confusing directives within the narrative. The following example is a portion of such an exercise.

EXERCISE 7-1: CREATING A FLOWER VASE

The key to creating this vase is to start from what you know. In this case, you know the heights and widths of the base and mouth, and the radii of the arcs. In addition, you know that the vase is symmetrical. Therefore, you can develop half the vase and use the Mirror tool to complete the project. However, you need to locate the curves that constitute the sides of the vase, as shown in the following illustration.

The challenge is locating the curves that constitute the sides of the vase. Note the tangency between the curves.

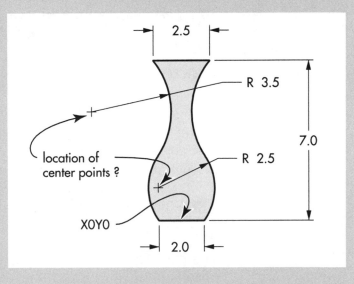

1 Open the design file *VASE.DGN*. This is an empty design file, so you will need to establish a starting point for this project.

2 Draw some lines establishing the base and mouth of the vase. Use the bottom, center of the vase as the origin point for the base (X0Y0).

3 Select the Place Line tool (Main toolbox > Linear Elements > Place Line) and place two lines: one for the base (1 unit long), and one for the mouth of the vase (1.25 units long). Use AccuDraw to accurately locate the two lines, as shown in the following illustration.

Try-It!

This element suggests exercises related to processes you might want to try in addition to the numbered exercises.

Special Icons

Several other features you will find in this book are notes, tips, warnings, timeouts, "try-its," and V8 notes. Each of these elements is designated for quick access by its own icon. The following explain these features.

 NOTE: *Notes present important information or concepts that might otherwise be overlooked.*

 TIP: *Tips convey shortcuts and hints that help you to be more productive.*

 WARNING: *Warnings point out functions and procedures that could get you in trouble if you are not careful.*

 TIMEOUT:

This is used to elaborate on a fact that may not be germane to the discussion in progress but is nonetheless important.

V8: V8 notes point out significant differences in functionality and other aspects of the software you can expect in MicroStation version 8.

How to Use This Book

The following are suggestions on several ways this book can be used.

The New User: Reading from Beginning to End

This is highly recommended for those who are new to MicroStation. *Inside MicroStation* is an accessible introduction to CAD. More than just a compendium of command descriptions, it shows how to work through a design, from where to place the first point, to detailing, and finally to plotting. Reading this book will turn the complete beginner into a real MicroStation user.

The Intermediate User: Looking Up Commands

Not only are most of MicroStation's tools and commands in this book indexed, but exercises and illustrations are indexed by the tools they represent. This makes *Inside MicroStation* a powerful reference for the intermediate user who still needs to look up facts about MicroStation's operation. In that tool illustrations cover all MicroStation platforms, this book will be useful to anyone working in a multiplatform environment.

The Advanced User: A Training Guide

The advanced user who has any responsibility for training others will find *Inside MicroStation* an excellent training guide. Its practical, "working design" will help you instill good working habits from the outset. See also the publisher's contact information on the copyright page of this book for other OnWord MicroStation and related CAD titles.

What Is Not Covered in This Book

Because *Inside MicroStation* is an introductory-level book, some important concepts have necessarily been covered only lightly. The focus of this book is on showing new users how to start with a blank screen and complete a usable drawing.

Companion CD-ROM and Web Site

The companion web site, found at the following address, contains all data files necessary to complete exercises found in the book. (See also the section "Exercises" in this introduction.)

www.onwordpress.com/microstation

PART ONE

1

GETTING STARTED

WELCOME TO THE MICROSTATION COMMUNITY

NO MATTER WHAT YOUR FIELD OF WORK—whether architecture, civil, facilities management, electrical, mapping, mechanical, industrial plant design, or another—MicroStation packs the power to meet your needs. If you have a vision, MicroStation can help you give it reality. And, as you will discover during the course of reading this book, MicroStation is not at all difficult to master.

Compared to many of its contemporaries, MicroStation has consistently been deemed one of the most capable CAD products, and the most consistent in overall operation. This comes from years of product refinement by Bentley Systems Corporation of Exton, Pennsylvania, USA.

A Bit of MicroStation History

MicroStation is markedly not a Johnny-come-lately product of the dot-com. In fact, its origins predate that of the personal computer. However, its story does not start in the 1980s at Bentley in Exton but in the late 1960s and in Huntsville, Alabama, in the late 1970s.

Intergraph: Turnkey CAD to Technical Desktop

MicroStation's origin dates back to the late 1960s, when Intergraph Corporation, then known as M&S Computing, got its start during the Apollo moon mission years by developing real-time software for the space program. During these early days, the company developed a number of software programs for NASA, which included, among other things, a printed circuit board design software package, and probably most important, a mapping graphics package that would later evolve into IGDS. IGDS, an acronym for Interactive Graphics Design System, was a minicomputer-based computer-aided design (CAD) system designed to support the engineering professional.

IGDS was a true "system" because Intergraph provided everything: software, support, and all associated hardware. At that time, personal computer state-of-the-art was typified by the Apple computer, an 8-bit microprocessor machine with limited graphics capabilities and no hard drive. Meanwhile, an Intergraph system would consist of several high-performance graphics terminals (monochrome at first, color later), a Digital Equipment Corporation PDP-11 minicomputer (later replaced by the DEC VAX super-minicomputer), large disk drives the size of washing machines, and huge digitizing tables and equally large plotters.

Despite the very high entry costs of such systems and the initial concentration on the cartographic (mapping) world, IGDS was quickly adopted as a general-purpose CAD system capable of supporting almost any engineering discipline. From electronic and mechanical design through architectural and civil engineering, Intergraph's IGDS clientele represented the total spectrum of engineering worldwide.

In the early 1980s, the IGDS system migrated to the 32-bit DEC VAX super-minicomputer, and its more primitive graphics workstations were refined to the point that Intergraph's systems were recognized for their very ergonomically designed workstations and their nearly universal use of dual graphics displays (see following illustration). (At the time, many CAD companies offered two displays, but one was always dedicated to text-only information.)

Dual graphics terminals, typified by the Interact workstation, were the hallmark of Intergraph's turnkey systems from the beginning.

It goes without saying that Intergraph's IGDS system, at $75,000 a seat, was expensive. Only large firms could afford to employ them, and only on large projects. In those days, it was quite common to utilize the workstations with three shifts of designers to maximize productivity and generate a favorable return on investment. By this time, the IBM PC and PC-XT had come on the scene. Still only 8-bit computers (versus the VAX 32-bit architecture), the PC was still considered underpowered for most design applications.

During the early to mid 1980s, computer hardware began evolving very quickly, as did Intergraph's systems. At the same time, the first purely PC-based CAD products came on the scene. IGDS running on the DEC VAX computer continued to be refined. Intergraph developed several new products that utilized IGDS as its core CAD "engine," but with tools and capabilities oriented to a specific engineering discipline. Examples included APDP for architects, EPDP for civil engineers, and PLRMS and TFIDS for telephone companies.

At the same time, Intergraph started developing its own standalone CAD hardware platform to replace the centralized VAX system architecture. Based on the Fairchild (later bought by National Semiconductor) Clipper microprocessor, the Interpro graphics workstations used a variant of the UNIX operating system called CLIX (short for CLipper unIX). Presumably, Intergraph's strategy was to port IGDS and all of its related vertical applications to this platform. However, a small upstart company in Pennsylvania would forever change those plans.

Bentley Develops MicroStation

During the early 1980s, Dupont, a large chemical product manufacturer, had begun the process of standardizing their design practices around Intergraph's IGDS system. At that time, Keith Bentley, who was employed in Dupont's CAD support group, made the observation that not all engineers needed the full capabilities of the then very expensive Intergraph graphics workstations to do their work.

Keith began to explore the use of relatively inexpensive single-screen graphics terminals (such as the Digital Equipment Corporation VT100 alphanumeric terminal equipped with a Regis graphics option) as a viable alternative to the dedicated graphics workstation. At one-tenth the cost of a fully equipped IGDS graphics workstation, this design file access software, originally called PseudoStation, marked the very beginnings of MicroStation.

Acquiring the sales rights to PseudoStation from Dupont, Keith moved to California and with his brother Barry began Bentley Systems in 1982. Soon, they had more than a hundred customers initially paying $7,945.50 a copy for PseudoStation (the author worked for one such customer at the time and was even able to interface his Mac 512 to the VAX using PseudoStation, but that is another story).

However, as good as PseudoStation was, it still represented only half the story. It replaced the dedicated graphics workstation but not the VAX computer, where IGDS was still the host program. With the early success of PseudoStation, and the continued performance improvements in the personal computer market, Bentley Systems (now relocated to Philadelphia, Pennsylvania) began the task of developing a CAD program that ran entirely on a personal computer. Needless to say, they succeeded with the initial release of MicroStation!

Considering that the IGDS design file format had been fine-tuned for the 32-bit minicomputer environment, developing MicroStation on a circa 1983-84 personal computer, essentially an 8-bit hardware platform, was no small task. That Bentley Systems was able to create a software program that not only could read and write IGDS-compatible files but would retain the refined operational feel of the Intergraph system is a testimony to the Bentleys' software development prowess (by this time, there were three Bentley brothers involved in the business).

MicroStation 1.0, which was never commercially released, was a read-only review product that ran entirely on a personal computer. It did, however, provide simple plotting capabilities (the author still retains his original MicroStation 1 manual and its *single* 1.2-MB floppy install disk).

It was not long before MicroStation, version 2 (the first fully interactive IGDS-compatible CAD program) was available. As opposed to other CAD packages that only supported 2D graphics, MicroStation included all of the high-end features associated with IGDS right from the start. This included reference files support for workgroup computing on a network, external database manipulation for managing associated nongraphic data, and 3D graphics for sophisticated modeling.

At this time, Intergraph purchased half of Bentley Systems and became the marketing and support channel for MicroStation. This included a version of the product compiled to run on Intergraph's new line of Clipper graphics workstations. Ultimately, MicroStation was ported to several UNIX-based graphics workstations, including Sun SPARCstations, Hewlett Packard's Apollo-based workstations, Silicon Graphics, and IBM's RS6000 workstations.

During the mid to late 1980s, MicroStation evolved quickly through version 3—which introduced screen menus in 1987, and the first release of the Apple Macintosh version—to version 4, which introduced the first modern graphic user interface of any DOS-based CAD program. Version 4 was also significant because it marked the introduction of Bentley's MDL programming environment, which supported the development of seamlessly integrated engineering-specific enhancements. MDL is still with us today, in the form of Bentley's own engineering specific ECs (Engineering Configurations).

The release of version 5 in 1993 greatly enhanced the software, with special emphasis on modularity and usability, with a host of new features, such as multi-lines, custom line styles, workspaces, and realistic rendering with material mapping capabilities. MicroStation 95, released in 1995 (of course!), departed from the version numbering scheme of the previous releases but was very notable for the introduction of AccuDraw, undoubtedly the single most profound "feature" released to date.

AccuDraw introduced an entirely new way of interacting with MicroStation, with a combination of keyboard-entered data and mouse movements that dramatically improved nearly every user's drawing productivity, especially 3D work. MicroStation 95 was also notable for its introduction of SmartLine, a streamlined method of creating and manipulating composite shapes and strings, as well as the MicroStation Basic macro language (designed for the end user, complete with the Macro Record feature). The illustrations that follow catalog the version development of MicroStation.

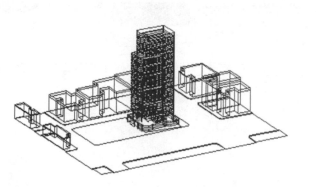

In MicroStation 2, released in 1986, the user interaction was bare, with only keyboard-entered commands, status messages, and no menus.

```
Locks=GR,SN,              LVL=1,SOLID,WT=0,LC=SOL,PRI,CO=0
(1) uSTN> _
```

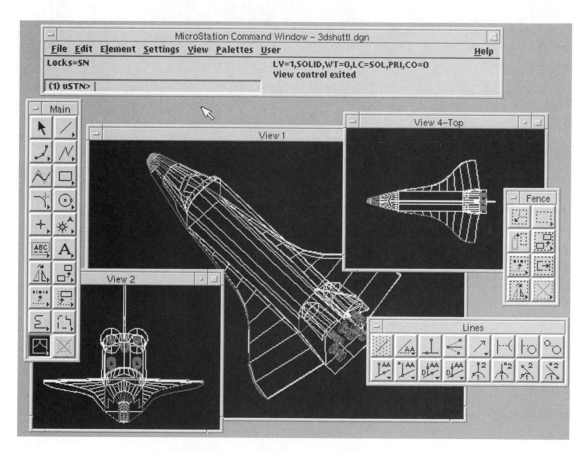

MicroStation 4 was a dramatic change in the look and feel of MicroStation, when it gained a true motif-based graphical interface on all supported platforms.

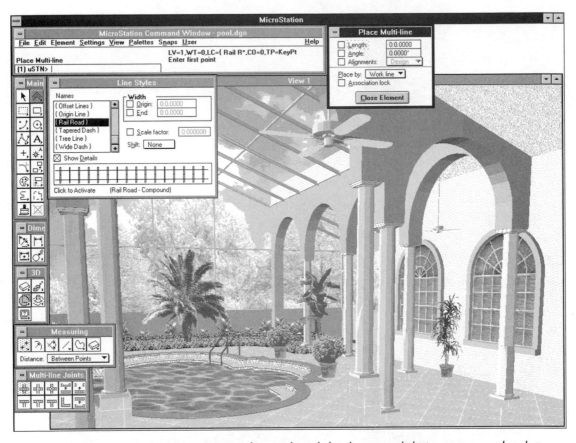

MicroStation 5 enhanced usability by consolidating command palettes and increasing functionality by adding multi-lines, workspaces, and custom line styles.

Despite its evolution over the years, MicroStation has not forgotten its roots. It continues to include features to allow the interchange of design data with the older IGDS system. This point is crucial to understanding the MicroStation philosophy. Bentley has gone through great pains to maintain software and graphics file compatibility across all computer platforms.

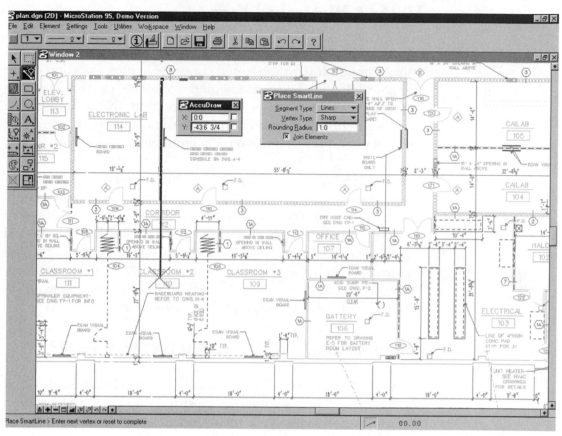

MicroStation 95 adheres to version 5's file format, adopts the popular Windows interface, and improves usability through innovative tools such as AccuDraw.

Platform compatibility now extends to everything from the old VAX systems, to PC compatibles, to modern workstations. When Bentley Systems, Inc., or simply Bentley, as it is now known, finished work on MicroStation in 1986, its founders, the Bentley brothers, got in touch with Intergraph for support. Intergraph bought half the shares in the company, and entered into an exclusive distribution agreement with Bentley. MicroStation would be developed and enhanced by Bentley, and Intergraph would market it. This arrangement remained in place until December of 1994.

1995: A Watershed Year

In 1995, Bentley also underwent a major business-related transition. Although Intergraph would continue to own a portion of the company, Bentley Systems assumed complete control over the sales, marketing, and support of MicroStation. By this time, there were five Bentley brothers involved in the company (Barry, Greg, Keith, Ray, and Scott) and over 500 employees worldwide.

The year 1995 also marked the year Bentley shed its image as a single-product vendor. Although MicroStation is still the core product, with over 300,000 users worldwide, Bentley has continued to not only improve MicroStation (SE, /J) itself but has added numerous engineering-specific products, such as Triforma (architecture specific), Geographics (mapping), PlantSpace (piping), Modeler (mechanical), Geopak (civil/transportation design), Descartes (aerial/satellite imaging), and many more.

MicroStation/J Introduces a New Objective

In 1997, Bentley introduced its next major milestone release of MicroStation. MicroStation/J is most notable for the integration of Sun Microsystems' Java virtual machine technology into the core of MicroStation.

Java introduced the first true object-oriented programming into the MicroStation development environment. By integrating OOP into the heart of MicroStation, Bentley set the stage for the release of ProjectBank DGN, Bentley's groundbreaking workgroup information tracking "product." ProjectBank is actually not a product but a technology at the center of Bentley's move to the true distributed processing environment, exemplified by the success of the Internet and the World Wide Web.

Over the years, Bentley was always quick on the uptake of new technology into both their products and their internal operations. From its inception, the Internet has played an important role at Bentley, both as a major method of communication with their users (all Bentley software is downloadable via the Internet). In

2000, Bentley expanded its Internet business by launching Viecon, its B2B (business-to-business) portal.

Hardware and Operating System Considerations

Over the years, MicroStation has run on a wide variety of computers and under several different operating systems. Prior to MicroStation/J, you could get MicroStation for everything from an Apple Macintosh to an HP-UX system. However, as a result of several business factors (including the dominance of "Wintel" in the marketplace), and for technical reasons, MicroStation/J has been released only for Windows 95/98/2000 and NT running on Intel processor based platforms. This focus on one platform type and operating system has simplified the system requirements for the average user. The sections that follow briefly review the hardware and operating system environment within which MicroStation/J operates.

Video Screen

One aspect of a MicroStation workstation that dates back to the IGDS days is the use of dual video graphics monitors. The current generation of the Windows operating system (Windows 2000, NT, 98) support multiple video cards within the same machine. MicroStation can take full advantage of millions of colors, but most users, performing everyday types of operations, can usually get by with 256- or 16,000-color environments.

TIP: *If your video card has only limited memory, it is better to trade color depth for resolution. The more "desktop" you have, the more tools and other windows you can have open at one time.*

When you work in 3D, it is often helpful to display your design in a shaded or rendered view, as opposed to the wireframe rendering most CAD products use as the default display environment (see following illustration). If your computer uses a video card equipped with an OpenGL-compliant 3D accelerator processor, MicroStation can take full advantage of its faster rendering speeds. The result is smoother dynamic view navigation in all three axes (X, Y, and Z), as well as quicker display of texture-mapped objects (e.g., a wood grain finish on a bookshelf).

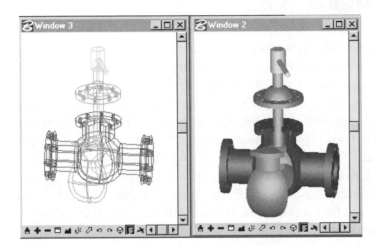

More and more users are opting for rendered views (right) when working in 3D designs.

With Windows 2000, NT, and 98's integration of OpenGL support into the operating system, the price of OpenGL video cards has dramatically decreased over the past several months. Look for the OpenGL compliance logo when considering a new video card.

Of Mice and Tablets

If video is how MicroStation presents its information to you, the mouse is the primary method for you to communicate with it. Of course, there is still the ever-present keyboard, where you enter commands and parameters, but the fun stuff (entering locations in your designs, selecting elements, selecting tools, and so on) is performed with the mouse (see the following illustration) and its on-screen pointer or *cursor*. The latter is a term originally used with the Intergraph IGDS system, and is another name for the mouse pointer.

An example of a mouse.

Using the Mouse

MicroStation utilizes two to three buttons on the mouse. Because mice come in configurations of two, three, or more buttons, MicroStation allows you to configure the button settings to meet your needs.

Prior to the introduction of the "wheel" mouse (a mouse with a small rolling wheel positioned between the two buttons), the decision was usually between a two-button chord setup (where you push both buttons simultaneously for one function, each button individually for two other functions) and a three-button setup (each button assigned a separate operation). These days, the wheel mouse is definitely the preferred input device, because not only do you get the three-button operation (the wheel is pushed down as a button) but automatic scrolling of the current view when you roll the mouse wheel.

Graphics Tablet

The graphics table (or digitizer, as it is also known)—a reference to its use as a tool to digitally "trace" drawings as input into MicroStation (or IGDS)—differs slightly from the mouse in its use and operation with MicroStation. The graphics tablet consists of a flat, book-like (thus the term *tablet*) box and a mouse-like "puck." Electronic circuitry in the tablet senses the position of the puck as it is moved over the tablet's surface, and transmits its absolute location to the computer. This differs from the relative position reporting of the typical mouse.

By knowing precisely where the puck is with respect to a reference point in the coordinate system of your drawing, you can see how it would be possible to trace the lines and other features of a paper drawing directly into MicroStation. In addition, most pucks come equipped with additional buttons that can be programmed for the most common MicroStation operations. Graphics tablets are also used to select additional commands from special tablet menus as an alternative to the more contemporary toolbox and menu user interface used in most Windows programs. An example of a graphics tablet is shown in the following illustration.

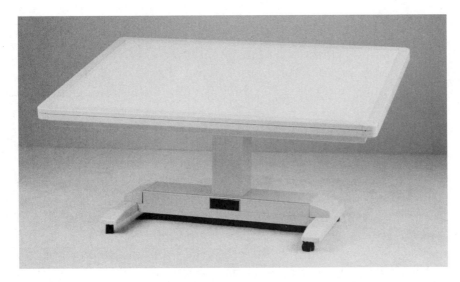

An example of a graphics tablet you might use with MicroStation. (Courtesy of Intergraph Corporation.)

On the puck there is a series of buttons. The use of these buttons plays an important role in how MicroStation interprets your input. For this reason, MicroStation requires the use of a puck with a minimum of four buttons. Because there are other pucks with fewer buttons (and indeed one style of puck that looks like a pen or pencil), you should read the instructions about which graphics tablets are supported by MicroStation.

There are two major advantages to tablets over mice. First, you can use a tablet to trace an existing drawing into your MicroStation design. Second, because of its puck location technique (it absolutely knows where the puck is at all times), a graphics tablet can be used with tablet menus.

The major disadvantage of the tablet is its size and cost. Even a small tablet takes up significant space on your desk and costs many times more than a mouse. It should also be noted that in certain configurations you may actually desire both devices; for instance, a mouse for use with Windows and a tablet for use with MicroStation.

Keyboards

Your keyboard is still a very active device for your input. Most of the time, however, it is relegated to entering short numeric entries for coordinates and the occasional text string. Most of MicroStation's operations are invoked using the now familiar Windows graphic user interface features such as pull-down menus (henceforth referred to simply as menus) and iconic tools organized in movable graphic toolboxes (more on this in a later chapter).

With that said, however, there is a powerful behind-the-scenes keyboard interface that allows you to directly enter graphics commands. Called the Key-in window, it is normally not present on the screen (you invoke it by selecting Utilities menu > Key-in Window). The term *key-in* comes from the original Intergraph IGDS system. A "key-in" is anything you type into MicroStation using the keyboard. MicroStation also provides full support for your keyboard's function keys (labeled F1 through F12), including Ctrl–, Shift–, and Alt–function-key combinations. This gives you 48 possible key combinations.

Plotters

Just as every office has a photocopier, all MicroStation-equipped design groups include some sort of hardcopy or plotting device. This is necessary to get that all-important "paper" copy of your design information to the client. MicroStation supports an impressive array of plotter types, including everything from the Windows system printer (the default and preferred output device) to the entire line of Hewlett Packard wide-format plotters (the latest being the DesignJet family of inkjet plotters), an example of which is shown in the following illustration.

Hewlett Packard's wide-format plotter.

It is not uncommon to have laser printers serving both as document printers and plotters for smaller drawings. Larger-format drawings are usually created on dedicated plotting devices. A discussion on the variety and capabilities of plotters can, and does, take up an entire chapter later in the book; just keep in mind that MicroStation supports all of the industry standard plotters available today. In other words, if you can draw it, MicroStation can plot it.

NOTE: *In the past, plotters tended to be large, designed to support 36-inch (1-meter) tall E-size prints. These days, with most engineers relegated to cubicles and workstations, the trend has definitely been toward the more compact C-size (or smaller) page. For this reason, many shops now utilize the more mainstream workgroup laser printers, such as Hewlett Packard's Deskjet 5si (capable of generating 11 x 17 prints).*

Installing MicroStation

To get started with MicroStation, you need to first install it on your system. MicroStation's installation procedure is very straightforward, and familiar to anyone who has ever installed a Windows application before. Bentley uses the same Installshield installer found in Microsoft's own line of products, as well as the lion's share of other applications. An example of a MicroStation installation screen is shown in the following illustration.

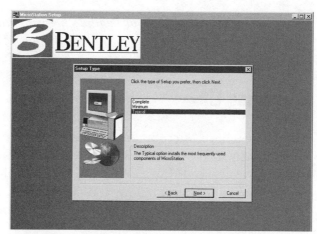

A snapshot of one of the installation screens you will see when you install MicroStation.

Most users will choose to install the "typical" setup, which includes the full MicroStation software product, as well as some example files and a limited set of database operations. In addition, you can opt to install other MicroStation components, which can require upward of 200+ MB of disk space (the "typical" installation requires about 100 MB to install).

Licensing MicroStation

After the initial installation of MicroStation, one of the first items of business is to install some sort of product license. Failure to do so will not prevent you from using MicroStation, but it will limit your design session to 15 minutes maximum before it will automatically shut down the application (do not worry, you do not lose your work; it is automatically saved). There are two major licensing models used with MicroStation and other Bentley products:

- Node licensing
- SELECT Server license manager

Node licensing simply refers to an encoded license key that is provided by Bentley to you and placed in a special directory within the product, which MicroStation then reads to unlock the product. SELECT Server refers to a separate network service application that issues temporary licenses to users on an as-needed basis.

NOTE: *More details about Bentley's licensing are beyond the scope of this introductory text. Questions about the subject should be directed to your system administrator or Bentley Technical Support.*

Starting Up

Once you have MicroStation installed and the license configured, to actually start MicroStation is very easy. Simply invoke MicroStation as you would any other Windows application (by default, Start menu > MicroStation > MicroStation). A desktop shortcut can also be set up, with the following definition:

```
Target: <drv>:
  \Bentey\Program\MicroStation\ustation.exe

Start in: <drv>:\Bentley\Program\MicroStation
```

where *<drv>:\Bentley\Program\MicroStation* is the location where the MicroStation product is installed. The illustration that follows shows a typical definition for the MicroStation shortcut.

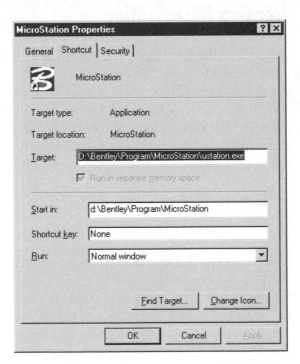

The "properties" window for the MicroStation shortcut.

You can invoke MicroStation from a Command Prompt window, provided your current directory is MicroStation's installed directory or is part of your system's PATH statement. In addition, the *ustation.exe* program will interpret additional parameters you provide after its name (referred to as command line parameters) to further tune its initialization. Your system administrator is normally responsible for the "care and feeding" of these parameters. However, the following are some of the more commonly used parameters.

- `-wu<workspacename>`

Sets a specific workspace as the initial environment.

- `-wr<\\sysname\sharename\workspacedir>`

A shared initialization workspace root that allows a single network location to serve all MicroStation users for common project parameters and resources.

- `-ws<configvarname=value>`

Sets a specific configuration variable to an initial value.

- `-r`

Open the file in read-only mode.

- `<filename>`

If no dash precedes the parameter, MicroStation interprets it as the target *dgn* file you wish to open in the graphics environment.

 NOTE: *It is not recommended that you play around with command line parameters until you have more experience with the MicroStation application. Most users should simply invoke MicroStation via the provided shortcut.*

MicroStation Manager

Unless you specify a file name at the time you start the MicroStation application (the *ustation.exe* portion of the shortcut), the first thing you will see when MicroStation initializes is the MicroStation Manager window. Similar in function to the standard Windows file open dialog box, the MicroStation Manager, shown in the following illustration, provides you with an option to perform several useful file maintenance functions, as well as set several additional configuration options prior to actually entering MicroStation's design environment.

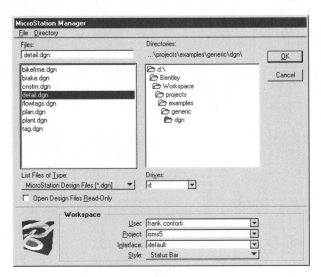

The MicroStation Manager dialog box is your gateway to MicroStation's design environment.

You can create new files, or copy, rename, delete, merge, or compress existing files (compress is simply a way of eliminating wasted disk space being occupied by your design file). You can also manage directories, select the type of supported file to open, and specify the workspace you wish to use when opening a file. In other words, it is a lot more like Windows Explorer than the Windows file open box it resembles.

NOTE 1: *At the time of this writing, MicroStation still displayed the older Windows file open style of directory and file navigation rather than the Windows Explorer-like interface found in most Windows applications. However, it is expected that newer releases of MicroStation will provide full support for the Windows Explorer in keeping with the Windows 98 and 2000 look and feel.*

NOTE 2: *If your software has not been properly registered, you will be presented with the Registration Information dialog box (see following illustration) to complete the registration process. If you opt not to register your copy, you can choose the "15 min. tryout" button. This allows you to enter MicroStation, but the software will terminate after 15 minutes.*

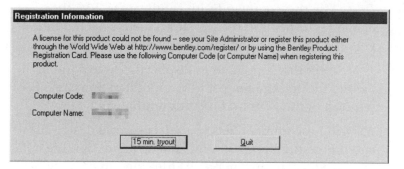

If you see this dialog box, your software has not been properly registered.

Think of the MicroStation Manager dialog box as the entrance door to MicroStation. Its primary function is to let you select a design file (a name given to drawing files you create with MicroStation) to open and work on in MicroStation's design environment. There are, however, several other functions you perform from the MicroStation Manager window. These are described in the following sections.

MicroStation Manager's File and Directory Menus

The MicroStation File menu provides commands for performing common file maintenance operations you are probably already very familiar with. The following describe these options.

- *New:* Select this command to create a new design file. This is the first of two commands under the File menu that do not require you to first select a file in the Files list box.

- *Copy:* Select this command to copy the highlighted file to another location. If copying the file to the same directory, you must specify a different name for the file. However, if you copy it to another location, such as another drive or directory, you can maintain the same name. If no file is selected in the Files list box, this option will be grayed.

- *Rename:* Use this command to give another name to an existing design file. You may choose to change just its file name extension, or the entire file name. If no file is selected in the Files list box, this option will be grayed.

- *Delete:* Use this command with caution, as a file once deleted is difficult to recover. If no file is selected in the Files list box, this option will be grayed.

- *Info:* Select this command to display the size of the high-lighted file and the time of its last modification. If no file is selected in the Files list box, this option will be grayed.

- *Merge:* This command lets you select several design files and merge their content into an existing design file. Of course, you can only merge 3D files into an existing 3D file. This is the second of the two commands under the File menu that do not require you to first select a file in the Files list box.

- *Compress:* As you edit elements in a design file, a copy of the edited elements is maintained for the purpose of undoing edits if necessary. The more editing you do, the larger this base of reserve elements. The Compress option under the File menu, like the pack operation in a database software, actually removes elements that had been marked earlier for deletion. If no file is selected in the Files list box, this option will be grayed.

The Directory menu provides commands for performing directory or folder maintenance.

- *New:* Use this command to create a new subdirectory under the current directory.

- *Copy:* This command copies all *dgn* files in the current directory to another location you type in.

- *Compress:* You can compress all *dgn* files in the current directory with this option.

- *Select Configuration Variable:* Use this command to set the current directory to the value specified by a MicroStation configuration variable.

As with many Windows applications, a list of your last four most recently accessed files is provided on the bottom of the Files menu for quick selection.

NOTE: *Just like many Windows applications, MicroStation is capable of opening many different types of files, including IGES, CGM, DWG/DXF, and its native DGN file format. However, non-MicroStation data is actually translated into a temporary MicroStation DGN file whenever you open it in this manner (this is not always desirable), so use it with caution.*

MicroStation with ProjectBank DGN

No discussion about MicroStation Manager is complete without a word or two about ProjectBank, Bentley's object-oriented data technology. Initially released as an add-on to MicroStation, ProjectBank is likely to be integrated into the core of the product in future releases. ProjectBank essentially replaces the file-centric approach to storing design data with a project-centric server model, in which all design sessions are managed from a central site. When you install ProjectBank into your MicroStation product, an additional shortcut is created in MicroStation's Start menu entry, as shown in the following illustration.

MicroStation
with Proje...

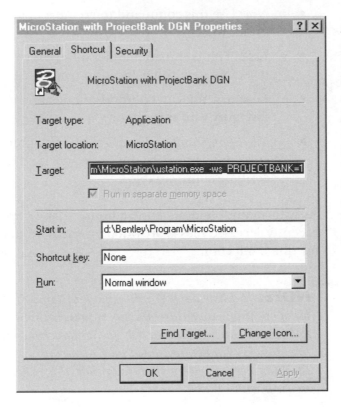

MicroStation with ProjectBank shortcut definition showing the additional -ws parameter. Note also the special ProjectBank "pyramid" in the lower right corner of the MicroStation icon graphic.

If you study the properties of this shortcut, you will see that it is exactly the same as the MicroStation shortcut previously described, with the following additional parameter:

```
-ws_projectbank=1
```

This innocuous parameter dramatically changes how MicroStation stores your engineer design session data. Instead of simply recording the changes you make to your drawing to a single design file (the name of the file type used to hold MicroStation design information), your design "session" is recorded and subsequently posted (in ProjectBank terminology, "committed") to a central project server, which permanently stores those changes as a single transaction against the project.

If this sounds a bit like a financial transaction at your local bank, you would not be far off the mark. In the financial industry, it is very important to track not only the current state of individual accounts but all prior transactions against that account (your monthly checking account or credit card statement is an example of a transaction report). ProjectBank essentially provides this sort of accountability in the design environment. It remembers *who* did *what*, and *when*—information that is not normally associated with any CAD product in use today.

NOTE: *In the opinion of this author, who has over two decades of CAD experience, ProjectBank DGN represents a revolutionary way of working with your design data—one that is bound to change the way design information is captured. As a new user, you are in a position to more readily adopt this technology than some of your more experienced workmates.*

When you invoke the MicroStation with ProjectBank shortcut, you are presented with a different MicroStation window. Instead of the MicroStation Manager dialog box, you will see the Project-Bank Explorer, shown in the following illustration.

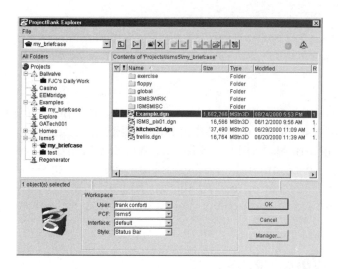

The ProjectBank Explorer is the primary interface for working with ProjectBank-enabled project servers, as shown on the left side of this window. The project Isms5 is the one used throughout this book by the author.

A definitely more Windows 98/2000-like window, ProjectBank Explorer provides the same fundamental operation: the selection of the "design file" you will open in the MicroStation design environment. However, the "design file" is actually initially stored under the control of the ProjectBank project server, as identified in the left-hand pane of the window.

The list of projects you actually see will depend on your office's current project load and configuration. To actually open a design file in ProjectBank, you must first create a briefcase against that project. This is simply a handy organization name you specify for holding project information on your local machine (see following illustration). In this way, should the server connection be lost (for instance, you take your laptop to a work site), you can continue to work unimpeded and still submit your changes to the server when it is available. To create a briefcase, just select the project you are going to work with and select the New Briefcase icon. This command is also available as a right-mouse menu command and from the File menu.

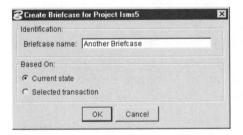

The name of the briefcase is not that important, but you should use something that makes sense to you.

NOTE: *You can only create a briefcase against a project that is currently active (as indicated by the blue triangle icon next to the project's name).*

Once you have created a briefcase, the content of the project is presented in the traditional folder/file name manner in the right pane of the ProjectBank Explorer window. As with MicroStation Manager, there is a command (available from the icon tool bar and File menu) for creating a new design file or folder or for deleting existing files in the project (more on this type of operation later).

Introducing Workspaces

Whether you are using the file-based MicroStation Manager dialog box or the ProjectBank Explorer, you will also need to review and possibly set several design environment parameters prior to opening the MicroStation design environment. One of these parameters deals with a concept called the *workspace*. A workspace is essentially a customized MicroStation environment for a particular user, project, site, or application.

Suppose you need to work on two projects today, the first architectural and the second civil. When creating the architectural drawing, you need access to your company's architectural fonts, line styles, and symbol library. For the civil drawing, you need the Civil Symbols and Settings groups. By implementing two workspaces, one architectural and the other civil, you can simplify the selection of various workspace *modules* (a term that refers to various MicroStation support files such as fonts, cells, seed, data, glossary, and others) as easily as selecting a menu option.

The workspace parameters or options are always presented on the bottom of the MicroStation Manager or ProjectBank Explorer window as a series of separate option menus. The first of these is the User option menu, which is fairly straightforward. You specify the user you wish to have your individual user configuration information stored under. By default, the user is, strangely enough, Default.

NOTE: *Many system administrators configure the initial User parameter to recognize your Windows log-in account name, and will display it as the current User value automatically.*

The Project option menu in MicroStation Manager and the PCF (short for Project Configuration File) in ProjectBank Explorer are used to set the initial project-level settings for MicroStation. The Interface option menu is a very interesting one indeed! Depending on what additional software products you have installed on your system, this list may be as short as one entry (default) or several entries, each one corresponding to a specific *vertical application* (a fancy term for an engineering-discipline-specific configuration). In fact, Bentley has adopted the term *Engineering Configuration* to reflect the concept that you can reconfigure your engineering environment for different engineering disciplines as needed. For this book, the default MicroStation interface should be used.

The final workspace parameter, Interface, is a holdover from the days when MicroStation was transitioning from version 4 to version 5. In version 4 of MicroStation, the user interface utilized a separate command window containing all of the menus and prompt fields associated with user communication. In version 5, this was replaced with the now familiar Windows application window, with its menu bar along the top edge and a status bar on its lower edge. Users who still want to use the version 4 Command Window style of user interface can select it via the Style option menu. For the remainder of this book, it is recommended that you use the default status bar style.

Touring the MicroStation Design Environment

Okay, so you have identified your design file and have set the workspace options. Now all that remains is to click on the OK button. MicroStation will open the selected design file in its design environment.

Finally, it is time to enter MicroStation's drawing environment. Just call up any DGN design file from any directory, or create a new one if you so desire. For the purpose of this tour of MicroStation's interface, the file chosen is not critical.

Dialog Boxes and Settings Boxes

Before moving on, take a moment to distinguish between a dialog box and a settings box. As far as MicroStation is concerned, a dialog box is modal, which means you cannot proceed with any operation until you click on the dialog box's OK or Cancel button. MicroStation Manager and ProjectBank Explorer are examples of modal dialog boxes, meaning that you cannot proceed into MicroStation's design environment without first selecting a design file and clicking on OK. The Dimension Settings window is an example of a non-modal settings box.

In contrast, a settings box, though similar in appearance to a dialog box, is non-modal in operation. This means you can open a settings box, make some changes to its settings, and simply return to your design activity while the settings box remains open. All changes made in the settings box will immediately be interpreted by MicroStation and any tools you are using.

In practice, however, most people do not make a distinction between what is a settings box and a dialog box, as the differences are subtle. The terms are here used interchangeably when discussing settings boxes, but true dialog box windows will always be identified as such. When you do finally open a design file in MicroStation, you are presented with the MicroStation design environment, shown in the following illustration.

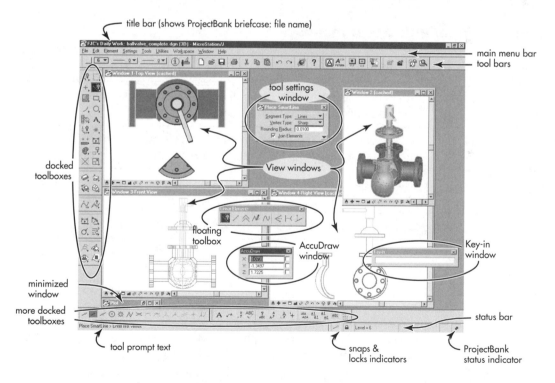

MicroStation's various interface elements. The exact layout of your screen may be different as a result of a previous user session. MicroStation is normally configured to save each session's layout at exit time.

Title Bar and Menu Bar

Along the top edge of the MicroStation application window is the title bar. It displays the name of your active design file (and brief-case name if you have ProjectBank active), whether the file is 2D or 3D in nature, and the name of the software, MicroStation. This last item may seem rather redundant until you remember that MicroStation supports several engineering configurations, each of which will identify itself as the software name (Triforma, Geographics, and so on).

MicroStation's main pull-down menu is where you would expect it, just below the title bar. The menus are organized logically based on function. Table 1-1, which follows, describes the various main pull-down menus.

Table 1-1: Main Pull-down Menus

Menu	Description
File	Commands affecting the entire active file. Includes import/export functions, print/plot functions, and related commands.
Edit	Undo/redo commands, element selection commands, and the standard cut, copy, and paste to/from the Windows clipboard operations.
Element	Commands for invoking settings/dialog boxes for specific types of design elements.
Settings	Commands for invoking a variety of MicroStation global settings.
Tools	Main menu for directly invoking the rich set of design tools for creating or manipulating design elements within the active design.
Utilities	Wide variety of useful utilities. This could be considered the "catch-all" menu for commands and features that do not fit into the other categories.
Workspace	Commands for the general working environment of MicroStation, including customization, mouse button assignments, and function key assignments.
Window	Commands for manipulating the various windows used within MicroStation.
Help	Options associated with MicroStation's online help facility.

Many of these menu names and their associated commands and operations will be familiar to the Windows user. Many closely resemble similar commands found in other Windows applications, such as Microsoft Word, Excel, and Outlook.

Many of MicroStation's utilities that present their own windows within MicroStation are often equipped with pull-down menus for accessing additional commands associated with a respective utility's operation. For instance, when you have the Cell window open (Element menu > Cells), you will have access to its menu bar containing its own File menu. From this menu you can open call libraries, a special type of file for storing repetitive symbols (discussed in detail later in the book). The Cell Library window is shown in the following illustration.

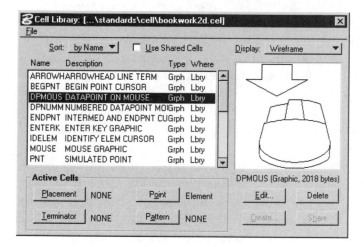

The Cell Library window is invoked from the Elements menu. Note the names of the cells contained in the library and their descriptions, as well as the image of the selected cell on the right side of the window.

Tool Bars

Another set of tools is normally found just below MicroStation's main menu bar. These tool bars, shown in the following illustration, should also be familiar to most Windows users, as they are another common feature of many Windows applications. They differ from the menu bar in one distinct area, their icons. These pictogram icons are so prevalent throughout the Windows operating system that they need no further definition. Suffice it to say, you click on the picture of the tool to invoke its related function.

This might include a pop-up submenu of options or the appearance of a dialog window.

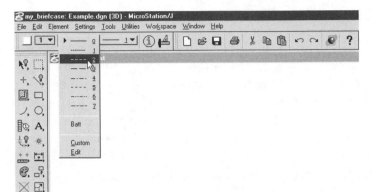

The Primary (shown with Line Classes menu active) and the Standard tool bars are normally "docked" in the upper left corner of the MicroStation application window.

MicroStation includes two tool bars (Standard and Primary), which most users leave docked just beneath the main menu bar. Tool bars do not have to stay docked, however, at the top of the MicroStation application window. If you click on their "handles" (those double lines on the left edge of each tool bar) and drag them into the middle of MicroStation's application window, they will reappear in their own floating window. Again, this is in keeping with the overall look and feel of most Windows applications.

You can also "dock" these two tool bars on the bottom edge of the MicroStation application window, but not along the left or right edges. Finally, you can close these tool bars so that they do not appear on your screen at all! To reinstate them, simply go to the Tools menu and select them again (Tools menu > Primary/Standard).

Primary Tools

As you draw various elements—such as lines, circles, and arcs—you will want to assign them a certain symbology, depending on your need to organize and differentiate the real-world objects they represent. For instance, on a site layout plan, existing buildings may be shown red, and new buildings may be shown yellow and with a thicker line style. The Primary toolbox is designed to let you easily switch active symbology settings. It also includes the Analyze Element and Start AccuDraw tools.

NOTE: *The term* symbology *is often used to refer to element attributes, such as color, style, weight, and level of a drawing element.*

The Primary tool bar, shown in the following illustration, is used to adjust the active element attributes. These attributes are automatically applied to the new elements as they are created by various tools. This is also where you go to start AccuDraw.

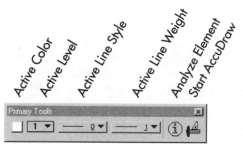

Primary tool bar.

V8: In MicroStation version 8, because of changes to the Level Manager, the Primary tool bar's appearance will be quite different from that displayed in MicroStation/J. However, there is no fundamental difference in the functionality of the two versions.

Standard Tools

No matter what software packages you use, regardless of their category—such as word processing, spreadsheet, database, or CAD—they all share many standard functions. Each will offer a command for creating a new file, opening an existing file, or calling up help. MicroStation organizes commands for standard file and clipboard operations in its Standard tool bar.

The eleven icons on the Standard tool bar (shown in the following illustration) are very similar to the icons found on most Windows applications. You may recognize them right away. In any case, from left to right, the icons represent these commands: New File, Open File, Save Design, Print, Cut, Copy, Paste, Undo, Redo, and Help. As a default, the Standard tool bar comes up docked to the right of the Primary tool bar under the menu bar. If it is not there on your screen, you can activate it from the Tools menu.

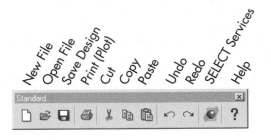

The Standard tool bar contains the most often used file and clipboard operations, as well as access to MicroStation's online Help facility.

NOTE: *Most users leave the Primary and Standard tool bars open and docked just beneath the main menu bar in MicroStation. Presenting a consistent location for accessing specific functions greatly enhances your drawing productivity.*

Toolboxes and Tool Frames

With very few exceptions, the standard MicroStation pull-down menu and tool bars do not contain tools for actually creating or modifying the components that make up your drawing. Instead, drawing tools are organized in toolboxes, much like a mechanic's toolbox. The Main toolbox is home to most of MicroStation's drawing, modification, and annotation tools.

Most users "dock" this master toolbox on the left edge of the MicroStation application window, just beneath the Primary tool bar. In the event you do not see the Main tool bar (docked or undocked), it is easy to re-open (Tools menu > Main > Main). To dock the Main toolbox once it appears, simply drag it to the left or right side of the MicroStation application window.

The Main toolbox (in strictly technical terms, it is a tool*frame*, not a tool*box*) is organized by tool function. All of the Line creation tools are grouped together into the Linear toolbox, which has its own entry position in the Main toolbox (frame). The circles have their own toolbox, text has its own, and so forth.

To select a specific tool, you click on the Main tool frame icon for that tool group and hold the button down (left mouse button, better known as the *data point*). The toolbox associated with that particular tool grouping will temporarily appear on the screen adjacent to its position on the Main toolbox. Sliding the cursor over the appropriate tool and releasing the data point button will activate that specific tool. This sequence is depicted in the following illustrations.

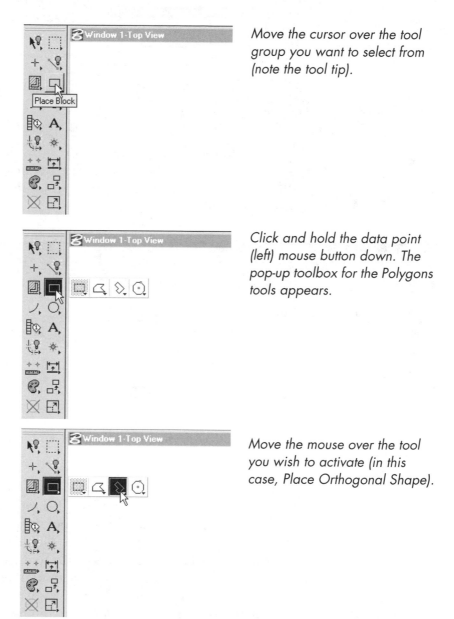

Move the cursor over the tool group you want to select from (note the tool tip).

Click and hold the data point (left) mouse button down. The pop-up toolbox for the Polygons tools appears.

Move the mouse over the tool you wish to activate (in this case, Place Orthogonal Shape).

You will note that whatever tool you select, a line of text will appear in the status bar prompting you for your first action, as shown in the following illustration.

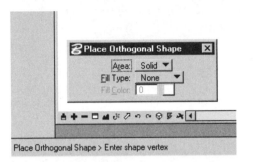

Upon clicking a tool, its name, along with feedback on sequence of data input, appears on the status bar. Note the "Place Orthogonal Shape > Enter shape vertex" prompt on the status bar.

If you continue to slide the cursor over the popped-up toolbox, you "tear off" the toolbox, which then appears in its own window within the MicroStation application window, as shown in the following illustration.

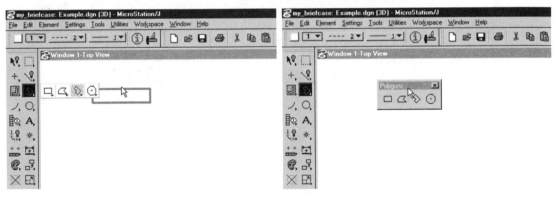

Dragging past the end of the pop-up toolbox results in the windowed version of the toolbox.

As with most Windows applications, when you slide the cursor over a tool icon (whether in a toolbox, tool bar, or frame), the name of the tool will temporarily appear as a tool tip (small black text in a yellow box). This tip is very helpful for the first-time user who has not associated specific icons with specific tools. Tool tips can be disabled by deselecting the Tool Tips option from the Windows menu, but who would want to do that? Even today, the author still uses tool tips to help identify seldom-used but vital tools.

Tool Settings Window

One reason for MicroStation's continued popularity with its users is its comprehensive set of design tools. Virtually any drawing element can be created or modified with very different tools, depending on the design circumstances. Just as the automobile mechanic may have several sizes of wrench, MicroStation provides many tools that, on the surface, look as though they do the same function. However, as you learn how to apply MicroStation to your design process, you will learn to appreciate its varied array of tools.

Quite often, MicroStation implements several variations in a single tool's operation through the use of tool settings specific to that tool. These are set via the Tool Settings window, a window that appears whenever you select a tool. Depending on the options chosen, MicroStation will adjust the operation of the current tool. In addition to looking at the status bar for feedback on what MicroStation expects by way of data input, you will always want to keep an eye on the Tool Settings box, shown in the following illustration. Here you will be able to select from among the variety of settings a tool might provide.

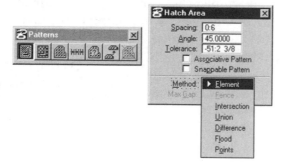

The title of the Tool Settings box, shown to the right of the Patterns toolbox, changes to reflect the name of the active command (Hatch Area) and displays available command options.

Status Bar

At the bottom of the MicroStation application window you will find the ever-present status bar. This non-dismissable part of MicroStation's interface is the primary portal for MicroStation-to-user communication. All MicroStation tools display special prompting messages to this status bar at each step of their operation, as indicated in the following illustration. Experienced users are always aware of the information displayed in the status bar as a matter of course during their design session, as a way of staying "in sync" with MicroStation.

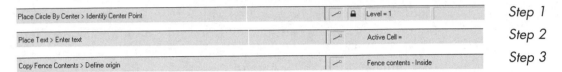

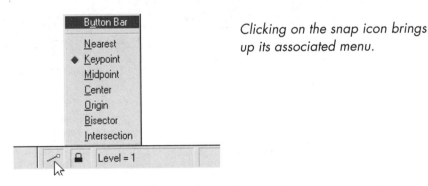

Here you can see how three different tools affect what information gets displayed in MicroStation's status bar.

The status bar is actually a fairly dynamic area of MicroStation. On the far left it is used almost exclusively to display text output from the current tool or command, usually in the form of prompt strings. These text strings "prompt" you to perform a particular activity such as place a data point, enter text, or otherwise provide critical data to the active command or tool program.

In the center of the status bar you will see two icons. The one on the left shows the current snap mode in effect (snaps are discussed later in the book). Depending on the snap mode, you may see one of several icons occupying this space. To change the snap, you can simply click on the snap icon and an option menu will appear, as shown in the following illustration.

Clicking on the snap icon brings up its associated menu.

The other icon near the middle of the status bar deals with MicroStation's locks feature. You can constrain MicroStation's drawing operations by enabling or disabling specific types of locks. Clicking on this icon brings up an option menu of the most-often set lock options and their current status, as indicated in the following illustration. A check mark next to a lock name indicates that the lock is enabled.

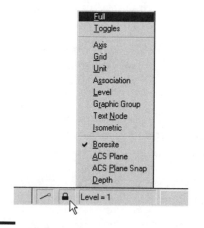

Clicking on the lock icon brings up its associated menu.

NOTE: *There are two dedicated dialog boxes also available to access the full range of MicroStation locks. Both are available from MicroStation's Settings menu (Locks > Full).*

The right half of MicroStation's status bar is the part that dynamically changes in content and display depending on the current situation within your design environment. By default, it displays your active level, but it can also be called upon to display other critical information, such as the number of elements you currently have selected (using the Element Selection or Power Selector tool) or whether you have a fence set.

On the far right edge of MicroStation's status bar you will find a small icon pictogram area will change in appearance depending on whether you are accessing a ProjectBank-enabled project or a local file. If you are accessing a local file and you have write access to the file, a small diskette icon appears. If you are accessing a local file in read-only mode, this disk will have a red X drawn over it, as indicated in the following illustration.

A red disk with an X through it informs you that the active design file cannot be modified in any way (read-only).

Under ProjectBank, the file status indicator provides a visual cue to your current file's status with respect to the master ProjectBank project server. The various colored-arrow coding and combinations are shown in the following illustrations.

If you have made edits to your file, a blue "up" arrow appears, which indicates you will eventually need to send those edits to the ProjectBank server (better known as "committing your changes").

It the server has received changes to your current tile since you last synchronized with it, a green "down" arrow will appear, indicating you will eventually need to go "get" those new changes from the server (better known as a "synchronize").

You can also have the situation in which you have made local changes and the server has accepted other changes; in which case, the indicator will display both arrows.

If you see a yellow exclamation point in the file status indicator, you have some potential conflicts between the work you have done and that found on the server.

The exclamation point will only appear after you have performed the synchronize operation. When ProjectBank is enabled, MicroStation opens a special toolbox containing the ProjectBank-related tools, as indicated in the following illustration. However, you can also access many of these tools by simply clicking on the status indicator. The pop-up menu includes the Commit and Synchronize commands, as well as options for configuring key ProjectBank behavior.

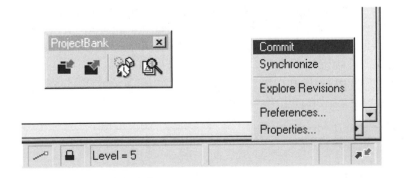

The ProjectBank toolbox and the pop-up menu are only available when working with a ProjectBank-enabled design project.

V8: A red minus sign superimposed on the disk icon will indicate that MicroStation version 8 is running in DWG mode. In this mode, any tools that would result in non-DWG elements are locked, preventing their use.

Customizing MicroStation's Design Environment

Although you do not realize it yet, every time you enter and exit MicroStation, a certain amount of customization to your working environment is occurring. Called User Settings, MicroStation remembers what toolboxes you have open, and other key parameter settings. However, you can move from this passive to a truly active customization using the Customize dialog box (Workspace > Customize), shown in the following illustration. This powerful dialog box allows you to change MicroStation's menus and toolboxes using a point-and-click approach.

NOTE: *Many companies discourage the use of the Customize feature within MicroStation, for purposes of consistency within a project or work environment. Before experimenting with this feature of MicroStation, check with your system or project administrator.*

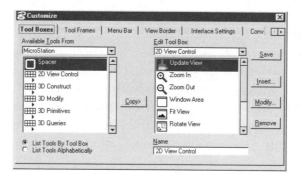

The Customize option under the Workspace menu allows you to customize MicroStation's menu and other interface components.

View Windows

Thus far you have looked at all interface components on your screen except for the one feature that dominates the entire design environment. Called "view windows," these are the primary portals into your drawing and are where you interact with the components of your design. MicroStation can display up to eight of these windows, labeled Window 1 through Window 8. Commands—such as open, close, tile, and cascade—that operate on view windows are located under the Window menu. Within the view windows, the various instructions you enter will construct your design.

As it progresses, this design will start to fill up the window, and eventually you will need to "zoom out" to get a better view of it. In fact, the commands associated with this and other screen control commands are collectively known as the "view control commands." Most of the common view control commands are located to the left and along the horizontal scroll bar of the view window, as shown in the following illustration. They may also be invoked from a separate toolbox. See the option View Control under the Tools menu.

Most frequently used view control commands are located to the left of the horizontal scroll bar of a view window.

Each view can be resized by clicking on and dragging its surrounding border. In addition, you can move, minimize, or maximize it. Located along the top edge of all view windows is a series of push buttons that control these functions. The upper left corner of each view window also includes the window control menu for access to these and other functions. The title bar that contains the name of the window is used to move that window. What each view displays (or does not display) is controlled by various "view-dependent" commands. Note the View Attributes option under the Window Control menu. View manipulation is discussed at length later on.

MicroStation and the Mouse

MicroStation makes maximum use of the "pointing device" (i.e., the mouse) by not only using it to generate coordinate information but to convey what you intend to do with that coordinate entry. This is accomplished by using two or more of the mouse buttons for specific coordinate-related operations. Whether or not the mouse is used with a graphics tablet (where it is commonly referred to as a "puck"), MicroStation expects three types of input from the mouse: *data point*, *tentative point*, or *reset*. Each button on your mouse (or the simultaneous pressing of two buttons, called a chord) identifies your intent. Table 1-2, which follows, summarizes MicroStation mouse button assignments.

Table 1-2: Mouse Button Assignments

Function	Assignment	Description
Datapoint	Left button	Used to select tools, pull-down menus, and buttons, and to identify coordinate locations in your active-design place and manipulate elements
Tentative Point	Middle button (on three-button mouse), left and right button chord (on two-button mouse)	Used to temporarily locate a point in space or an element of a drawing in your design file
Reset	Right button	Releases the current operation or rejects a highlighted element

The button assignments listed in table 1-2 are the default values and are almost universal in use today. Sometimes, however, left-handed users will mirror these functions, with left and right buttons switching functions. You can reassign these functions very easily using the Button Assignments dialog box (Workspace > Button Assignments). The illustration that follows shows the button assignments used in MicroStation.

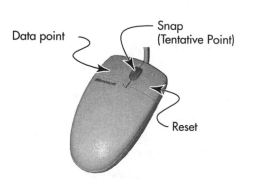

Data point

Snap
(Tentative Point)

Reset

The mouse button assignments.

Puck Buttons

In addition to the three mouse button functions, the graphics tablet puck also has a Command Point button. With this button, a command can be selected from the tablet menu attached to the graphics tablet. Tablet menus have been around for a long time (on Intergraph IGDS systems, they were the *only* method for invoking drawing commands), and MicroStation continues to support them. However, most users select their tools and functions from toolboxes and menus displayed on the video display. A four-button puck is shown in the following illustration.

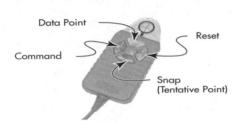

A typical four-button puck showing the Datapoint, Reset, Tentative, and Command buttons.

Data Point Function

This button is used to select tools and other functions both within MicroStation and within the Windows operating system environment. In addition, the data point is used to identify coordinate locations on the screen. When you press this button with the cursor over one of the view windows, you pass the present coordinates of the screen's cursor to MicroStation. These coordinates are used by the active tool or command to perform an operation. Data points are also used to identify specific views with many of the view control tools.

Reset Function

Right after the data point, Reset is the most often used function. Pressing Reset instructs MicroStation to release the present element selected or to exit out of the active operation. With few exceptions, this is the only function provided that does not alter the design file when activated.

Tentative Point Function

The tentative point is one of MicroStation's most powerful, and often most misunderstood, features for the new user. Think of the tentative point as a non-reproducible pencil (bluish colored for photocopy-type reproduction equipment, or the older purple hue for the older blueprint equipment). The tentative point allows you to select a coordinate location in your drawing to consider but not yet commit to the current tool operation. In other words, the tentative point lets you adjust the coordinate location before actually committing it as your data point. Tentative points also have the ability to "snap" to other elements for selecting that location as a data point.

This snap-to feature is one of the real strengths of the tentative point. It is always available, regardless of which drawing tool you have active. There is even a separate button bar, invoked from the Snaps option under the Settings menu (and elsewhere), to let you modify the behavior of this button.

Command Point Function

This button is used only with graphics tablets (digitizers). It is used to identify "commands" from a previously initialized graphics tablet menu. This button is necessary because MicroStation shares the surface of the tablet with the screen tracking cursor and any tablet menus you initialize.

Other CAD products reserve portions of the digitizer surface to map to the display screen, which greatly decreases the amount of space available for any graphics tablet menus. The downside is that you must remember to press the Command button when selecting a command from the tablet menu, and press the data point when selecting on-screen tools and menus.

Most first-time users of MicroStation who work with graphics tablets will find themselves inadvertently selecting a command from the tablet, instead of placing a data point, because they put their finger on the Command button. A good rule of thumb to remember is that the Command button only works when you are looking "down" at the graphics tablet, and the data point only works when you are looking "up" at the screen.

Key-in Window

As you will learn later, MicroStation has a very powerful coordinate input system for those times when you need to directly enter key coordinate information rather than pointing and clicking (data-pointing). AccuDraw has revolutionized the way most users interact with MicroStation when it comes to inputting coordinate data via the keyboard.

However, there is an older method, still available, that allowed the user to directly enter coordinate information into the current tool or command operation. Called the Key-in window, this command line interface involved the entering of special two-character "key-ins" to identify the type of coordinate value you wanted to enter (X,Y versus angle and distance). In addition, the Key-in window (or the Key-in field, as it was once known) can be used to directly enter tool names, and even some parameter data.

By default, MicroStation does not display the Key-in window; you have to invoke it (Utilities menu > Key-in). When you do, a multi-paneled window appears, as shown in the following illustration.

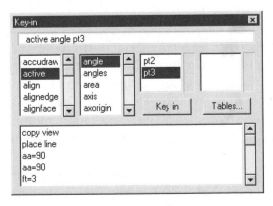

The Key-in option under the Utilities menu brings up MicroStation's Key-in window.

The topmost field is the actual key-in field, where you can directly type in valid commands and parameter settings. The middle field is called the command browser. It allows you to explore the currently valid command names and options by simply pointing and clicking on each key word. You can also type in the first few letters of any valid command and MicroStation will complete its name (a shortcut system, as it were).

At the bottom of the Key-in window is the key-in history. Every command you type into the key-in field is recorded here for instant recall (just click on the entry using the data point or scroll through them using the up/down keyboard arrows).

Most important, the Key-in window can be resized to selectively close each of the three components in inverse order. You essentially "squeeze out" the part of the window you do not need by shrinking its size. Because the key-in history and the key-in browser take up so much screen real estate, most experienced users will resize the Key-in window until it is only a fraction of its default size (most type-in commands are rather brief), and then dock it with the main MicroStation application window, as shown in the following illustration.

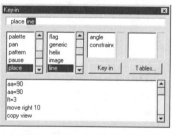

The Key-in window can be resized to selectively display its various components. It can also be docked to the bottom or top edge of the MicroStation application window.

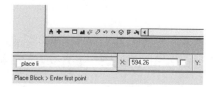

MicroStation's Help, Including Tool Tracking

Before concluding this chapter, a few words need to be said about MicroStation's help facility. Although most programs running within the Microsoft Windows environment include some sort of help file, MicroStation's includes a complete online reference and user guide that can be an invaluable source of information, especially when first learning about the product. Once opened (Help menu > Contents), you simply navigate through the help materials as you would a browser (see following illustration).

NOTE: *Bentley Systems is as of this writing just introducing a new help "engine" using the Microsoft Windows HTML-based help subsystem. The following material describes this new help, rather than the older, Dynaweb-based help.*

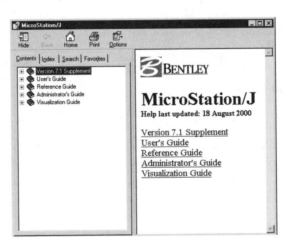

MicroStation Help begins with the Contents page. Here you will see a list of the "books" available via the online help engine.

First-time users will want to tour the *User's Guide*. Here you will find information on the fundamentals of MicroStation, which includes a very nice online glossary of terms used in the MicroStation environment.

One of the more powerful features of MicroStation's help system is Tool Tracking (Help menu > Tool Tracking), shown in the following illustration. This is an option you selectively turn on and off via the Help menu. When enabled, you will get the specific help page on any newly selected tool in MicroStation. For instance, if you want to learn about how the Place Line tool works, you can simply turn on the Tool Tracking feature and click on the Place Line tool (located in the Linear Elements toolbox) and MicroStation will respond by showing you the Place Line tool information.

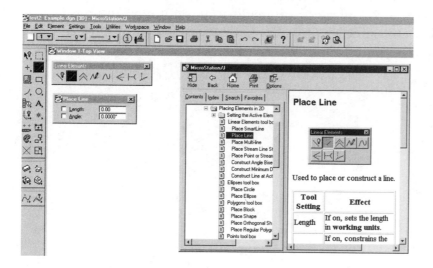

MicroStation's Tracking option provides you a quick means of finding out how a specific tool works within MicroStation.

TIP: *Be sure to turn off the Tracking option when you are done studying a particular tool or feature. Because of its design, the Help window will always reappear on top of the MicroStation application window, which can be very annoying at times.*

Exiting MicroStation

No discussion about MicroStation is complete without telling you how to close a MicroStation session. To exit MicroStation, go to the File menu and select the Exit command. You can also key in *EXIT* or *QUIT* in the Key-in window. Other ways to exit? Double click on the Control menu, or press Alt-F4.

Summary

You have just been given the briefest of tours of the MicroStation design environment. In the next chapter you will get an opportunity to actually use MicroStation to draw something substantial. After that, you will learn about the individual features and capabilities of MicroStation chapter by chapter. Once again, congratulations on choosing such a powerful tool, and do not worry, it is not nearly as complicated as it might seem right now. In no time you will be creating fantastic designs and ideas that would make any manager proud. Good luck!

A SAMPLE DESIGN SESSION

IN THE PREVIOUS CHAPTER YOU RECEIVED A BRIEF INTRODUCTION to the MicroStation design environment. There is nothing like a real-world problem to help you really understand how MicroStation works and how to apply it to your own design work.

Prairie-style Art Glass Layout

Even if you are not an architect, there is no doubt you have heard of Frank Lloyd Wright, architect. To say he has had an influence on American architecture is like saying that putting a man on the moon was just another business trip. From his involvement in the Arts and Crafts design movement at the turn of the *last* century through his Usonian design philosophy, there is no question Mr. Wright was a Visionary, with a capital V. (In case you could not tell, the author is a big fan of Mr. Wright.)

So, what better way to learn about a twenty-first-century product than to use it to adapt a design from the beginning of the last century? The Tree of Life art glass (shown in the following illustration) is arguably one of Mr. Wright's most recognizable stained glass window designs. Part of his Prairie home design period (late 1800s to early 1900s), this design is most notable for its use of heavy lines and repeating geometric patterns. It is also one of the author's favorite Wright art glass designs.

The Martin house Tree of Life window is immediately recognized as a Frank Lloyd Wright hallmark design. (Image reproduced by permission of the Frank Lloyd Wright Foundation.)

This art glass makes the perfect subject for an introduction to MicroStation. It is both simple in design (just a few lines) yet complex in execution (many repeating and mirrored patterns), which are two features of many design projects at which MicroStation excels.

Pre-design Analysis

Before you begin "slinging lines around," let's take a closer look at the subject. First, you will need to define the project. The project goal in this case is to take you through the entire process of applying MicroStation to a specific design situation, which in this case is the creative application of Wright's Tree of Life design to a window of a given size. This process will start with the creation of a new drawing, or *design file* in MicroStation parlance (design files can contain one or more drawings), and end with a completed window layout showing the finished design. There are several steps along the way:

- Lay out the initial window dimensions.
- Create the pattern that constitutes the essence of the Tree of Life design.
- Apply the basic design layout to the window in question and adjust as required.

Again, the purpose of this chapter is to show you how MicroStation works, so the design aspect for this project is kept very simple. Fortunately, simple in this case also means elegant and inspirational.

NOTE: *Although the Frank Lloyd Wright Foundation has granted permission to use the Tree of Life image as the inspiration for this design project, it does not grant any further rights to reproduce this design beyond the purely educational purposes of learning MicroStation. If you decide to pursue the use of this design beyond this context, make sure you get the appropriate Frank Lloyd Wright Foundation permission before proceeding.*

As you can see in the reproduction of Mr. Wright's original sketch for the window, shown in the following illustration, there was a fair amount of adjusting of the design's individual details to arrive at the final results. However, you can see that the overall idea of repeating one pattern several times across the window was an important aspect of the design from its outset.

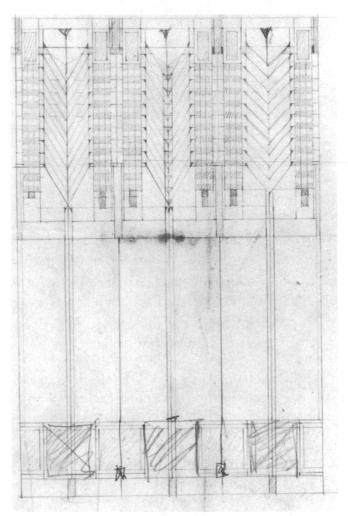

Tree of Life sketch. (Image reproduced with the permission of the Frank Lloyd Wright Foundation.)

Although it would be fun to generate an exact replica of this design, such an endeavor is beyond the scope of this chapter. The idea is to show you how to apply MicroStation to the design process in a hands-on approach to a simplified example of the window, such as that shown in the following illustration.

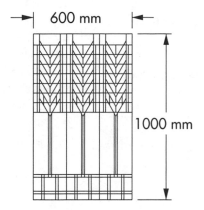

An illustration of one possible design. Note the use of repeating patterns throughout the design.

Even the simplified version contains several interesting features. From a CAD user's perspective, this design is interesting for a number of reasons. It would be interesting to return to this design after you have spent some time with MicroStation and try refining it to more accurately reproduce the original design.

The layout of the window has several repeating elements. First, there is the individual repeating of the tree's branches. Next, there is a mirror copy of the tree itself (left and right halves of the tree) and, finally, there are three complete copies of the tree pattern to fill out the design. MicroStation can be used to good effect to develop this design by applying its element manipulation tools such as Mirror and Copy Element to quickly replicate the "simple" pattern into the full, rich pattern of the final product.

Analysis of the Window's Pattern

As evident in the following illustration, the repeated pattern of the window is easily divided into the upper "branches" and the lower "roots and trunk" portions. It is also apparent that the pattern can be distilled to half an image to be mirrored about the vertical axis later. From a production perspective, this means that once you develop half of one tree pattern, the rest of the design process is a matter of mirroring and copying the appropriate design elements, tasks easily accomplished within MicroStation.

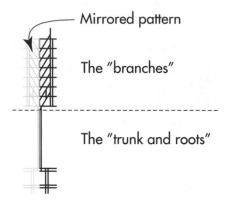

Mirrored pattern

The "branches"

The "trunk and roots"

The upper portion of the design is dominated by a small repeating pattern of angled lines, whereas the lower portion is a simple box-like structure. Note how the design lends itself well to a mirroring process (a common operation in MicroStation).

Remember, the goal here is to fit this overall design to a predetermined window size. This is very similar to a typical design scenario most users are likely to encounter. To get started, you will need to first divide the window into three sections (overall pattern repeats three times on the window). Next, you will need to divide one of the three sections into quarters to both define the upper and lower half of the design as well as the axis for the mirroring process.

Creating a New Design File

The first step in beginning a new project is to create a design file. The illustrations that follow show how to start MicroStation and use a seed file to make a new design file named *window.dgn*.

TIP: *You will find a completed version of this exercise in the data set on the companion CD-ROM (file name* windowcompleted.dgn*), which allows you to review the outcome at any time during this exercise.*

Creating the New Drawing

To get started, you need to "launch" MicroStation. This is accomplished with a simple click on the MicroStation shortcut (normally, Start menu > Programs > MicroStation V7.1 > MicroStation).

NOTE: *If you have been instructed to develop this exercise under ProjectBank, use the MicroStation with ProjectBank shortcut instead of the
regular MicroStation shortcut. You will need the name of the project
server (and folder optionally contained within this project) to proceed.*

With MicroStation started and the MicroStation Manager (or ProjectBank Explorer) dialog box displayed, you will need to create a
new design file to house your design data. Exercise 2-1, which follows, takes you through this process using MicroStation Manager.
Exercise 2-2 takes you through this process using ProjectBank.

EXERCISE 2-1: CREATING A NEW DESIGN FILE USING MICROSTATION MANAGER

To create a new design file using MicroStation Manager, perform the following
steps.

1 Select the New command (File menu > New). The Create Design File
 dialog box appears, which is shown in the following illustration. This is
 a familiar dialog box used to define a new design file. It uses the same
 file navigation feature found in many Windows applications.

*Create Design File
dialog box. Note the
seed file at the bottom
of the dialog.*

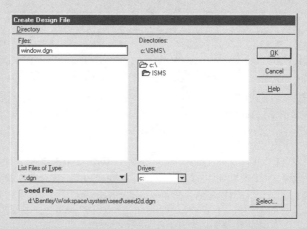

2 In the text field under the label Files, key in *WINDOW.DGN* but do not click
 on OK yet! Look at the Seed File section of the dialog box. MicroStation
 creates new design files by referencing settings from an existing seed file.
 MicroStation is delivered with dozens of seed files. In addition, most companies have their own set of seed files for use in specific types of projects.

Because of this overabundance of seed files from which you could potentially create your initial design file, you will need to pay close attention to this selection process. For this exercise, you will use the default seed file named *SEED2D.DGN*, as displayed by default in the Seed File section.

3   Click on the OK button on the Create Design File dialog box. *SEED2D.DGN* is the seed file for *WINDOW.DGN*.

4   Select *Window.dgn* if not already selected, and click on the OK button. MicroStation opens the selected file in the design environment.

In exercise 2-2, which follows, you will create a new design file using ProjectBank Explorer. This is an alternative to the previous exercise, which used MicroStation Manager to create a new design file.

EXERCISE 2-2:   CREATING A NEW DESIGN FILE USING PROJECTBANK EXPLORER

To create a new design file using ProjectBank Explorer, perform the following steps.

1   Select the project name as provided by your instructor, project manager, or system administrator (depends on your organization and/or the environment within which you are currently performing this exercise). The illustrations here use the project *Isms5* as the project, with the exercise files located in the folder named *exercise*.

2   Create a new briefcase against this project (File menu > New > Briefcase). The briefcase name is not important. The following illustration shows the creation of the briefcase *my_briefcase*.

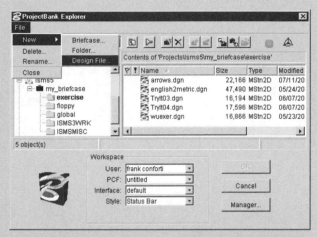

*New briefcase
my_briefcase.*

3   Select the new briefcase *my_briefcase* (the folder *exercise* is further opened in
the steps that follow), as shown in the following illustration.

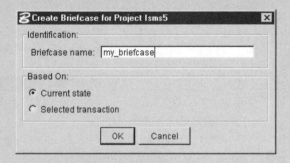

*Selecting the new
briefcase.*

4   Create a new design file in the briefcase (File menu > New > Design File), as
shown in the following illustration. The New Design File dialog box
appears. Here, you enter the name of the new design file and its associated
seed file.

*Creating a new design
file.*

5   Enter *window.dgn* as the design file name. Use *seed2d.dgn* as your seed file.
Click on OK to create the file. At this point, you need to commit your
change to the server (the initial creation of the design file is considered a
change), as shown in the following illustration.

*Committing the
design file to the
server.*

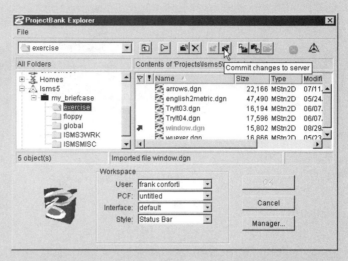

6   Select the Commit command (PB Explorer toolbar > Commit Changes
icon). The Commit Changes dialog box appears.

7   Enter the description *created initial window design file* and click on the Commit button, as shown in the following illustration.

*Entering a description
for the design file.*

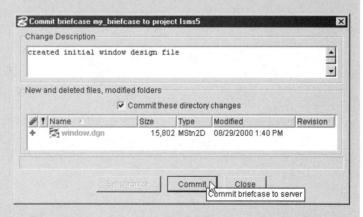

At this point, the file is now posted to the ProjectBank server. Therefore, you can go ahead and enter MicroStation's design environment.

8   From the ProjectBank Explorer dialog box, select the file *window.dgn* and
click on OK. MicroStation's design environment appears, ready for your
design work.

Setting the Measurement System (Working Units)

Congratulations! You have now successfully created and opened your first new design file. Regardless of which method you used to create the design file, you are now presented with an empty design file, except for several initial design parameters the *window.dgn* file inherited from the seed file. Before you can begin the design proper, you will need to set the working units of your new design. This is a step usually taken care of by the selection of a purpose-built seed file, but because you used the default *seed2d.dgn* you will need to manually adjust this critical setting. You do this by changing the working units in the Design File Settings dialog box. Exercise 2-3, which follows, takes you through this process.

 NOTE: *The working units of your new file will be the same as those of the seed file you initially selected. You will want to maintain a library of seed files, with appropriate working units, for the types of drawings you expect to create.*

EXERCISE 2-3: ADJUSTING WORKING UNITS

To adjust working units, perform the following steps.

1   Open the Design File Settings dialog box (Settings menu > Design File). Select the category Working Units. Here, you will set the system of measurement you want to use during the design process. Although it does not really matter what values you use for this exercise, you should probably stick with the metric measurement system, as used here.

2   Set the Master Units label to M (for meters), the Sub Units label to mm, and the subunit-to-master unit ratio to 1000 (1000 mm to 1M).

3   Set the Pos Units option to 1000 as well.

4   Click on OK and accept the change to your working units definition. A dialog box appears, warning you of this fundamental change. This is normal.

>>>> >>

Defining the Window's "Box"

At this point, you are ready to draw something in the design file; namely, the rectangle that defines the size (or *extents*) of the window. The window is 600 mm wide by 1000 mm tall. In exercise 2-4, which follows, you will use the Place Block tool to draw the rectangle representing the finished size of the window.

EXERCISE 2-4: PLACING THE WINDOW BOUNDARY RECTANGLE

To place the window boundary rectangle, perform the following steps.

1   Select the Place Block tool (Main tool frame > Polygons toolbox > Place Block).

2   Activate MicroStation's AccuDraw feature (Primary tool bar > Start Accu-Draw). The AccuDraw window will appear, with X and Y fields. There is no Z field, as you are working with a simple 2D design.

3   Place a data point (left mouse button) in the lower left corner of the empty View window 1. The AccuDraw "compass" appears at your first data point location. Note how a dynamic rectangle tracks the movement of the cursor as you move it.

4   Move your mouse to the right until the X field of the AccuDraw window is highlighted. Enter the distance *:600* via the keyboard. Do not forget the colon character! This tells MicroStation you want to enter 600 millimeters, not 600 meters! You could also enter *.6* for 6/10 of a meter. Note how the X field of the AccuDraw window "locks" to this entered value, as does the X length of the yet-to-be-created rectangle.

5   Move your mouse up from its current location until the Y field of the AccuDraw window is highlighted. Enter the distance *1* (for 1 meter) via the keyboard. You can also directly select the X field of the AccuDraw window and enter your value, but most MicroStation users prefer the gesture control demonstrated here. Note also how the Y field now locks to the value you entered. At this point, you need to set the rectangle.

6   Place a data point to create the fully defined rectangle, as shown in the following illustration.

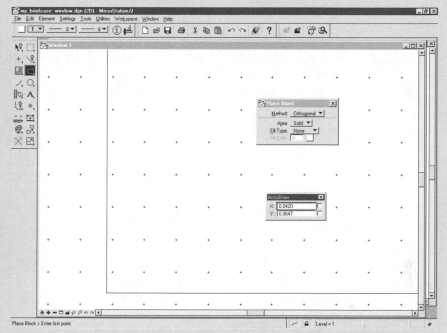

Placing a data point to establish the rectangle.

At this point, you have a rectangle that represents the extents of the window but only part of it is probably visible in the View window. You need to fit the view's scale to see the entire rectangle.

7 Select the Fit Active Window tool from the View window 1 view control tool-box (lower left corner of the view). The rectangle scales to fit in the view's extents, as shown in the following illustration.

Scaling the rectangle to fit within the view.

Adjust the magnification of a view so all elements are visible

8 Press the Reset button on the mouse to release the Fit Active tool and return to the Place Block tool.

It is time to review the original Tree of Life sketch, as shown in the illustrations that follow. Note how the window is divided into three equally spaced areas, with a repeating pattern in each (that is a hint on how you are going to develop this drawing). You need to

subdivide the window into three equal spaces. Exercise 2-5, which follows, takes you through this process.

EXERCISE 2-5: DIVIDING THE WINDOW INTO THREE PANELS

To divide the window into three panels, perform the following steps.

1   Set the active color to blue (Primary tool bar > Color selector > Blue), and the line style to dashed (Primary tool bar > Line Style selector > 2), as shown in the following illustration.

Establishing active color and line style.

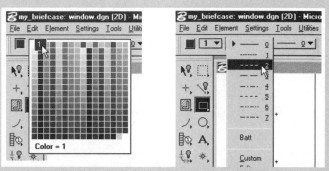

2   Select the Place Line tool (Main tool frame > Linear Elements > Place Line), as shown in the following illustration.

Selecting the Place Line tool.

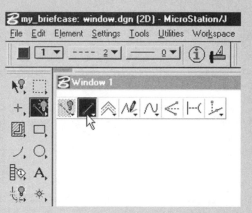

At this point, MicroStation focuses on the Place Line tool's tool settings window. For this part of the exercise, however, you need to change that focus back to AccuDraw so that you can accurately position the lines that will subdivide the window. This can be done either by clicking on the AccuDraw window with the cursor or by pressing the Esc key on your keyboard.

3   Set the input focus on the AccuDraw window and press the K key on your keyboard. This brings up the Keypoint Snap Divisor window. Here, you are going to tell MicroStation how to divide existing elements for locating snap points. You need to adjust the window so that you can click on thirds of the rectangle.

4   Type in *3* in the Keypoint Snap Divisor window for three equal spaces on the window. You can dismiss this Keypoint Snap Divisor window by pressing Enter, or you can leave it open.

5   Using the Tentative Point button, snap to the lower left third of the rectangle to start drawing the vertical line. The Tentative point snap is normally the middle button on a three-button mouse, or the simultaneous pressing of both mouse buttons on a two-button mouse. Snap to the bottom-most line of the rectangle.

6   Click the data point button to define the starting point of the new line. A dynamic line appears that starts at the one-third distance along the bottom edge of the rectangle.

7   Using Tentative Point (again) and a data point, snap to the same location on the top edge of the rectangle. A dashed line appears that rises vertically from the bottom of the rectangle to the top of the rectangle, delineating one-third of the rectangle's area, as shown in the following illustration.

Delineated one-third of the rectangle's area.

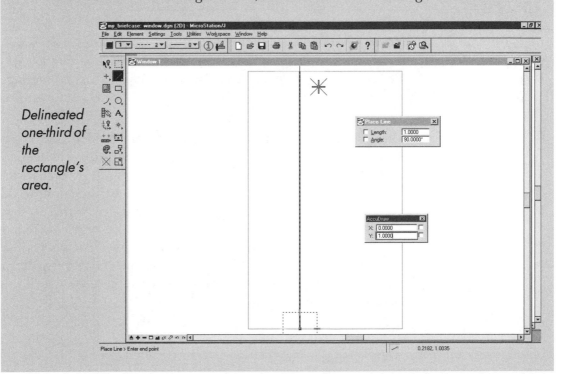

8   Press the Reset button to terminate the current line drawing mode.

9   Using the tentative point snap/data point technique, create the other vertical line to complete the division of the window into three vertical segments, as shown in the following illustration.

Division of the window into three areas.

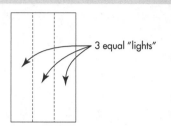

NOTE: *Do not forget to press the Reset button between each line segment placement!*

ProjectBank Users: *Now would be a good time to commit your changes made so far. Select the Commit command from the File menu, enter a description (Initial window layout for three lights), and commit your changes.*

At this point, you have a single window frame divided into three equal areas, as shown in the following illustration. Now, you need to focus in on the area where you will be developing the primary window pattern. This will be the rightmost third of the window. As mentioned earlier, you need to subdivide this space in half so that you will be able to mirror your pattern around a central axis. To aid in visualizing the design, you should place a line that bisects the working area where you will be developing the repeating pattern (henceforth referred to as the *design area*). Exercise 2-6, which follows, takes you through this process.

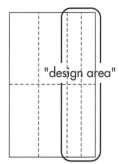

Divided window frame.

"design area"

EXERCISE 2-6: TEMPORARILY REDEFINING THE KEYPOINT SNAP

To temporarily redefine the keypoint snap, perform the following steps.

1   With input focus on the AccuDraw window, press the K key again.

    This returns the Keypoint snap window to the forefront. You need to place that vertical bisector in the design area by splitting its space in half. The quick-and-dirty method for doing this is to simply manipulate the keypoint snap (in other words, snap to 1/2 of 1/3 of the overall window's dimension). It sounds more difficult than it really is!

2   Enter *6* for the new keypoint snap value (1/3 divided by 1/2…). Next, you should differentiate the color of the center line from your previous work.

3   Set your active color to green [Primary toolbar > Color selector > Green (3)].

4   With the Place Line tool still active from the last exercise, snap a tentative point to the center of the bottom edge. Data point to accept. A dynamic dashed, green line appears.

5   Snap a tentative point and data point on the top edge of the design area. You now have your design area bisected vertically down its center.

6   Press the Reset button to release the dynamic line, still following your cursor movement.

7   With input focus on the AccuDraw window, set the keypoint snap to 2. This results in the ability to snap to the endpoints and the midpoint of any element, a nice neutral value to use.

At this point, you have a nicely divided window containing three vertical lights. The rightmost light acts as the design area, with its single vertical bisector. Now is a good time to use MicroStation's view control tools to focus in on the design area, as shown in the following illustration.

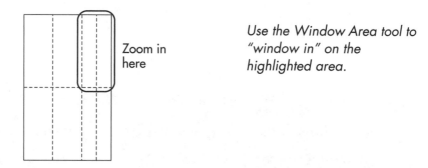

Zoom in here

Use the Window Area tool to "window in" on the highlighted area.

Creating the Upper Portion of the Tree of Life Pattern

Reviewing the art glass photograph or Mr. Wright's original sketch, it is apparent that the upper portion of the pattern is itself a repeating pattern of the angled line and selected horizontal members. Further investigation shows that there are seven "cells" of repeating geometry consisting of an angled line (about 60 degrees from horizontal, give or take 5 degrees), a series of vertical and horizontal lines that intersect this angled line in an interesting pattern. To either side of each pattern is a smaller checkerboard pattern that appears to fill about a third of the repeating pattern's horizontal dimension, as shown in the following illustration.

The exact dimensions of the individual pattern elements are not as important as how you arrive at the final pattern. If you look at Mr. Wright's original sketch, you will see that he experimented with several different angles before arriving at the 60-degree angle. You will need to keep this in mind as you develop the first few lines, as this is the aesthetic aspect of design that MicroStation can assist you with but cannot perform on its own.

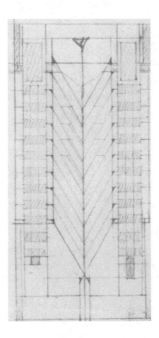

Close-up showing the repeating pattern.

For this reason, the placement of the line work from this point on is not going to involve precise dimensions. Instead, you will need to try your hand at "sketching" the location of the various elements with an eye on how they look in the overall pattern.

NOTE: *Do not worry if your first placement does not look right. You can always select MicroStation's Undo function (Edit menu > Undo, or Ctrl-Z) and try again.*

Next, you need to place that angled line, which helps establish the "branches" of the pattern. Exercise 2-7, which follows, takes you through this process.

EXERCISE 2-7: PLACING THE ANGLED LINE

To place the angled line, perform the following steps.

1   Set the active color to red (Primary tool bar > Color > Red or 3), and the line style to solid (Primary tool bar > Line Style > Solid or 0). MicroStation users often use color and line styles as a differentiation between different design elements within the overall design. In this

case, red is used to display the stained glass cames (not canes!) that run between the window's individual glass elements.

2   Returning to the Place Line tool, select the tool settings option Angle, and set its value to 60 (degrees), as shown in the following illustration. This forces all lines to be placed at 60 degrees from the horizontal, and is an example of a specific tool setting unique to only the active tool.

Angle option set to 60 degrees.

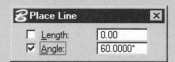

3   Snap to the midpoint of the green dashed bisector line, and data point. A dynamic line locked at 60 degrees appears. Because you do not know the length of this line yet, just draw it out a distance (to the edge of the window light if you want).

4   Data point to set the other end of the line. This is a sketch element you will refine a bit later.

At this point, you have your first solid element of the design (pat yourself on the back!). Next, you need to establish a few more of the vertical and horizontal elements of the pattern, as shown in the following illustration. These involve more of an aesthetic eye than a literal "CAD" eye on the design.

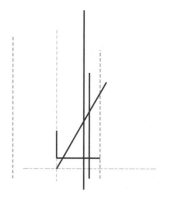

You will place the lines shown to set up the first of the repeating patterns.

Before you begin placement, a short explanation about Accu-Draw's axis indexing feature is in order. A moment ago, when you placed that 60-degree line, it was locked to the value you entered in the Place Line tool's settings window. You can probably guess that placing lines at 0 degrees and 90 degrees could be accomplished by setting this parameter to those values. However, a lot of design work involves working with horizontal and vertical lines. Constantly returning to the tool settings window to set these values can be very tedious.

Fortunately, AccuDraw incorporates a helpful feature that allows you to temporarily lock your current drawing element to either the X or Y axis by simply positioning the cursor close to either axis of the AccuDraw compass. This works like a "sticky" lock-to-axis feature. If you move within a few pixels of either axis, the element snaps to that axis and will remain locked to that axis as long as you keep the cursor within the prescribed few-pixel distance. If you move too far off the axis, it "unsticks" and you are free to place your data point anywhere. Move it close to the axis again and, viola!, it sticks again. You will use this feature to place the horizontal and vertical components of the pattern. Exercise 2-8, which follows, takes you through this process.

EXERCISE 2-8: PLACING MORE OF THE PATTERN ELEMENTS

To place more of the pattern elements, perform the following steps.

1   Select Place Line if it is not still active (remember, a tool stays active until you replace it with another tool). Turn off the Angle option in the Place Line tool settings window (deselect its checkbox).

2   Set the input focus back to AccuDraw (hint: pressing the Esc key will usually return focus to AccuDraw), and press the N key. This temporarily sets the Tentative point snap to Nearest mode. This is necessary so that you can snap to the bisector at the nearest point, not just at its midpoint or endpoints.

3   Snap to a location on the vertical green bisector a bit above the angled line, as shown in the following illustration. Keep in mind that later you will want a small triangle of color between the three lines.

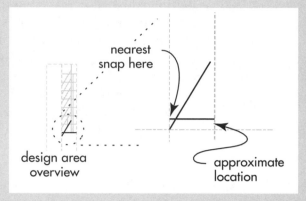

*Snap location on the
bisector.*

4   Data point to start the horizontal line. Moving your cursor directly to the
right will keep the new line indexed to the X axis. As long as you keep the
line indexed to the axis, the resulting line will be precisely aligned in the
horizontal plane.

TIP: *You can lock the line to the axis by pressing the Enter key while
the line is indexed to the X axis. This invokes AccuDraw's Smart
Lock feature, which locks whichever axis is currently selected in its
window.*

5   Moving the cursor directly to the right of the first data point, place another
data point toward the edge of the window light. Again, you are simply
sketching so that precise endpoints are not absolutely necessary…yet.

6   Press the Reset button to release the dynamic line still active from your last
data point location.

7   At this point, you have one of the orthogonal lines that make up the branch
pattern. You need to place a few more lines roughly in the locations shown
in the following illustration. Once you are happy with your sketch, you can
proceed to the next exercise.

Placement of lines for the branch pattern.

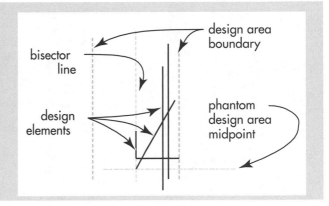

Now that you have your single branch pattern in place, you need to clean it up a bit before you can replicate it. This involves the use of a MicroStation element modification tool (found in Main toolbox > Modify Element toolbox). Exercise 2-9, which follows, takes you through this process.

EXERCISE 2-9: CLEANING UP THE BRANCHES

The first item of business is to "trim back" the angled line you placed earlier, so that it intersects the leftmost vertical line (the one that defines the left edge of the checkerboard pattern), as shown in the following illustration. This is accomplished using an Extend Element tool.

Trimming the angled line.

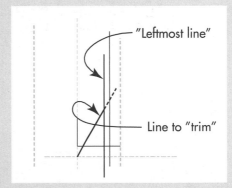

1   Select the Extend Element to Intersection tool (Main tool frame > Modify Element > Extend Element to Intersection), shown in the following illustration.

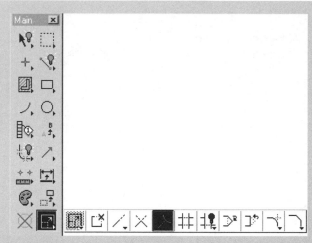

Extend Element to Intersection tool location.

2   Data point on the angled line to identify it as the element to be modified.

3   Select the leftmost long vertical line you placed in the previous exercise. You want the angled line to terminate exactly at the point of intersection with that vertical line.

4   Data point once more anywhere on the view to accept the modification, preferably away from other elements. This is sometimes called *accepting the change* or just *accept*. To optimize data point usage, you can also use this "accept" data point to select the next element you want to be modified. In this way you can quickly use one tool on several elements.

5   Using this same identify target, identify the element to be intersected, and modify the two vertical lines so that they touch the design area box on its top edge and the horizontal line you placed in the previous exercise. At this point, the design should look as shown in the following illustration.

Modifying the two vertical lines.

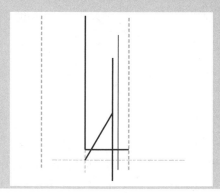

Now that you have cleaned up things, it is time to replicate select elements that constitute the rest of the tree pattern's canopy (branches). Although there is a Copy Element tool you could use to simply select each element individually and copy them several times over, it is much more efficient to select several elements at once and perform a single copy operation. A combination of the PowerSelector tool and the Copy Element tool (shown in the following illustration) makes this possible.

The PowerSelector and Copy Element tools.

PowerSelector lets you to choose the exact elements you want to copy. Once you have identified your target elements, you select the Copy tool and proceed with the copy operation. Exercise 2-10, which follows, takes you through this process.

EXERCISE 2-10: CREATING THE TREE CANOPY USING POWERSELECTOR AND COPY ELEMENT

To create the tree canopy using PowerSelector and the Copy Element tool, perform the following steps.

1   Select the PowerSelector tool (Main toolbox > Element Selection toolbox > PowerSelector).

2   Using the "Individual" method (first icon on the left of the PowerSelector's tool settings options), identify the elements shown in the following illustration.

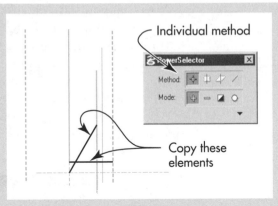

Identifying elements.

3 Select the Copy Element tool (Main toolbox > Manipulate Element > Copy Element). You have to give the Copy Element tool a "from" location so that it knows where to start the copy offset. Because you want to have a very accurate offset, use the Tentative point to snap to the bottom of the angled line.

4 Tentative point snap/data point to the lower left end of the angled line.

5 Move the cursor up along the Y axis, keeping it indexed to the axis. Input focus should be in AccuDraw's Y field.

6 Enter a distance of *.06* (for 60 mm). Note how the second copy is now locked at 60 mm along the Y axis.

7 Data point once to make one copy of your line work.

8 Move your cursor up the Y axis again. Note how, as you get close to the same 60-mm distance, MicroStation temporarily locks to 60 mm. This is another feature of AccuDraw, the Previous Distance lock, shown in the following illustration. If you data point while it is locked on 60 mm, you can be assured that the second copy is offset from the first copy by the same 60 mm.

Previous Distance lock in effect.

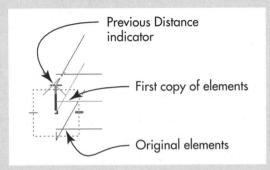

9 Move your cursor along the Y axis again, and data point when the previous distance marker appears. Do this four more times.

10 Select PowerSelector and click on the Clear button (lower rightmost option) or press the Space key. You should have a total of seven branches defined using this technique. To finish up the top part of the design, place a line between the first two branches, just to the left of the leftmost line in the previous exercise, as shown in the following illustration.

Finishing the top part of the design.

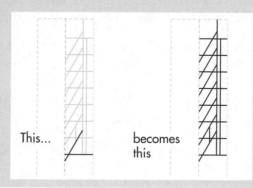

In exercise 2-11, which follows, you will create the root system and trunk of the tree.

Exercise 2-11: Creating the Root System and Trunk

The next step is to create the trunk and roots. In the following illustration you can see that this part of the design is very simple, consisting of eight lines total. As before, this part of the design is very subjective. However, a dimension or two is provided to help you get a feel for placing the various elements.

Creating the trunk and roots.

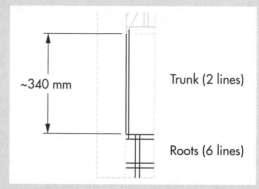

1 Select the Place SmartLine tool (Main toolbox > Linear Elements > Place SmartLine).

2   Snap to the bottom of the branches pattern (left end of angled line) and data point. You are drawing the centerline of the tree's trunk.

3   With AccuDraw indexed to the Y axis, move down and enter a fixed distance (0.34), or eyeball it for the best aesthetic appearance. Data point to accept the length. Press the Reset button to release the line placement operation (and place your one line segment).

4   Place a horizontal line from the end of the just-placed line to the right edge of the bounding (dashed) rectangle. Use the Y lock to "lock" the AccuDraw compass to the X axis, and snap to the bounding (dashed) rectangle. Alternatively, you can use the Extend Element to Intersection tool introduced in the "branches" exercise to set the line to intersect with the bounding box.

5   Place the vertical, bottom line that defines the root ball of the tree. You will use a new tool to create the parallel lines that make up the rest of the root design.

6   Select the Move Parallel tool (Main toolbox > Manipulate > Move Parallel), shown in the following illustration. Select the Make Copy option.

The Move Parallel tool.

7   Set the Distance field in the Move/Copy Parallel tool settings to 0.02 (for 20 mm). This is the distance you will be offsetting the parallel line from the original.

8   Data point once on the vertical "trunk" line, and once to its immediate right. A copy of the line appears exactly 20 mm to the right of the original.

9   Press the Reset button to release this line.

10 Using the same data point/data point/reset procedure, select each of the three lines that make up the root ball and create a line parallel with them.

At this point, you have a complete half of a tree. In exercise 2-12, which follows, you will use the Mirror tool to create a symmetrical copy of the tree to complete one-third of your final design. The copy of the tree is shown in the following illustration.

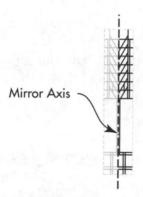

Mirrored half of the tree.

Mirror Axis

EXERCISE 2-12: MIRRORING THE TREE PATTERN

To mirror the tree pattern, perform the following steps.

1 Select the PowerSelector tool. Set the Method to Block with Overlap by clicking on the PowerSelector Block icon in the tool settings until it appears with the dotted outline, as shown in the following illustration. This indicates that any elements you identify with the imaginary block, defined by your next two data points, will be added to the selection set whether those elements are contained entirely within the selection block or simply touch it.

Selecting Block with Overlap.

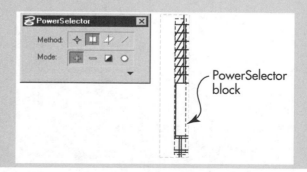

PowerSelector block

2 Using two data points, identify the "half tree" elements shown in the previous illustration. Note that you do not have to select those small vertical lines between each of the branches, nor the vertical line that runs through the tree's trunk. You will notice that the horizontal line you placed earlier to bisect the bounding rectangle is highlighted as part of the selection set. You will need to release it before you perform the Mirror copy operation.

3 Set the PowerSelector Method back to Individual, and the Mode to Subtract. Identify the horizontal dashed line that bisects the bounding rectangle.

4 Select the Mirror tool (Main toolbox > Manipulate toolbox > Mirror Element). Select the Make Copy option and set the Mirror About option to Vertical, as shown in the following illustration. This tells the Mirror tool you want a copy of the elements to be mirrored and that it should be mirrored about an axis parallel to the Y (vertical) axis.

Selecting mirror operation parameters.

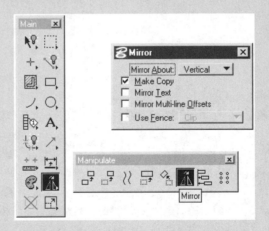

5 Using the Tentative Point button, snap to the vertical dashed line that defines the centerline of the tree. A mirrored copy of the tree appears, as shown in the following illustration.

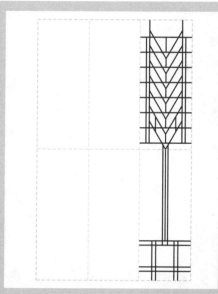

Mirrored copy of the tree.

6 Press the Reset button to release the Mirror operation. If you accidentally data point again, another copy of the tree will be mirrored. If you do this, simply select the Undo command (Edit menu > Undo, or press Ctrl-Z).

At this point, you have one completed tree. Pat yourself on the back! The only operation remaining to complete the window design is to copy the entire tree two more times, in the remaining two "windowpanes" to the left of the one you just created. Now you see why you were originally instructed to divide the overall window into three separate windowpanes. Exercise 2-13, which follows, takes you through the process of completing the window pattern.

EXERCISE 2-13: COMPLETING THE WINDOW PATTERN

If you just completed the Mirror Copy operation from the last exercise, you should still have the left half of the tree in your selection set (highlighted). Before you can copy the tree to the other two windowpanes, you will need to first use the PowerSelector to add the original right half of the tree to your selection set, including those small vertical lines between the branches you did not mirror-copy in the last exercise (see the following illustration).

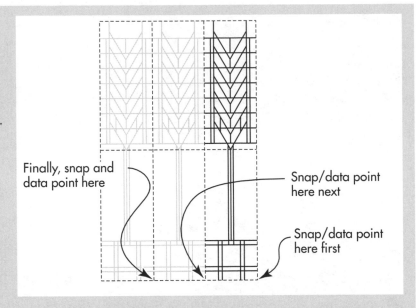

*Original right
half of the tree,
to be added to
the selection set.*

Finally, snap and
data point here

Snap/data point
here next

Snap/data point
here first

1   Select the PowerSelector tool. Set the Method to Block, and the Mode
    to Add.

2   With two data points, identify the elements that make up the right half of
    the tree. If you carefully select the data points that define the PowerSelec-
    tor block inside the tree's edges, you will avoid grabbing the outline that
    makes up the entire window. If you do inadvertently grab any of the dashed
    lines, simply set the Mode to Subtract, set the Method to Individual, and
    deselect the unwanted elements. This is a nice aspect of the selection set
    and PowerSelector. Being able to fine tune which elements you want to
    operate on before selecting the appropriate design tool makes for a very
    efficient work flow.

3   Select the Copy Element tool (Main toolbox > Manipulate toolbox > Copy
    Element), as shown in the following illustration.

Selecting the Copy Element tool.

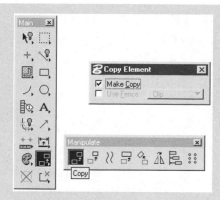

4 Using the Tentative Point button, snap to the rightmost, bottom edge of the bounding rectangle to establish the point to copy from, and data point.

5 Snap to the right, bottom edge of the first empty windowpane (the one to the left of your tree design), and data point. The second complete copy of the tree appears next to the first copy. Note how the second copy is now your selection set. This allows you to select the next location, where you want the selection set copied.

6 Snap to the right, bottom edge of the last empty windowpane and data point. You now have your three copies of the tree. This completes your Tree of Life design layout, shown in the following illustration.

The completed exercise!

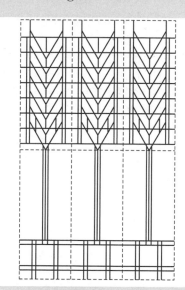

7   Press the Reset button to release the Copy tool, and click on the Clear button in the PowerSelector tool settings to release the selection set.

Summary

You have successfully completed the creation of the basic window layout. There are many other operations you could proceed with from here, including the generation of individual colored window segments. However, the purpose of this chapter was to show you how you use various combinations of MicroStation's design features to go from a blank design to an almost finished product.

You were introduced to AccuDraw, Bentley's patented precision drawing aid, as well as its very powerful element selection tool, the PowerSelector. Combining these two features with a relatively simple set of basic drawing and manipulation tools led you from a few lines you "sketched" to a complete design ready to be refined and used for manufacturing. This "place-and-refine" methodology works well with many design flows, and is one you can easily master while trying out tool combinations.

If you did not follow every step as described in this chapter, do not worry! The lion's share of this book will help you through the learning process, as each of the major features of the MicroStation design environment is explained in detail. Once you have completed a few more chapters, you might want to return to this design and try your hand at it again. In addition, you can further refine it using other tools and techniques introduced later in the book.

WORKING WITH WORKING UNITS

How MicroStation Models the Real World

WHEN YOU SIT DOWN IN FRONT of a computer-aided design (CAD) system for the first time, chances are you feel a little intimidated and even a little nervous. Do not worry; this is natural. Unlike the pencil you have been using all of your life, the computer may seem an awkward tool for drawing pictures. Fear of the unknown may have something to do with this. After all, how many of us really know how a computer "thinks"?

The point is, with a little understanding of how MicroStation works, the mystique of CAD may be alleviated enough to allow you to go about your business designing things. Let's take a look at what is inside MicroStation.

Accuracy of Design Information

First and foremost, MicroStation is a computer program, and as a program it has limitations in how it does things. For one, it cannot resolve ambiguous information into usable drawings; that is, "garbage in, garbage out."

MicroStation's literalness is actually an important feature of the product. When you provide accurate information to MicroStation, it will be remembered in exact detail. If you tell MicroStation to create a line 22.3435 meters long, it will store that line as exactly 22.3435 meters long. The lesson is that accuracy of design information is all important. This is an important aspect—a law, as it were—of CAD.

NOTE: *The First Law of CAD: All spatial data (XYZ coordinates) will be stored to the maximum accuracy possible.*

This is a fundamental difference between CAD and manual drafting. Even though we are taught in drafting classes to be as accurate as possible, the underlying technology (pencil lead) is just not up to the challenge. It is for this reason we always double check our dimensions with a calculator.

If you were to think of manual drafting as people-aided design ("PAD"), you would find that there are only two commands at your disposal, as indicated in the following illustration. You draw lines on the paper to represent a particular idea or concept using the Place Lead command. When you make changes to the drawing, you remove some of the pencil lead(with the Erase Lead command, as it were) and replace it with some new lead (again, the Place Lead command).

On the other hand, CAD provides literally hundreds of commands and functions that assist you in developing your idea into a final product. In addition, unlike manual drafting, where you thought you were drawing a line 22 inches long, but were merely laying down clay and graphite powder, CAD really does create a line 22 inches long.

"Aha!" you say. "But I can't really trust CAD any more than I can my trusty drafting board. How do I know the line is 22 inches long and not 21 and 63/64?" Easy! Ask MicroStation to measure its length and report it (the Measure Distance tool). MicroStation will report that the line is 22 inches long.

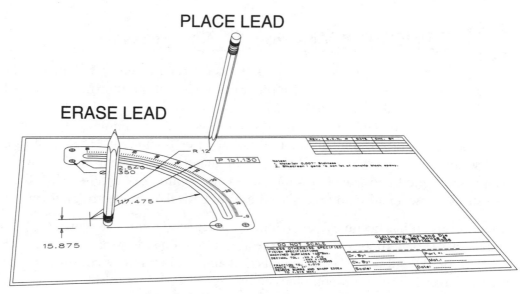

In people-aided design (manual drafting), the only two commands at your disposal are Place Lead and Erase Lead.

Modeling Versus Drawing

Whether it is the layout of a printed circuit board, or the plans for a new power plant, the drafting or designing process is used to convey an engineering concept in a graphically intensive manner. These concepts are usually expressed on paper as a series of diagrams with enough text annotation to eliminate any ambiguity in the design. This annotation normally takes the form of dimensions and various notes and callouts. In most cases, the final "product" of the design process is a rich set of such drawings that are then used in the manufacturing process (yes, even a road could be considered a manufactured product in this context).

The process of creating these drawings is commonly referred to as the "drawing process." In most engineering operations, a CAD system is used as the primary tool in this process. The total focus of the CAD effort is to generate a finished set of plans. In most cases, CAD mimics the methodology used in traditional manual drafting, with a few additional functions. In this case, CAD stands for "computer-assisted drafting." Today, however, many companies have started to use CAD in computer modeling.

TIMEOUT: A View from the New Millennium

Since the advent of CAD in the early 1970s, its focus has been on enhancing the production of paper drawings. With the introduction of the Internet in the early 1990s, a very real alternative to this paper product as an end product has emerged. Just as e-mail has supplanted traditional mail in volume, so too may engineering e-output supplant the traditional paper document. However, this is still some time in the future, as there are still many issues to be ironed out with respect to signed and sealed drawings, and court rulings on such issues as what constitutes "ownership" and "fair/legal use" of CAD data in electronic form.

In any event, it has become a very common practice to require as a project deliverable all CAD data that went into the creation of the final drawing submittals. This has led to CAD submittal requirements that almost always stress coordinate accuracy as a paramount condition of acceptance.

Engineering document publishing to the Web is becoming more commonplace. Several products are available to automatically generate web-compatible output from CAD drawings. An example of such a product is Bentley's Model Server Publisher, shown in the following illustration.

Bentley's Model Server Publisher product.

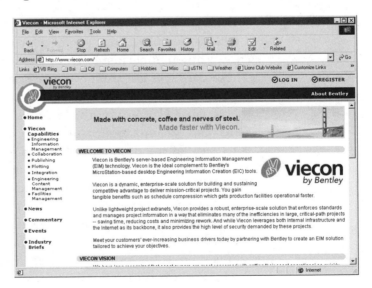

By modeling the underlying objects of the active design project, CAD has much more to offer to the design professional. Now, instead of relying on approximations of the design on paper, the engineer/designer can construct accurate, arithmetically driven models of a design. As a result, the CAD software becomes an active component of the design process.

Most times when you think of modeling, some sort of physical scale model (such as a clay or foam-core model) comes to mind. When you use a CAD system to construct an idea, you are essentially creating an electronic or virtual model.

Unlike manual drafting or physical modeling, where you are working at some sort of scale (for instance, 1 inch = 8 feet, a common North American architectural drawing scale), CAD allows you to engineer at the full size of the product. This is true if you are designing the longest bridge or the smallest electronic device.

This is not to say that scale plays no role in CAD, but that the only time that scale normally comes into play is when defining a paper drawing feature such as plotted text height or hatch pattern spacing. After all, seldom do you see text stenciled on the side of a building from the construction plans!

NOTE: *The Second Law of CAD: Work in real-world units and true size wherever possible.*

Measurement Systems and MicroStation

Of course, if you had to design a road exclusively in inches or create a printed circuit board using kilometers, you would probably consider CAD an expensive nuisance. In reality, any CAD system worth its salt must be able to work with the user in the measurement system he or she is most comfortable with. This leads to another "law."

NOTE: *The Third Law of CAD: The user must be able to express his or her ideas in any measurement system available. It should also be possible to convert from one measurement system to another without loss of geometric accuracy.*

To this end, MicroStation supports a variety of measurement systems. In fact, its method of defining units of measure is fully user-customizable. This means that if you were inclined to design a widget in rods and chains, MicroStation would be able to accommodate you. If, on the other hand, you decided to use feet and miles, you could do that as well. Or, meters and millimeters, parsecs and light-years—you name it.

There is a minor cost to this flexibility. MicroStation provides user-selectable units of measure by fixing the size of the "design plane." Picture a sheet of blank paper. On this paper you can draw anything you like; however, you can never go beyond the edge of the paper. Does this sound a little like Columbus sailing west? It would be if it were not for the sheer size of this paper. As indicated in the following illustrations, if you were to set up MicroStation to work in miles and feet, for instance, the paper size would be large enough to map an entire continent, down to about a tenth of an inch!

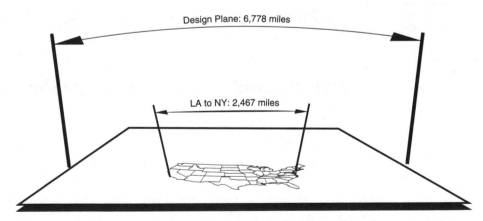

Using miles and feet for measurement units, conceptually you could "map" the United States down to a tenth of an inch.

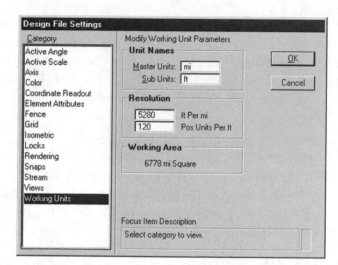

A working units setup that allows you to map the United States to a tenth of an inch.

Getting Started with MicroStation

When you start MicroStation (whether from a command line via Windows' Start menu, or double clicking on a shortcut on the desktop), you are presented with one of two initial dialog boxes. If you are working on a standalone configuration of MicroStation, you will be presented with the MicroStation Manager dialog box. If your copy of MicroStation has been ProjectBank-enabled, you will be presented with the ProjectBank Explorer dialog box. In either case, the displayed dialog box is the "front porch" to the MicroStation design environment. At this point, you need to identify the "document" (design file) to be worked on.

Creating a New Design File

What if you are just getting started on a project or need to create a new design file? With MicroStation, you can create a new design file with either the MicroStation Manager or ProjectBank Explorer.

NOTE: *Although you can configure MicroStation to start up with an untitled document (design file), most companies use the "create the document (design file) first, and then open it" approach. This is the default setting for MicroStation and is the only method used for ProjectBank DGN enabled MicroStation.*

So, how do you create a new design file? If you have the MicroStation Manager dialog box displayed, select the New command from the File menu. This opens the Create Design File dialog box, shown in the following illustration. From this dialog box you identify the name and location of the new design file and, almost as important, the seed file used to configure the initial content of the design file.

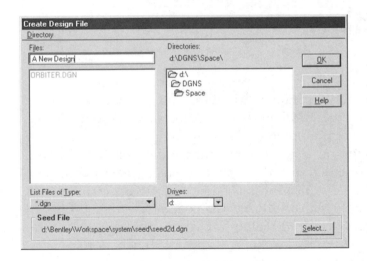

Using regular directory navigation techniques, you specify the location and the type of design file (via the Seed File section of the Create Design File dialog box) you want to create.

Seed File

To complete the creation of your new drawing, MicroStation requires a seed file. This file is nothing more than a regular design file that contains all of the information you need to get started on your design, which in essence is a design "template." In most cases, the seed file contains preconfigured settings such as working units, view settings, and even the "dimension" (2D or 3D) of the new drawing. In addition, a seed file can contain actual graphic elements and even reference file definitions, the latter often used to pre-attach drawing sheet borders to a design.

Seed files are also closely associated with specific engineering disciplines. It is not unusual for a company to have a separate seed file defined for each of a project's engineering phases or departments.

You identify the seed file in the Seed File section of the Create Design File dialog box. The seed file last selected is displayed in this section. For instance, the previous illustration shows *d:\Bentley \Workspace\system\seed\seed2d.dgn* as the seed file for the new design file. This is the default seed file, delivered with MicroStation, which contains the most basic of settings associated with a seed file (more on this later). The *2d* part of *seed2d.dgn* refers to the 2D orientation (XY, no Z values) of this seed file. Choosing *2d* or *3d* is discussed later in this chapter.

To choose a seed file other than the one listed by default, click on the Select button. This opens the Select Seed File dialog box, shown in the following illustration, where you can use standard directory navigation to select whatever seed file you need. By default, it starts in the same directory as your current seed file. The standard seed files delivered with MicroStation can be found in the *\Bentley\Workspace\system\seed* directory. The list of seed files displayed will vary depending on other factors such as specific system configuration data, the workspace selected, or a particular software application you may be running with MicroStation (for instance, MicroStation Triforma).

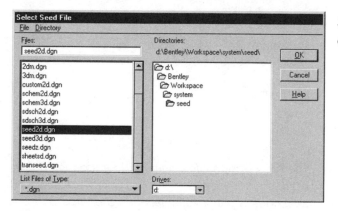

Select Seed File dialog box.

MicroStation is delivered with only the most basic of seed files used to select such rudimentary settings as metric (*2dm.dgn/3dm.dgn*) and generic 2D or 3D. There is also a special seed file for use with translated drawing data, such as is created when translating a DWG file from AutoCAD. Once you have selected the proper seed file and specified the new design file's name, MicroStation will generate the new file with the seed file's settings and data.

Creating a New Design File Under ProjectBank DGN

Because ProjectBank is a project-centric working environment, it provides a slightly different procedure for creating a new design file. With a local briefcase opened against an active ProjectBank project server, you simply click on the briefcase (or a folder contained therein) and right mouse click. One of the options on the right mouse menu is New Design File, shown in the following illustration. When New Design File is selected, you are prompted for the name of the new design and its seed file.

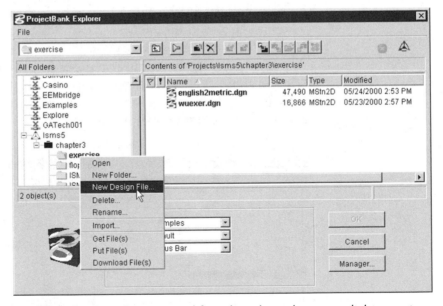

The New Design File command found on the right-mouse-click menu is used to create new designs in an active ProjectBank project. This command is only available when a briefcase or a folder within a briefcase is selected.

You cannot "browse" for your seed file as you can in MicroStation Manager's New Design File dialog box. Because seed files are closely associated with a specific project, only those seed files residing within the current project are presented to the user. Once you have created your new design file, you must still commit the newly created design to the ProjectBank server. You do this using the Commit tool located in the ProjectBank Explorer tool bar.

Opening a Design File

Once you have created your new design file, it is a simple matter of making sure it is selected in MicroStation Manager (or in ProjectBank Explorer) and clicking on the OK button. This starts MicroStation's design environment with the selected file as your current design. The term *active design* is used to identify the design file currently open for read and write by MicroStation. You will see this term used throughout the MicroStation documentation and in several dialog boxes (most notably with respect to reference files described later in this book).

Returning to the "mapping the United States to a tenth of an inch" discussion from earlier in the chapter, the area in which you perform actual design work is referred to as the *design plane*. This "space" in which you work consists of a network of absolute positional units. These units are indivisible; that is, they cannot be split. This is also referred to as an integer coordinate system.

So, how limiting is this absoluteness of design size? Not very! There are over 4.2 billion of these positional units (4,294,967,295, to be exact) in each direction of a 2D drawing, not to mention another 4.2 billion units along the Z-axis of a 3D drawing.

This means you have a total of 4.2 billion^2 points with which to create your design; a truly awesome number. These "dots" are also referred to as *units of resolution*, which is a unit of measure the computer understands. The following illustration shows a design plane containing positional units.

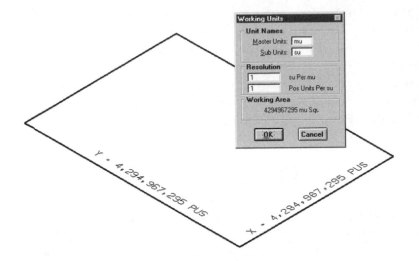

The design plane's 4.2 billion positional units can be pictured as a 2D or 3D matrix with 4.2 billion "points" on each axis (2D plane shown here).

If you have ever attended an introductory data processing class, you probably remember that all computers think in binary, or base 2. This is true of MicroStation. The 4.2 billion is not a random number somebody just thought up. It is the decimal (base 10) equivalent of 2^{32}. This is an especially significant number when you consider that all computers on the market today use 32-bit processors. Thus, each coordinate axis value (X, Y, and Z) takes up one computer storage location, or "word."

NOTE: *The binary nature of MicroStation is also very evident in other places. These include the number of standard line styles, 8 (2^3); levels, 64 (2^6); reference files, 256 (2^8); and many others.*

As a mental exercise, try to spot all of the places where this base 2 nature of MicroStation reveals itself.

MicroStation's Working Units

If all measurement systems were decimal in nature, MicroStation's job would be a lot simpler. Unfortunately, this is not the case. You have feet and inches, miles and feet, and meters and centimeters. Indeed, you even have microns and angstroms. The sections that follow explore MicroStation's working units, which incorporate this level of complexity.

A Look at MicroStation's Units of Measure

To accommodate these widely varying measurement systems, MicroStation supports the concept of "working units." By collecting equal numbers of units of resolution (UORs) into a single unit, assigning that unit a name (say, *inch*), and in turn collecting several of these inches into a foot, you begin to see how the working unit system works.

The inch-to-foot relationship is a good example of how working units accommodate the designer's needs. There are times when a designer will want to work in feet (say, for the length of a wall), and other times when inches are more appropriate (as with the thickness of a wall). In either case, MicroStation keeps track of the relationship of feet to inches.

There are two relationships a user defines to create working units. These are the number of *subunits* in a *master unit* (the inch-to-foot relationship in the previous example), and the number of *positional units* (as termed by MicroStation) in a subunit.

 V8: MicroStation version 8 will replace the 32-bit integer numeric representation with 64-bit floating point numeric values. This results in a 4 million-fold or better degree of improved accuracy in addition to MicroStation's existing phenomenal capacity to store coordinate values!

The Positional Unit

The relationship of feet to inches as an example of the relationship of master units to subunits is relatively easy to follow. However, the positional unit is a little trickier to understand. MicroStation's standard definition of a positional unit is that it is the smallest unit of measure you can use to accurately describe a distance in your design. As mentioned earlier, you cannot split a positional unit into finer units of resolution. A positional unit is the smallest possible unit of resolution, displaying identical qualities with every other positional unit. A subunit is defined as a set number of positional units. The relationship of master units, subunits, and positional units is collectively known as a MicroStation design file's *working units*.

NOTE: *Experienced MicroStation users commonly refer to a design file's working units by its master unit/subunit labels. For instance, an American architectural design file might be described as using feet and inches working units.*

Thinking back to the 4.2-billion-unit limitation of the design plane, there must be somewhere you "add up the leftovers and divide evenly." This is the primary purpose of the SU (subunit)-to-PU (positional units) definition.

As you increase the number of positional units each subunit contains, the overall "size" of the design plane shrinks. This is an unequivocal fact. The relationship among the three parts of the working unit can be expressed mathematically, as follows.

```
Master units x #subunits per master unit x #positional units per subunit
  = 4,294,967,295 (4.2 billion)
```

Restated, the formula makes more sense.

```
4.2 billion + #subunits per master unit + #positional units per subunit
  = total number of master units available
```

This representation is how MicroStation displays its working units. Exercise 3-1, which follows, takes you through the process of setting working units in MicroStation.

V8: The importance of evaluating the number of UORs to master units/subunits will be all but eliminated in MicroStation version 8.

EXERCISE 3-1:   SETTING WORKING UNITS

1   Start MicroStation. From MicroStation Manager, open the design file *wuexer.dgn* located in the *INSIDE MicroStation* exercise folder and click on OK. This opens the selected file in the MicroStation graphic work environment, as shown in the following illustration.

File opened in the MicroStation graphic work environment.

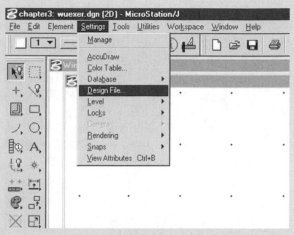

2 Select the Design File option from the Settings menu (Settings > Design File). The Design File Settings dialog box appears.

3 Select Working Units from the Category list in the left pane of the dialog box. Looking at the right side of the dialog box, you should note the three sections Unit Names, Resolution, and Working Area.

The Unit Names and Resolution sections contain data fields in which you adjust the various parameters associated with working units. The Working Area section contains the calculated size of your design plane.

4 Enter under Unit Names something you can relate to (FT for feet and IN for inches), as shown in the following illustration. Under Master Units, enter *FT*. Under Sub Units, enter *IN*. You can use the Tab key or arrow key to move from field to field.

Entering Unit Names information.

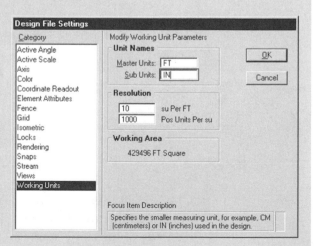

5   You should now be in the Resolution section of the dialog box. The first
    field sets the subunit-to-master unit relationship. In this case, you want to
    select 12 INches Per FooT. Key in *12* and press the Tab key. This sets your
    master unit-to-subunit relationship to 1:12. The total working area changes
    as soon as you press the Tab key. If your Pos Units Per IN is 1000, the Work-
    ing Area now reads *357,913 FT Square.*

6   In the Pos field, shown in the following illustration, enter *8000.* This sets
    the positional unit-to-subunit relationship. Note that the Working Area
    number is reduced to 44,739 feet.

Entering Pos field information.

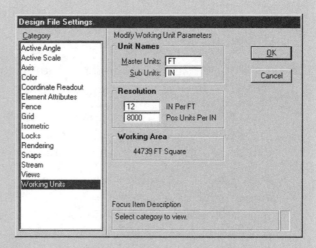

7   Click on the OK button to close the Working Units dialog box. As shown in
    the following illustration, MicroStation warns you that changing the work-
    ing units will change the size of any existing elements. Go ahead and click
    on OK.

MicroStation warning about the
effect of changing working units.

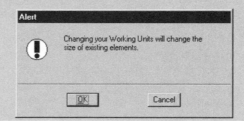

    That is all there is to establishing working units. If you decide to
    change any value of the working units, you can go back and click
    on any of the fields and enter a new value. As a review, the follow-

ing illustration depicts the relationship of the components of MicroStation's working units structure.

NOTE: *Be sure to select Save Settings from the File menu to save your working units after you change them.*

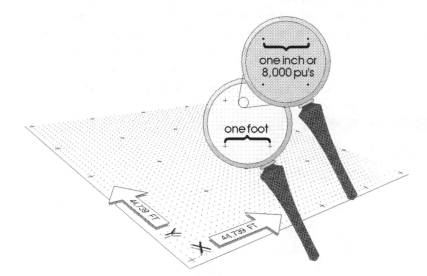

Design plane showing the relationship among master units, subunits, and positional units.

Now that you understand working units, do not worry about them. You should understand this discussion about working units and selecting the right relationship among master units, subunits, and positional units, but you probably will not have to worry about setting them up. In most companies, the working units used for a given project are set up beforehand, as a matter of course in the "seed files" used as the starting point for all new designs.

Enhanced Precision Option

Starting with MicroStation SE (released in 1997), an optional feature called Enhanced Precision began providing an additional 15 bits of accuracy to every coordinate value stored in a design file. With MicroStation/J (released in 1998), Enhanced Precision is active by default, and most users are not even aware that it is there. This feature dramatically increases the accuracy of MicroStation, especially when dealing with non-rational components such as circles and arcs. It also led the way for the introduction of

true 64-bit floating point coordinate support, which came with
ProjectBank DGN in January of 2000.

V8: MicroStation version 8 will also support 64-bit floating point
coordinates.

As indicated in the following illustration, you can verify that
Enhanced Precision is active by viewing a key configuration variable (*MS_ENHANCEDPRECISION*) in the Configuration dialog
box (Workspace menu > Configuration > Operation category >
Enhanced Precision). When set to 1, Enhanced Precision is
enabled (now the default setting starting with MicroStation/J.)

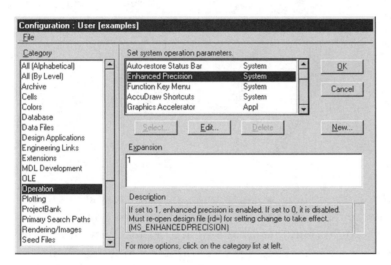

*Verifying that Enhanced
Precision is active.*

WARNING: *It is not a good idea to change the Enhanced Precision
value without first consulting your CAD administrator or project manager.*

Converting between Metric and English Measurements

In these days of international cooperation, one area of special
concern is the reconciliation of the metric and English measurement systems. Due to the uneven conversion factor, it is a good
idea to incorporate the 25.4 mm to 1 inch ratio into the positional
unit-to-subunit definition in your design file. In most cases, main-

taining this ratio will not negatively impact the design process or the final design results. Exercise 3-2, which follows, shows you how to incorporate the English/metric conversion ratio into your design file and how to easily move from English units to metric units and back.

EXERCISE 3-2: CONVERTING BETWEEN ENGLISH AND METRIC MEASURING SYSTEMS

1   Open the design file *English2Metric.dgn.* You are presented with a detail drawing of a meter assembly. Note the strange dimensional values.

2   Open up the Design File Settings dialog box (Settings > Design File). The values for this design have been preset to the metric measurement system, as shown in the following illustration. Let's see what happens when you set the working units to the English measurement system.

Preset metric system values.

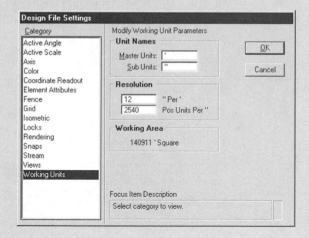

3   Use the following settings for the Master Units and Sub Units values.

   • Enter ′ for the Master Units unit name (the ′ symbol is a common label for the English foot measurement unit).

   • Enter ″ for the Sub Units unit name (the ″ symbol is a common label for the English inch unit).

   • Enter the value *12* for the subunit-to-master unit (Per) relationship.

   • Enter the value *2540* for Pos Units Per ″.

4   Click on OK to close the Design File Settings dialog box. Select Update View (from Window 1 view controls) to update the graphic view. Note the change in the dimensional values (values are in feet and decimal feet), as shown in the following illustration.

Dimensional values reflect conversion to English units.

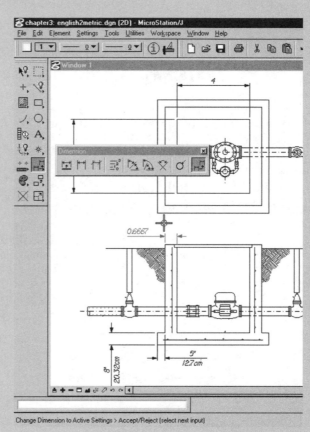

5   (*Optional Step*): Change the dimensional settings to show both the English and metric values, as follows.

- Select the Update Dimension tool (Main toolbox > Dimensions > Change Dimensions to Active Settings).

- Data point on each dimension.

The dimensions now show feet and inches, as well as the equivalent metric values in centimeters. By setting up your working units as demonstrated in the previous exercise, you can easily switch back and forth from metric to English. A neat feature of this approach is seen when you begin dimensioning your design. By selecting the appropriate working units, you can set up a complete set of dimensions in both English and metric units.

The Organization of the Design File

To understand what is going on behind the scenes of MicroStation, it is handy to know how MicroStation stores the information you place in the design file. A convenient analogy is a grocery shopping list, as depicted in the following illustration. As you add elements to the design, you are in essence adding items to the design file "list." As you add elements to this list, more space is consumed on your computer's hard disk or network directory.

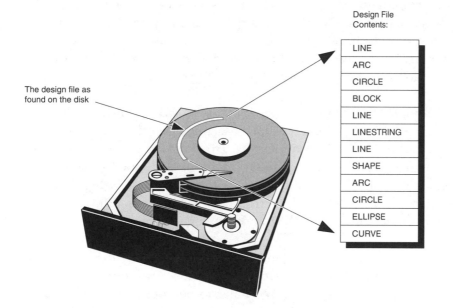

Design File Contents:

| LINE |
| ARC |
| CIRCLE |
| BLOCK |
| LINE |
| LINESTRING |
| LINE |
| SHAPE |
| ARC |
| CIRCLE |
| ELLIPSE |
| CURVE |

The design file as found on the disk

If you lifted the computer file off the disk, it would look like a shopping list with the elements that make up the design listed in the order they were added to the design file.

When MicroStation needs to locate a specific element, say at the location of your cursor, it scans this list for an element matching the cursor's coordinates. As you create your design, MicroStation is constantly working through the list, looking for elements, modifying elements, and yes, even deleting them.

Part of the information MicroStation stores with each element placed is its XYZ coordinates. These coordinate numbers are always stored as a count of positional units (2^32.) In a previous exercise, you will recall there was a message warning you about changing your working units whenever you exited the Working Units dialog box.

This message alerts you to the fact that you may have redefined how the positional units are divided. However, the actual numbers stored with each element stay the same. Thus, an element may appear to grow or shrink after you change the working units, but in fact it remains the same in the design file. The only thing that changes is how MicroStation divides these positional units and displays them.

The following is a little bit of experienced user insight. Some of the commands you use, such as Fence Copy, can dramatically increase the amount of storage space taken up by the design file. On the other hand, some commands, such as Move Element (or Fence Move), only affect the selected elements' coordinate values and do not change the overall size of the design file.

Although MicroStation is classified as a CAD application, it is in fact a highly sophisticated database manager, not that different from dBASE or Microsoft Access. However, its internal database structure, whether talking about the older DGN file format or the new ProjectBank component storage technology, is highly optimized for fast search and retrieval of a drawing's individual elements based on spatial (XYZ coordinates) information.

V8: The analogy of the "shopping list" previously discussed will still be valid for version 8. However, version 8 will employ a much more sophisticated, structured data storage design that will make possible smaller overall file sizes and faster program performance.

As you continue using MicroStation, you will gain a sense of how it organizes the graphic elements within the design file. For instance, when identifying elements within a design, watch how it always selects the elements in the order in which they were inserted into the design file.

2D Versus 3D: A Prime Consideration

One of the characteristics of the current release of MicroStation is the distinction it makes between 2D and 3D designs. This is intentional. Because most design efforts result in a set of paper drawings, it is often easier to execute the design in a 2D environment. Doing so eliminates many subtle user errors that can creep into a 2D design executed in a 3D environment (Z-skewed elements are the most prevalent). For this, and historical reasons, you must decide at the outset of a new design file whether it should be a 2D or 3D design, and select the appropriate seed file (*seed2D.dgn* and *seed3D.dgn* are delivered with MicroStation).

Keep in mind that MicroStation has incredible 3D capabilities, including solid modeling, rendering, and even animation, but as we are just starting out on our journey of MicroStation exploration we will stick to 2D for a while. Later, 3D will be introduced, and everything you have learned in the 2D environment will still apply in this 3D world.

Summary

In this chapter you learned that MicroStation can accommodate just about any coordinate measurement system used in the world today. It does this by varying the relationship of the design plane's master unit to subunit to positional unit values. In addition, you learned that it is important to "draw" your designs in real-world measurement values rather than produce a scaled drawing. This allows you to utilize the intrinsic accuracy of MicroStation to represent designs and concurrently arrive at the required final output. In the chapters that follow, you will explore MicroStation's design tools in detail and learn how to apply them to the design process.

2D Basics: Part 1

Learning the "Tools of the Trade"

YOU HAVE DONE THE EXERCISES. You have made your way through the "concepts," and now you are ready for the real thing. Welcome to the tools chapter. Here you are introduced to MicroStation's basic drawing commands. This chapter shows you how to create a drawing. It shows how to place elements graphically using the mouse, and with the keyboard using precision key-ins. All along the way you will see how MicroStation solves real problems.

Design File Elements

MicroStation is a full-function, full-featured CAD system. You can use it to tackle just about any design problem imaginable. MicroStation has been used to design everything from naval vessels to chemical plants. It is also used to generate extremely accurate maps, using just about every data source imaginable.

When you analyze most designs created in CAD, you find that they consist of no more than about ten different graphic elements. Most of these are already familiar to you: line, circle, text, ellipse, arc, polygon, and curve. When you become familiar with how to create and manipulate this handful of relatively "primitive" element types, shown in the following illustration, you will be a long way toward mastering MicroStation. Let's begin that familiarization.

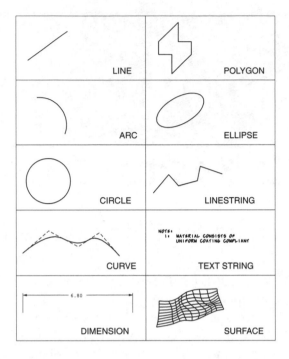

LINE	POLYGON
ARC	ELLIPSE
CIRCLE	LINESTRING
CURVE	TEXT STRING
DIMENSION	SURFACE

The basic drawing "elements" used in MicroStation.

Drawing Lines

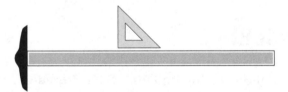

"A (straight) line is the shortest distance between two points."

This definition of a line is especially appropriate in CAD. You, the user, supply the two endpoints, and MicroStation provides the line. This is accomplished using the Place Line tool.

As with many of its tools, MicroStation provides several methods for accessing the Place Line tool. You can select it from the Tools menu (Tools > Main > Linear Elements > Place Line tool icon), from the Main toolbox itself (Main > Linear Elements pop-up toolbox > Place Line tool icon), or (heaven forbid!) you can type in the command *place line* via the Key-in window. The most common method for selecting this most basic of tools is from the Main toolbox normally docked on the left edge of the MicroStation application window.

Opening the Main Toolbox

Do not see the Main toolbox? No problem; here is how. From MicroStation's menu, navigate to Tools > Main > Main (yes, that is two Mains), but do not click on the item yet. If you see a check mark beside Main, the Main "toolbox" is already open, and will be visible somewhere within MicroStation's application window.

Should you select Main while it is open, it will close. Do not worry, this is simply a toggle action. Just select the command again (Tools menu > Main > Main) and you will be back in business. While accessing the Main toolbox, shown in the following illustrations, from MicroStation's Tools menu, you may have noticed there are additional toolboxes listed under the Main submenu selection. These are the toolboxes contained within the Main toolbox itself—sort of like tool trays within a bigger toolbox. (You will learn more about these shortly.)

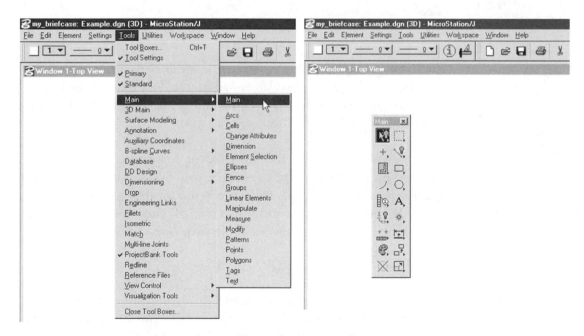

Selecting Main from the Tools menu will open the Main "toolbox." If there is a check mark (note Primary, Standard, and ProjectBank Tools), it means the Main toolbox is already open on your MicroStation desktop. The Main toolbox is displayed in the right-hand image.

With the Main toolbox open, you can directly access the content of its associated toolboxes by clicking and holding down the mouse button on one of the tool icons. When the pop-up toolbox appears, just drag your mouse over to the tool you want to activate. If you drag the mouse further, the popped-up toolbox tears off, which opens the selected toolbox in its own window, as indicated in the following illustration.

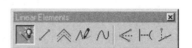

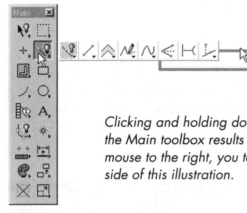

Clicking and holding down the left mouse button over a tool in the Main toolbox results in a pop-up toolbox. By dragging the mouse to the right, you tear off this toolbox, shown on the right side of this illustration.

Each of these pop-up toolboxes contains related tools. For instance, all of the line drawing tools are located in the Linear Elements toolbox, circle placement tools are in the Ellipses toolbox, and so forth.

Although you can slide the cursor over a specific tool from the popped-up toolbox, there are times you may want to work with several related commands from the same toolbox. In such a case, you will find it more convenient to tear off the toolbox so that all of its tools are immediately available with one mouse click.

Tool Settings Window

As mentioned in previous chapters, you will want to keep an eye on the Tool Settings window at all times to be aware of the options available to you for the selected command. However, the name Tool Settings does not actually appear anywhere on its window. Instead, the Tool Settings window takes on the name of the currently active tool, as seen in the following illustration, which shows two different tools selected.

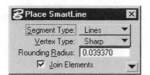

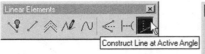

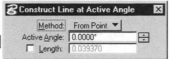

The "tool settings" window dynamically changes as you select different tools in MicroStation. Note how the Tool Settings window title box matches that of the tool tip text seen when you move the cursor over a tool.

If you close the Tool Settings window, MicroStation will automatically open it the next time you select another tool or reselect the same tool.

Place Line Continued

Place Line tool.

Now that you have your Main toolbox active, all that remains is to activate the Place Line tool, shown above. This is done by clicking and holding the cursor on the second icon in the second row of the Main tool frame to open the Linear Elements toolbox. Place Line is the second icon on the toolbox. When you select this icon and release the cursor, the command name with available options appears in the Tool Settings window. It also prompts you for input via the status bar at the bottom of the MicroStation window.

As the most basic of drawing tools, Place Line prompts you to draw individual lines by specifying their endpoints using the cursor and left mouse button. The first mouse click, better known in MicroStation parlance as a data point, specifies one end of the line; the second data point specifies the other end of the line. Between the two data points, MicroStation displays a dynamic or rubber-band line on the screen so that you can visually relate to how the line will appear when you do data point the second time. This method of drawing a line is shown in the following illustration.

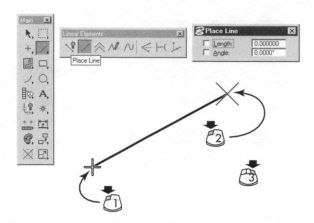

With no options selected in the Tool Settings window, the Place Line command requires only two data points to draw a line.

So much for placing your basic line. Let's look at the options associated with the Place Line tool.

Place Line Tool Settings

Placing a line by clicking data points at various locations in your design is a place to start; however, in the real world, you will want more control over how MicroStation creates your lines. That is the purpose of the tool options. By selecting combinations of these options, you enhance the capabilities of the selected tool. The sections that follow explore these options.

Length Option

This is a very simple but capable option. When you click on the Length option and enter a value in the adjacent text field, you are telling MicroStation that all lines placed from this point forward will be of the length entered, no matter how close or far apart the two data points are. For instance, selecting this option and entering 2 units (meters, inches, or whatever your current working units are) in the text field would result in a line exactly 2 units long, starting at the first data point and pointing toward the second data point.

Angle Option

When you select the Angle option and enter the direction in degrees in the adjacent data field, you are telling MicroStation you want to restrict the line to a specific direction. Think of it as striking a line with your triangle or protractor. The length of the resulting line is not affected by this option. That is the job of the Length option.

While you are still reviewing the Place Line command, you may want to try out the Length and Angle options. Try setting various values and combinations for these two options.

While on the subject of lines, you may have noticed other tool icons on the Linear Elements toolbox. These icons represent related commands that help you activate variations of the basic Place Line tool, or related linear element creation tools. These are covered later in the book. For now, just think of them as line-construction and special-duty line commands.

Try-It! Using Place Line's Length and Angle Options

1   Use Place Line's Length and Angle options to sketch the geo-metric design shown in the following illustration.

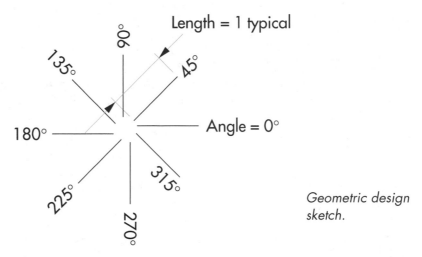

Geometric design sketch.

2   Select the Length option and set it to 1, and place a single data point. What happens? Next, select the Angle option and set it to 0. What happens? Data point. Set the Angle to 45

degrees and place the second line. Set the Angle to each value shown and place lines to complete this operation.

TIMEOUT: Teaching an Old Dog New Tricks

When you work with experienced users of MicroStation, chances are good you will hear them refer to tools as commands ("Use the Place Line command to draw that…"). With the earliest releases of MicroStation, drawing tools were called "commands," which referred to the command-line-only interface used to type in command instructions.

When MicroStation 4.0 introduced the concept of tools and toolboxes (originally called *palettes*), the line between what is a command and what is a tool got somewhat blurred. Technically, commands are actions you select from the pull-down menus, and tools are actions executed from a toolbox. However, the differences between these two terms are so minor that you may find yourself referring to tools as commands. Do not worry. Where it is important to make the distinction, this book refers to tools as tools and commands as commands.

Other Toolbox Tricks

By now you have noticed that MicroStation uses a lot of windows. Among these are toolboxes, view windows, the Tool Settings window, and other windows you will learn about later. To help you manage all of these windows, MicroStation provides special features to some of these windows. Chief among these is docking. When you drag an open toolbox to the edge of the MicroStation application window, it "docks" with that edge. The toolbox's title bar is replaced and the amount of room taken up by the toolbox is greatly reduced, as shown in the following illustration.

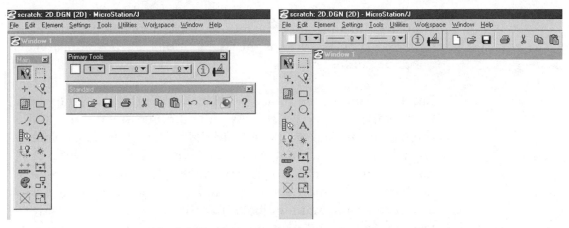

In the left-hand image you can see how the various windows take up a lot of space compared to when they are docked (right-hand image).

Toolboxes also have one other handy on-screen ability: resizability. If you move the cursor over the window border surrounding a toolbox (non-docked), it will change to a resize arrow. By pulling on the border of a toolbox you can change its shape for a more optimized layout, an example of which is shown in the following illustration.

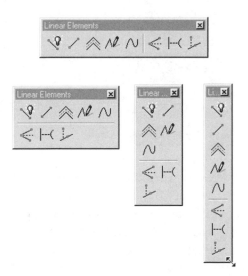

The four windows shown are the same Linear Elements toolbox resized to different window configurations. Note the resize cursor is still visible on the rightmost window.

There is one notable exception to the resize rule for toolboxes. Tool frames are a special type of toolbox that cannot be resized but have the extra ability of holding additional toolboxes (sound familiar?). In this chapter, you may have noticed that the Main toolbox was referred to as the Main "toolbox" (in quotes). This was done because Main is not really a toolbox but a tool frame.

Because Main is used to access toolboxes via the pop-up toolbox feature, it cannot be resized. As with all other tool frames, it also can only be docked on the left and right edge of the MicroStation application window (tool frames use a vertical orientation to optimize the pop-up menu feature). Although a minor differentiation, it is a common source of new user confusion ("Hey, why can't I resize the Main toolbox?") Fortunately, there are relatively few tool frames and many toolboxes. The identifying features of tool frames are shown in the following illustration.

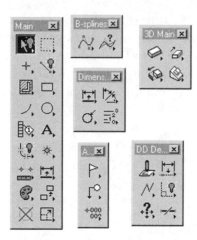

Tool frames are easily identifiable by their vertical orientation and the small pop-out arrow located in the lower right corner of each icon.

Introducing AccuDraw

Typing in values for various coordinates can be very monotonous (witness that last Try-It). However, precisely entering values for angles and coordinates is a very important part of all design work. You probably spend more than 80% of your time entering such information. As a result, you definitely want MicroStation to make it as easy to do as possible.

Enter AccuDraw. AccuDraw is Bentley's revolutionary (and patented) data input technology that takes the otherwise tedious entry of spatial (numeric) information and makes it an almost pleasant experience. Instead of entering "special" keyword values to command line prompts (e.g., "Enter the XY value..."), AccuDraw interprets your keyboard entry, your mouse movement, and what the currently active tool expects from you to generate the appropriate coordinate information. At times, it seems to anticipate your input!

Starting AccuDraw

By default, AccuDraw is not active. To activate AccuDraw, simply click on its icon located on the Primary tool bar, shown in the following illustration.

V8: In version 8, AccuDraw will be enabled by default. The Tee Square-and-Triangle icon shown in the following illustration is replaced by a compass icon in version 8.

Clicking on the Tee Square-and-Triangle icon activates MicroStation's AccuDraw feature.

Your only indication that AccuDraw is active is the sudden appearance of the AccuDraw window. Depending on whether you are working in a 2D or 3D file, this window will contain X, Y, and Z fields (obviously, a 2D file would not have a Z field). The identification of a 3D file is shown in the following illustration.

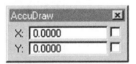

The appearance of the Z field (below) identifies your active design file as 3D.

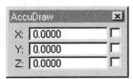

The real magic of AccuDraw does not happen until you start drawing things. For instance, with the Place Line tool (with the Length and Angle options deselected) active, start AccuDraw and place a data point anywhere in your active design. When you do, AccuDraw's other visual component appears, the compass, shown in the following illustration.

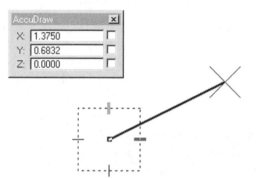

With AccuDraw active, the compass appears whenever you place a data point. The X and Y values in the AccuDraw dynamically change, reflecting the delta distance from the first data point to the current cursor location.

At first blush, the compass does not appear very exciting, but just wait. As you will see, it is key to the productive use of practically every drawing tool within MicroStation! With the first data point placed, AccuDraw's coordinate window begins to track the delta coordinate distance from that data point to the current cursor location. As you move the cursor, the X and Y values constantly change. If you try this and watch carefully, you note one other interesting response. As you move the cursor further along the X axis than the Y axis, the X field in the AccuDraw coordinate window highlights. You have to look closely because this highlight is a simple black frame around the X coordinate field.

More importantly, you will also see that a text insertion "I-beam" appears in the highlighted coordinate field. If you type in a number while this field is selected (the field has "input focus"), the axis locks to the value entered. This is immediately apparent by the way the current command (as with the Place Line tool) reacts to the input; the freely moving cursor "locks" at the value just entered along the AccuDraw compass axis, as indicated in the following illustrations.

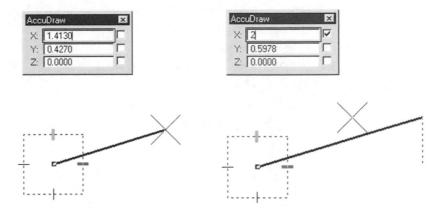

As soon as you begin entering a value on your keyboard, AccuDraw locks the selected axis.

You can still move the cursor along the unlocked axis, but the dynamic line maintains the locked value, indicated by a dashed line extending from the AccuDraw compass axis to the dynamic end of the line. What is even more amazing is that you can continue to type in numbers and the locked axis responds to your typing.

In other words, when you type in *2* in the X field, the line locks to 2 units. However, if you continue to type in a number, say 2.6789, the line changes in length as you type in each digit. This includes backspaces! The only caution is to not move the cursor while you are typing in your coordinate. When you do, AccuDraw assumes you are done typing in the locked value and locks the axis.

By moving the mouse up or down, the input focus automatically switches to the Y field, where you can simply type in the value for the distance along this axis. This is considered a form of gesture recognition, which helps minimize extra keystrokes, such as typing in the character X followed by a number, followed by Y, and then a number, as found in other CAD products. After a very short time you will find this gesture-type-gesture technique very easy and downright friendly.

AccuDraw's Context Sensitivity

When you place the second data point for your new line, the AccuDraw compass exhibits yet another interesting capability.

Instead of staying at the beginning point of the line, the compass jumps to the endpoint of the newly created line. It also rotates to follow the axis of the line just placed, as shown in the following illustration. This is called *context sensitivity*, another major feature of AccuDraw.

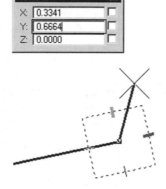

The AccuDraw compass automatically rotates to match the axis of the previously placed line segment. Note that the distances in the X and Y fields are measured from the AccuDraw compass axis, not the original line's starting point.

Context sensitivity simply refers to AccuDraw's ability to communicate with the currently active tool and respond to its needs. For the Place Line tool (and most other linear placement tools), it requests that AccuDraw rotate to the plane of the new line and to move to the last endpoint. Other tools affect AccuDraw differently depending on their unique situation (within their context, as it were).

AccuDraw's Shortcuts

Along the lines of context sensitivity, AccuDraw also responds to user requests. You direct AccuDraw by typing in single character commands or hot keys. With input focus on AccuDraw, typing any key other than a numeral or period is interpreted by AccuDraw as a shortcut—a one- or two-character input that, depending on what key you press, results in an instant action by AccuDraw. For instance, want to know what all the hot keys are? Type in a question mark (?). AccuDraw responds by opening the AccuDraw Shortcuts window, shown in the following illustration.

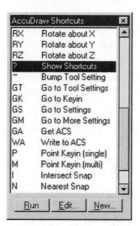

Typing a question mark in AccuDraw brings up AccuDraw's Shortcuts window. You can edit, add, or delete shortcuts to MicroStation via this window.

You will come to rely on these shortcuts to make AccuDraw and MicroStation do your bidding. For 2D work, they are conveniences; for 3D work, they are absolute necessities.

NOTE: *AccuDraw represents such a dramatic productivity enhancement that long-time MicroStation users often refer to their experiences as "before AccuDraw" and "since AccuDraw." Before AccuDraw, you had to know two-letter key-ins (DX=xval,yval, DI=dist,ang, and so on) entered via a command line interface. This resulted in a lot of extra typing, not to mention errors ("I meant up not left!"). For 3D, most "old-timers" do not even want to remember the days before AccuDraw.*

One of the first shortcut keys you learn is the one for switching between rectangular coordinate entry (XYZ) and polar coordinates (distance and angle). Pressing the space bar toggles between these two coordinate systems, as indicated in the following illustration. On-screen polar input is noted by the change of the AccuDraw compass to a circle. In addition, the coordinate input window changes to show the input fields Distance and Angle.

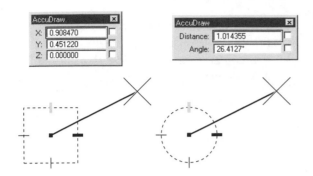

Pressing the space bar toggles between the rectangular coordinate system of input and the polar coordinate system.

As with rectangular input, typing numbers in a highlighted field locks that field. Before you go any further with AccuDraw, you need to learn about a few more drawing tools. As these tools are explored in the sections that follow, salient AccuDraw features are discussed.

SmartLine Tool

Place SmartLine tool.

Although most users think of lines as the primary drawing tool, most MicroStation users quickly abandon the Place Line tool in favor of the more powerful Place SmartLine tool, shown at left. The reason for this is simple: the SmartLine tool is much more productive. Instead of being limited to just single line segments, which are tedious to edit, Place Smart Line optimizes the element you create by monitoring your data points, as indicated in the following illustration.

For instance, when you place three data points (A, B, C) in your design using Place Line, you create two separate lines: one that goes from point A to point B, and another that goes from point B to point C. With SmartLine, a single line string is generated, which contains two endpoints and an intermediate vertex. Basically, a vertex is a "kink" in the line. Instead of calling it a line, MicroStation now refers to this entity as a line string (think of a string of vertices or line segments).

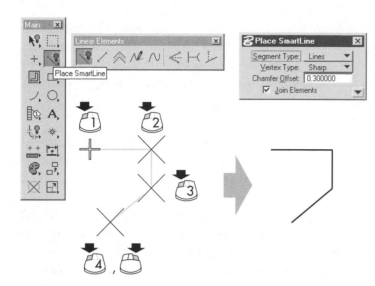

The Place SmartLine tool creates lines, line strings, and other element types based on its tool settings and how many data points are entered.

When creating a series of linked coordinates like this, it is almost always more desirable to generate a single line string rather than several discrete lines, as shown in the following illustration. When you begin editing or modify such elements, you come to really appreciate the difference between these two element types. A line string can have up to 101 vertices, including the start and endpoints.

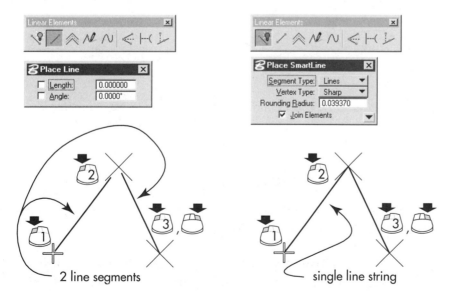

The Place Line tool always creates single line segments (left), whereas Place SmartLine optimizes its output based on all data points (right).

NOTE: *The SmartLine and its cousins the curve and multi-line are the only elements that require the use of the Reset mouse button as part of the element's creation process. Because MicroStation uses Reset to back out of most commands, this can be very disconcerting to the first-time CAD user.*

So, what do you do if you find yourself creating a SmartLine you do not want? Easy, you just undo it! Either press Ctrl+Z on the keyboard or select Edit > Undo from the menu bar.

SmartLine also supports special vertex treatments. By default, Place SmartLine generates sharp vertices, which means that the line segments literally come to a point. There are times, however, when you may want to transition a vertex to a tangent arc (known in drafting parlance as a fillet) or an intermediate angular line (a chamfer), as shown in the following illustration.

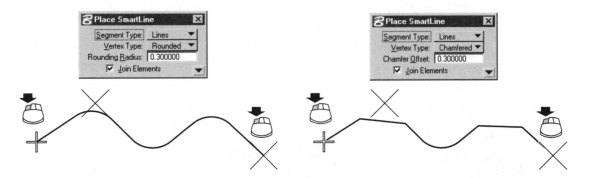

Setting the Vertex type to Round results in filleted corners, Chamfer in chamfered corners. The dimension of the fillet or chamfer is set in the appropriate Rounding Radius or Chamfer Offset fields.

SmartLine's Other Tool Settings

There are several other tool settings associated with Place Smart-Line. Descriptions of these follow.

- Segment Type: You have a choice between drawing line segments or arc segments. Placement of arc segments is similar to the Place Arc by Center tool, discussed later in this chapter. You can switch between line and arc segments at any time during the construction of a SmartLine while still keeping it all connected; just do not click on the Reset button.

- Join Elements: This checkbox is turned ON as a default. If you turn it off, the SmartLine segments will not be joined and MicroStation will treat them as individual elements.

- While constructing a SmartLine, if you snap to its beginning vertex, additional options appear in the Tool Settings window. These are shown in the following illustration.

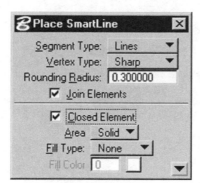

The Closed Element checkbox lets you decide whether or not to construct the SmartLine as a closed shape.

Drawing Circles

One of the other major drawing instruments you mastered after the straightedge was the compass. This handy device aided you in drawing precise circles and arcs by setting a radius and a center-point. MicroStation also has its own compass. Better known as the Place Circle tool, this tool draws circles from sets of data points and typed-in values, as follows.

- By clicking a data point to define the circle's center, and another on its circumference

- By predefining its radius and clicking a data point for its center

- By identifying three data points on the circle's circumference

- By predefining the circle radius and setting two data points on its circumference

- By setting two data points representing its diameter

The order of these circle methods is not an accident. When you select the Place Circle tool, this is the order in which the various options appear on the Tool Settings window. The Place Circle tool is found in the Ellipses toolbox, shown in the following illustration.

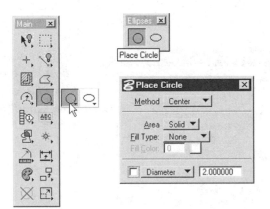

The Ellipses toolbox provides access to MicroStation's "other" compass.

Beneath MicroStation's friendly graphics interface there lies a powerful mathematical engine. To give you the accuracy and graphics performance necessary for CAD operations, all of the various elements are constructed using basic geometric formulas such as those you learned (and probably forgot) in high school geometry class. Just as there is a minimum number of variables that must be supplied to solve a geometric problem, so must MicroStation be supplied with key pieces of information before it can perform its task.

When MicroStation asks for certain information, you can be certain that it is important for solving the computation. The Place Circle command is an example. When the result of a given command does not appear correct to you, it is a good bet the information you supplied was either incorrect or in the wrong order.

MicroStation provides various circle placement options to solve the various drawing problems that incorporate circles. The following sections look at some of the ways you can place circles.

Place Circle by Center

Place Circle by Center is the most basic of the circle commands. It is familiar because it most closely mimics the compass instrument. To use this option, set Place Circle's Method tools settings option to Center. By setting a center point and selecting a corresponding point on the proposed circle's circumference, you generate a circle, as indicated in the following illustration.

TIP: *The Radius data field in the Tool Settings window dynamically indicates the radius after the first data point for the circle's center is placed. If AccuDraw is active, it will automatically switch to polar coordinates after you have entered the center point for the circle. This is another example of AccuDraw's context sensitivity.*

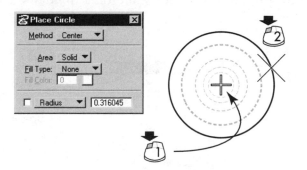

Place Circle by Center tool in action.

Key-in Radius or Diameter

In addition to freehand selection of your circle's proposed radius, you can also preset the radius value. This is accomplished by selecting the Radius checkbox in Tool Settings, shown in the following illustration. You supply a value for the radius in the Radius key-in field, and a single data point for the center point, and MicroStation draws the circle.

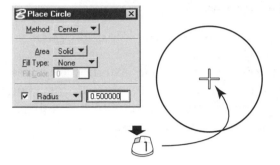

The Radius option associated with the Place Circle by Center command controls the size of the resulting circle.

You can also specify the size of your circle by diameter. The Radius option is actually a pull-down option menu. Clicking on it allows you to select either Radius or Diameter, as indicated in the following illustration. The data field associated with this option will be interpreted as a radius value or a diameter value depending on this option menu.

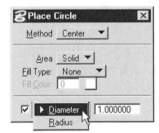 

You can select to enter a circle's size either by radius or diameter. Note how toggling this option doubles or halves the value in the adjacent data field (1.000000 to .500000).

Place Circle by Edge

One of the more subtle circle drawing options is Place Circle by Edge, shown in the following illustration. This tool places a circle through three given data points. The trick lies in how these three points are entered. Remembering that MicroStation is a mathematical "engine," and that there is only one mathematical solution to a circle intersecting three points, the result will always be the same. This is true no matter which order you enter the data points.

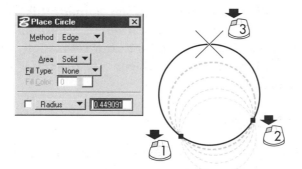

Place Circle by Edge. Note how the dynamically updated circle follows your cursor until you place the third data point. Note also how the radius value shown in the tool settings window is continually updated as you move the cursor around.

First-time users of CAD are sometimes confused by the behavior of the Place Circle by Edge tool. As you move the cursor around after the second data point, the dynamic circle frequently goes off the screen, especially if all three points are close together or along a common plane. Exercise 4-1, which follows, will provide you with practice in becoming comfortable with the use of the Place Circle by Edge tool.

EXERCISE 4-1: PLACING A CIRCLE BY EDGE

1   Open the exercise design file *Circle.dgn.*

2   Select the Place Circle by Center tool from the Ellipses toolbox. In the Tools Settings window, set Method to Edge. Set the Radius/Diameter option to off (checkbox not selected).

3   Place a data point anywhere in View 1. A small black-filled box appears on the screen at the location you just selected as a data point.

4   Place a second data point. A second black-filled box appears and you will see the dynamic (read: "rubber-band") circle attached to your cursor. As you move it about the screen, this circle will follow your cursor while still passing through your first two data points.

5   To finish the placement of your new circle, place a third data point.

6   Try placing a few more random circles using the Place Circle by Edge tool.

The use of the markers for your data points has "tamed" this command considerably over versions 4 and earlier of MicroStation. Because you can tell what points the circle is drawn through, you can predict how your final circle will appear. Try placing a few more circles in this manner.

Place Circle by Diameter

Another fine method for creating circles is "by diameter." Not to be confused with the Radius/Diameter checkbox option, this method lets you select two locations on the proposed circle's circumference and MicroStation will create a circle that passes through those data points. MicroStation assumes that the two points given are diametrically opposed (on opposite sides of the circle). The result is a circle "fitted" between the given points, as indicated in the following illustration.

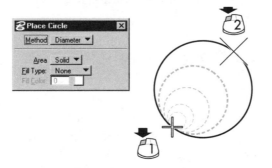

The Place Circle by Diameter tool lets you draw a circle by clicking two points that define its diameter. Note the lack of the Radius option for this tool.

Mixing and Matching Tool Settings

Earlier, you learned that you can set a method for placing a circle. You also learned that you can preset the radius or diameter of a circle and place it by center. However, did you know that you can both set the radius/diameter of a circle *and* place it by edge?

As you set the various options located in the Tool Settings window, MicroStation immediately interprets your changes and modifies the current tool actions to accommodate the tool settings you just changed. The most obvious of this immediate cause and effect action is the current tool's prompts displayed in MicroStation's status bar (bottom of the MicroStation application window). For instance, when you set the Method field from Center to Edge you will note that the prompt displayed in MicroStation's status bar changes from "Place Circle by Center > Identify Center Point" to "Place Circle by Edge > Identify Point on Circle." This is shown in the following illustration.

The prompt in the status bar changes in response to changes made on the Tool Settings window.

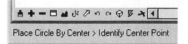

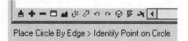

If you place the center point of a circle but have not completed the second point, and you switch methods from Center to Edge, MicroStation will automatically reset the tool operation to ask you for the first data point again. This makes sense because the types of input each of these Place Circle methods requires might be mutually exclusive. First-time MicroStation users are sometimes confused as to "Why it didn't remember my last data point" when switching key tool settings values.

The Art of Deleting Elements

If you performed the Place Circle by Edge exercise earlier, chances are good you are now staring at a display monitor full of circles. If you are, you are ready to perform your first element *manipulation* (as opposed to creation) operation with the Delete Element tool, shown in the following illustration at left. There is no question that Delete Element is *the* most popular tool used in the design process, and is probably one of the reasons it is always available directly from the Main toolbox (not found in a sub-toolbox). It is the only tool that holds this distinction.

To use the Delete Element tool, simply select it (from the Main toolbox) and data point on the "offending" element you want to delete. Once highlighted, the targeted element's demise is only a second data point away, as indicated in the following illustration.

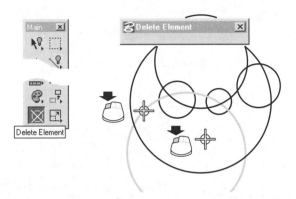

The first data point identifies the element to be deleted; the second data point confirms the deletion. You can also use the second data point to select an additional element for the next deletion operation (a third data point confirms).

Note how the Delete Element adds a small circle around the center of the normal MicroStation cursor's crosshairs. This tells you that the Delete Element tool expects you to identify an element. All of the element manipulation tools provide this same on-screen cue.

Of course, the success of Delete Element relies on identifying the correct element to begin with. Quite often, elements occupy common locations within a design. To make sure a potentially destructive tool such as Delete Element is going to affect the right element, you should always try to identify the target element at a point where no other elements occupy the same point.

However, if this is not possible (a not uncommon occurrence) or if the wrong element is highlighted with the first data point, clicking the Reset mouse button will release the currently selected element, and the Delete Element tool will attempt to locate the next element that occupies the same location. You can cycle through all of the elements that cohabit a given location by repeatedly pressing a Reset. Once the proper element is highlighted, entering a data point will delete the element.

NOTE: *Use of the Reset (also referred to as Reject or Reset/Reject) mouse button is a fundamental tenet of MicroStation. Instead of a separate "Are you Sure?" dialog, this second "Data point to accept or Reset to reject, Select next input" approach minimizes the number of user inputs required to perform a given operation. This leads to more productive design by not constantly requiring the user to confirm their actions or reselect the current tool.*

 ## TIMEOUT: Conservation of Data Points

This "accept/reject" (select next element) process is common throughout MicroStation and is meant to minimize the number of data points needed to perform your task. Although very efficient in the hands of an experienced user, this method will cause you some consternation (and yes, inadvertent loss of elements) until you get used to it.

To minimize unwanted deletions, always place a data point away from all surrounding elements (in a relatively empty part of the design) for the accept point. As you become more experienced with MicroStation, you will begin to conserve these acceptance data points and use them for the next selection.

TIP: *When you have finished with the Delete Element tool, it is a good idea to select a less, shall we say, deadly tool to "park" on while considering your next design move. This will keep you from making accidental deletions as a result of a misplaced data point. A "safe" tool to select is the Element Selection (or Power-Selector) tool, shown at right.*

Try-It!: Using the Delete Element Tool

Use the Delete Element tool to delete the circles you created in the last exercise.

Undoing and Redoing Your Work

Now that you have learned how to delete your work, it is a good idea to introduce the Undo and Redo commands. Nearly every current computer application in use today incorporates an Undo command. This command reverses one or more of the previous actions performed within the current document or file. Not surprising, MicroStation provides a very powerful undo facility.

As expected, Undo is located under MicroStation's Edit menu. A nice feature of MicroStation's undo command is the way it tells you what it is about to undo. This is done by appending the name of the previous action onto the menu label itself, as shown in the following illustration.

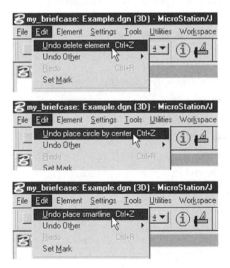

The Undo selection under the Edit menu always includes the name of the previous activity by tool, command name, or function.

You can also select Undo from the Standard tool bar (usually docked just under the MicroStation menu) or enter Ctrl-Z from the keyboard. Each time you select Undo, one previous operation is reversed. Another selection of Undo reverses the previous operation (it does not "undo" the Undo).

Although it may seem like MicroStation has an unlimited undo, there is a limit to how many previous operations you can reverse. The total number of steps you can undo depends greatly on the actions and quantity of elements affected by each operation. For instance, if you were to delete the entire content of a 20-MB design file, chances are good you would not be able to undo this operation. On the other hand, if you are performing most typical design operations, MicroStation is able to retain an undo buffer that probably will extend back to the moment you opened the file for edit. The undo buffer is cleared when you exit MicroStation (it is not persistent).

NOTE: *ProjectBank provides a true unlimited undo capability by providing you with the ability to restore the previous state of one, few, or several elements at once. This is accomplished through the Revision Explorer available from the ProjectBank Tools toolbox.*

"For every Undo there is the opposite and equal Redo." This means that in case you need to undo your Undo, MicroStation includes the equally capable Redo command. Located just a command down from Undo, Redo reinstates the previous operation undone by the Undo command. It is no problem if you did not quite follow that, as the Redo command is very intuitive in operation and is only available if you have just performed an Undo.

The Ellipse Element

When the Place Circle tool was introduced, you may have noticed that it resides in a Ellipses toolbox. The only other tool within this toolbox is Place Ellipse. The major difference between the circle and the ellipse is in the latter's treatment of the radial dimension. Unlike the circle, which has only one radial component, the ellipse has two. The ellipse consists of a single center point, with a primary radius length and a secondary radius length. In addition, the ellipse has an axis along which these two radii are situated (the two radii are always perpendicular to each other).

Sound complicated? If you think of an ellipse as a "squashed" circle, where the long side of the ellipse is the axis, you can picture how MicroStation treats the action of the Ellipse commands.

Place Ellipse by Center and Edge

Similar to the Place Circle by Center tool in its default condition (no options selected), this tool, shown in the following illustration, creates an ellipse from three data points. The first data point defines the center point, the second locates a point on the ellipse itself (and defines the axis tilt of the ellipse), and the third locates another point on the ellipse.

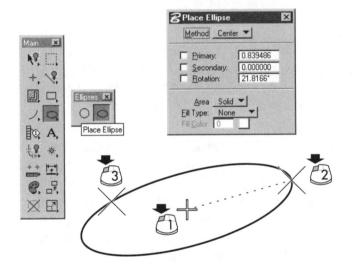

The Place Ellipse by Center and Edge tool. Note the additional option fields that provide the ability to enter the ellipse parameters (primary axis size, secondary axis size, and angular rotation or "tilt" of the resulting ellipse).

As with the circle, there is only one solution for creating an ellipse from three points. As a result, you do not have to know precisely where the "axis of rotation" is prior to creating an ellipse. However, there are times when you will want to specify this and other key data about your ellipse. This is done with option fields in the Tool Settings window.

With the Place Ellipse by Center and Edge command, you can control the two radii values and the rotation of the primary axis with respect to the X axis. If you specify all three values, you only need identify the location of the center to draw the ellipse. If you select the angle of rotation and the primary radius value, you will need to provide the center point, and the secondary radius via another data point.

Why would you want to control your ellipse axis? Simply put, to get the results you are looking for. For instance, you may be creating an isometric drawing for which "traditional" circles will not work. But knowing that all circles in an isometric diagram fall on either a 30-, 60-, or 90-degree axis means that you can control the direction of the resulting ellipse.

Place Ellipse by Edge Points

When you need to position an ellipse by identifying three points along its perimeter, the tool for you is Place Ellipse by Edge Points, shown in the following illustration. Similar in operation to the Place Circle by Edge tool, this tool requires two data points that serve to define the major axis of the ellipse, as well as the primary radius. A third data point generates the secondary axis. As with the previous ellipse tool, you have options that control all three aspects of the ellipse process.

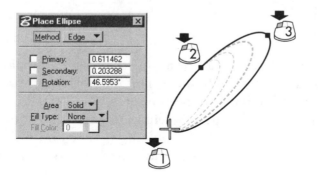

The Place Ellipse by Edge Points tool. Note how data points 1 and 3 lie along the major axis of the ellipse.

 TIMEOUT: A Bit of Trivia

Internally, MicroStation stores ellipses and circles as the same basic type of element. When you think about it, a circle is nothing more than an ellipse with equal primary and secondary radii. Later on, you will discover that this relationship is a key consideration in how a tool such as Scale Element can distort a circle into an ellipse with such ease.

Drawing Arcs

A close cousin to the circle is the arc. Sharing the same traits as the circle (a radius and a center point), the arc has one additional characteristic: *endpoints*. You might think of an arc as a partial circle. Drawing arcs in MicroStation requires use of the Arcs toolbox, shown in the following illustration.

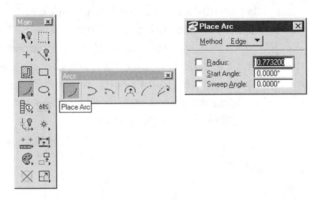

Arcs toolbox and its settings.

Earlier you learned that for any given three data points there is but one circle that can fit them. This is not true for arcs. For three given points you have two possible arcs, as indicated in the following illustration. The trick is telling MicroStation which arc segment you want. So, how do you tell MicroStation which piece of the circle you want? Through a little trick called the counterclockwise rule.

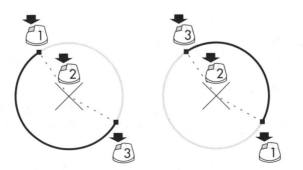

For any given three points there are two arcs: the one you want and the one you do not.

The Counterclockwise Rule

Just as no drafting tool set would be complete without a protractor, MicroStation must also be able to handle angular data. Arcs by their very nature use angular information. The "length" of an arc is typically measured in degrees of sweep. The problem is that there are two possible sweeps: the actual arc and the portion of the circle not occupied by the arc.

All radial input must be entered counterclockwise.

MicroStation "solves" the problem by requiring you to enter all radial information in a counterclockwise direction. When you give the first endpoint, the center point, and the final endpoint of an arc, MicroStation automatically strikes the arc from the first point counterclockwise to the final endpoint. This can be confusing to the first-time CAD user. For now, just remember that all radial input must be entered counterclockwise, as indicated at left.

V8: MicroStation version 8 will support both clockwise and counterclockwise input. When placing an arc, the direction you move the mouse after the second data point will determine clockwise or counterclockwise placement.

Place Arc by Center

Placing an arc with the Place Arc by Center tool is very straightforward. You select the first endpoint, the center point, and (following the counterclockwise rule) the final endpoint. The radius distance of the arc is set by the distance from the first endpoint to the center point. The final endpoint only sets the "sweep" angle of the arc. This is shown in the following illustration.

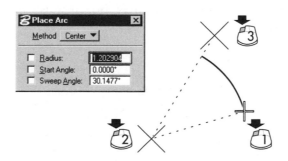

The Place Arc by Center tool relies on the counterclockwise rule.

Place Arc by Edge

After all of this talk about the counterclockwise rule, along comes an arc placement command that does not follow it! Instead, Place Arc by Edge uses the order in which you place data points to determine the final arc.

The first and third points still define the ends of the arc; however, the second data point is used to compute the radius. Think of "From-Through-To," an idea that will reappear in other commands. When you specify an arc by edge, you start from point 1, pass through point 2, and go to point 3, as shown in the following illustration.

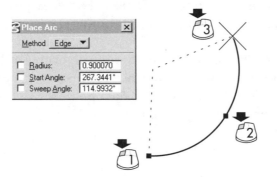

You switch the Method to Edge in the Tool Settings window to activate the Place Arc by Edge tool. An easy way to remember how the arc is drawn: From-Through-To.

Options When Placing Arcs

As with the Circle tools, the placement of arcs can be controlled by setting the appropriate options, as indicated in the following illustration. With both of the standard arc tools there are three specific options. These three options, as follows, match the three parts of an arc.

- Radius
- Start angle of the arc
- Sweep angle

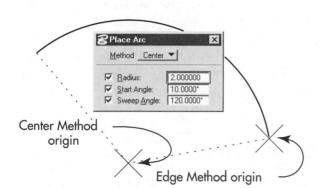

Center Method origin

Edge Method origin

The Radius option is easy to understand. With this option selected, you enter the radius you want the arc to have in the associated data field. Note the different origin points for the dynamic arc when you specify all three parameters. Which origin point you see will depend on the arc Method currently selected in the Tool Settings window.

The Start Angle option, on the other hand, is more difficult to understand. How the command operates depends on your selection of the drawing method, whether Center or Edge. Assuming you are placing the arc by center, and the start angle is 10 degrees, MicroStation locks your second data point to 10 degrees from your first data point. MicroStation responds by drawing a dynamic dashed line from the data point (noted by a small filled box) at the angle specified in the Start Angle data field. This gives you a visual reference for selecting your second and third data points.

In a similar fashion, when you select the Sweep Angle option and enter a value, MicroStation responds by displaying a dynamic arc attached to your cursor at the center point and tied to your first data point. In the following section, you will explore options for drawing curves.

Drawing Curves

So far, you have covered the triangle/T-square (Place Line), the compass (Place Circle), and the protractor (Place Arc). The next logical tool to discuss is the French curve found in every designer's drafting set. One of two of MicroStation's "French curves" is the curve string.

Place Point Curve

The Place Point Curve tool, shown in the following illustration, behaves similar to Place SmartLine; you place data points to define the shape of the curve with a final Reset to accept the curve. The curve string this tool generates is a gentle curving element that passes through each data point. The degree of curve is established by the angle between the data points. The more acute the angle, the sharper the curve. By keeping the angle very shallow, the result is a gently undulating curve.

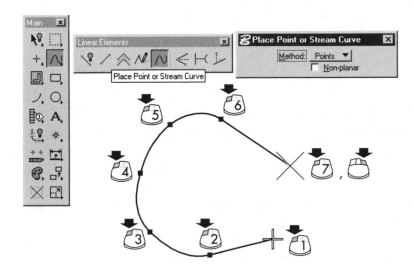

The Place Point Curve tool can be found in the Linear Elements toolbox. You must click on the Reset button to complete the construction of the curve.

Such curves constructed to pass through points are a favorite element for creating topographic contours on a site plan. As with SmartLines, you can modify the shape of your curves after you have placed them. (More on this later.)

Drawing Polygons

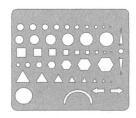

If you have ever used a general-purpose drafting template, you have no doubt drawn boxes, hexagons, and triangles. MicroStation, too, has its equivalent to these very useful shapes. Called polygons, these shapes share one important characteristic: area. All of these shapes are closed elements, in that they start and end on the same point. You have already seen one such element, the circle.

There are several tools to help you create polygons. The sections that follow examine some of them. All of the Polygon tools are located on the Polygons toolbox. There are tools here to create everything from simple shapes to specialized isometric ones, including the following.

- Place Block
- Place Shape
- Place Orthogonal Shape
- Place Regular Polygon

Place Block

The simplest of the polygon tools is Place Block, shown in the following illustration. This tool creates a four-sided rectangle based on the data points you supply. When you select this tool, you have a choice of methods for placing the block: Orthogonal or Rotated.

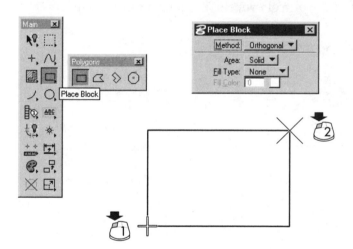

Place Block has only one option: Method. Seen here, the Orthogonal method results in a block aligned with the view.

With the Orthogonal method chosen (the default), the block is aligned with your view. To place this four-sided polygon, all you need to do is enter two data points for opposing corners of the box. It does not get much simpler than this.

With the Rotated method, you provide the axis along which the block will be placed. Your first two data points define this axis. A third data point provides the height and width of the new block, as indicated in the following illustration.

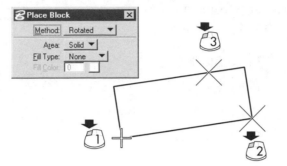

Place Block Rotated uses the first two data points to define the block's axis; the third data point defines the block's height.

Place Shape

There are, of course, other types of polygons. One of the more useful ones is the free-form shape, which you can make with the Place Shape tool, shown in the following illustration. By specifying every vertex, you have full control over the final form of the shape.

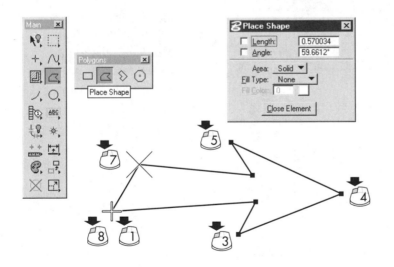

The Place Shape tool in action. The shape will close itself when you click a data point at the beginning point or click on the Close Element button.

The one tricky aspect of the shape is closure. Unlike the SmartLine, which needs a Reset for completion, the Place Shape tool relies on your returning to the beginning point of the shape. MicroStation provides a button on the bottom of the Tool Settings window (Close Element button) that forces closure of the shape.

TIP: *If your shape is small, or has closely spaced vertices, MicroStation may suddenly close on the beginning point before you have finished drawing the shape. The best defense against premature closure is to select your starting point along the longest side of your intended shape.*

There are a number of options associated with the Place Shape tool. The Length option allows you to specify the length of each shape segment. Entering a value in the Length data field will result in a dynamic segment, attached to your cursor, for which you data point at a given angle. The Angle option allows you to provide the angle at which the current segment should be drawn. One note of caution about the Angle option: you must change the value for each segment, especially when used in conjunction with the Length option, or else you will end up with a very flat shape!

Place Orthogonal Shape

Related in function to the previous tool, Place Orthogonal Shape, shown in the following illustration, sets the axis of the shape with the first two data points. The shape created from this "baseline" will be orthogonal in nature (i.e., at right angles to it).

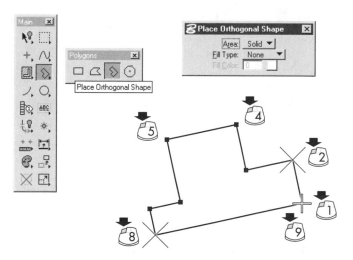

The Place Orthogonal Shape tool. Note the lack of a Close Element button on the Tool Settings window.

As each data point is entered, the resulting line will be set close to the data point. However, because it must be orthogonal to the axis set with the first two points, it will not be on the point itself. You close the shape by clicking a data point at its starting point.

NOTE: *Place Orthogonal Shape does not have a Close Element button.*

Place Regular Polygon

The tool that most closely matches the function of the general-purpose template is the Place Regular Polygon tool, shown in the following illustration. A very powerful shape maker, this tool allows you to set the number of equal-length sides (the Edges option) you want on your polygon. In addition, you have control over how your multisided polygon will be placed (the Method option). Finally, you can control the overall size of the polygon via the Radius option.

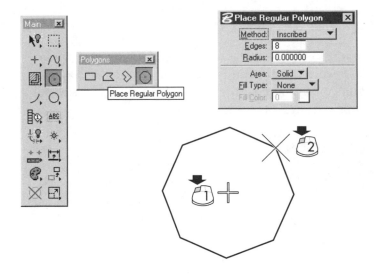

One of the fun tools, Place Regular Polygon gives you flexibility in creating your polygon.

There are three methods for placing a polygon. These are described in the following.

- *Inscribed:* Placement of the shape is by a center point and a point on one of the polygon's vertices. The vertices of the resulting shape fall on the radius of the phantom circle.

- *Circumscribed:* Placement of the shape is by a center point and the midpoint on one of the polygon's segments. The resulting shape's segments are tangent to the phantom circle.

- *By Edge:* Placement of the shape is by two adjacent vertices of the polygon. The radius value has no effect on this placement method.

Placing Isometric Elements

Isometric drawings, whether they are plumbing riser diagrams or PID heat-trace illustrations, use the 30/60/90 degree isometric drawing to establish the location of key elements with respect to an entire system in a 3D space. MicroStation supports isometric drawings with two special tools (Place Isometric Block, Place Isometric Circle). These tools, shown in the following illustrations, are accessed from the Isometric toolbox (Tools menu > Isometric).

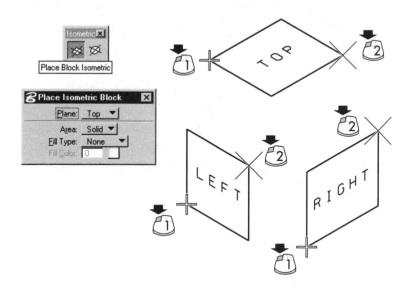

Place Isometric Block uses diametric data points to define the size of the block. The Plane option sets the isometric orientation.

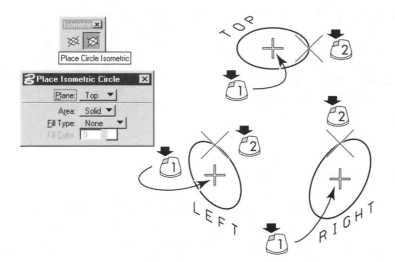

Place Isometric Circle requires a center point and a data point on the radius of the circle. The Plane option sets the isometric orientation.

Note that both isometric tools only simulate 3D when placing their respective element types. In most instances, these tools are used in 2D drawings, so this is not a problem. However, it could be very confusing were you to place isometric blocks and circles and then rotate the view!

Isometric Text

Although text manipulation has not yet been discussed, in regard to isometric drawings you should know that there are two text fonts delivered with MicroStation specifically designed to mimic the oriented text found on most isometric plans. These fonts follow. Both fonts are stick-figure-like, but do serve the purpose of creating realistic isometric plans.

- Font 30: Iso_fontleft
- Font 31: Iso_fontright

Selecting the Isometric Pointer

As a final aid in creating isometric drawings, MicroStation allows you to change the shape and orientation of the on-screen cursor. This is done via the Preferences dialog box (Workspace > Preferences). Selecting the Operation category provides you with two options regarding the pointer: Pointer Size and Pointer Type.

The defaults are normally Pointer Size/Normal and Pointer Type/Orthogonal, meaning the cursor/pointer is displayed as a small cross oriented along the X and Y axes. Changing the pointer size to Full View displays a large cursor that runs from one edge of your view to the other. Changing to Pointer Type/Isometric (see following illustration) results in a skewed cursor that visually shows the orientation of the current isometric plane value (set with the Plane option on the appropriate Tool Settings window).

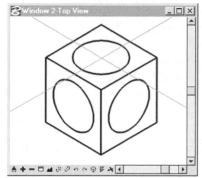

With the Pointer Size set to Full View and Pointer Type set to Isometric, this cursor is the result. Note how its orientation reflects the Plane setting shown in the Place Isometric Block tool.

Controlling Element Placement

Now that you have been introduced to a few drawing tools, you can address the subject of precision input. Although you can create reasonable drawings using the data point alone, the fact is most design work involves the accurate generation of coordinate information. MicroStation provides two major facilities for accurate coordinate input: element snaps and AccuDraw (introduced earlier in this chapter).

Snapping Along in Your Design

In MicroStation, the ability to define coordinate information from existing geometry is called *snaps* or *snapping to an element*. MicroStation supports the following snap functions.

- Snap to any point along an element
- Snap to the endpoints of an element
- Snap to the midpoint of an element
- Snap to the intersection between two elements
- Snap to a location of tangency along an element
- Snap to a point perpendicular to or from an element
- Force the creation of an element to be parallel, through a point or on a point, to an existing one
- Snap to the center of a shape
- Snap to the origin of a cell

Which snaps are available to you at any given time depends on the tool you have selected (also known as context sensitivity).

Snaps

Depending on what type of cursor control device you are using, the tentative point button may be a separate button (a typical example is a graphic tablet puck) or a combination of two buttons (a typical example is a Microsoft 2 button mouse). In all cases, the result is the same: the current snap function is invoked and an action appears on the display, as shown in the following illustration.

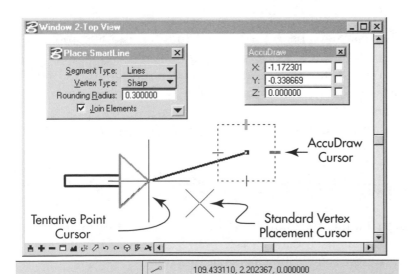

When the tentative point button is selected, the tentative point cursor appears. The normal object cursor is still free to move around the view but no longer has the dynamic element attached to it. Instead, the element is tentatively drawn at the current location of the tentative point cursor.

In the previous example, the snap mode set was called keypoint. In this mode, the tentative point cursor jumped to the nearest keypoint on the element closest to the cursor. One point to note about keypoints is their temporary nature. The action of generating a tentative point is, as the name implies, not permanent. In other words, as you are placing a line, the act of invoking the snap with the tentative point button does not, in itself, generate a line.

Instead, MicroStation requires you to accept the location of the tentative point snap by selecting the data point button. As long as you do not click on the data point button, you can continue to reselect tentative points, or even change the snap mode, without any effect on your drawing.

Mouse Button Assignments

Today, most people use a two-button mouse, a three-button mouse, or one of the new "wheel" mice (two buttons and with a small wheel sandwiched between), exemplified by the Microsoft IntelliMouse (the author's favorite input device). The Button Assignments dialog box (Workspace menu > Button Assignments), shown in the following illustration, is provided to configure the standard MicroStation mouse button functions for your particular mouse or input device.

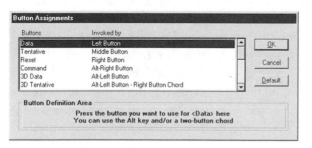

Data, Tentative, and Reset are the three most important button assignments found in the Button Assignments dialog box. In this example, Tentative has been assigned to the middle button, indicating the use of a three-button or wheel mouse (pressing down on the wheel is the same as a middle button click).

One point of confusion in using this dialog box is how you go about selecting the button to assign to a given command. First, select the command (for instance, Tentative). Next, move your cursor down onto the bottom section of the dialog box labeled Button Definition Area and press the mouse button (or graphics tablet puck button) you want to assign to that function. That is it.

However, many first-time users of MicroStation overlook the need to move the mouse into that specific area of the dialog box to actually set the button to the command and get nonplussed when the command does not update.

Note, also, that if you click a button that has already been assigned to another command (e.g., you press the left mouse button in response to setting Reset), the new assignment will override the existing one. However, you cannot accidentally assign no button to a command; MicroStation will simply swap the button assignments between your new one and the existing command assignment (i.e., the Reset function is controlled by the left mouse button, and the Data Point function is controlled by the right mouse button).

TIP: *Many left-handed users swap the Data and Reset buttons, which can be disconcerting for anyone who uses the same station. What makes this swap less obvious is the fact that you can still select menu items and other Windows operations using the left mouse button. If you suspect this is the case, simply visit the Button Assignment dialog box and verify the current mouse settings.*

Accessing the Tentative Point

Before leaving this discussion about mouse buttons, it is important for you to know how to set up and use the Tentative Point or Snap button. On a three-button or wheel mouse, Snap (also called Tentative Point) is normally assigned to the middle button. In the case of a two-button mouse, it is assigned to a mouse button chord. A chord is the action of pressing both mouse buttons simultaneously. It takes some practice, but if all you have is a two-button mouse it is still possible to use this all-important function in MicroStation.

Overview of the Tentative Point or Snap

Before exploring individual snaps and their operation, it is a good idea to understand how MicroStation uses snaps in its operation. Unlike other CAD products, MicroStation provides access to most snap operations during all drawing operations. When you select a tool, such as Place Arc, and want to locate an existing element's end point as the start for this element, you simply "snap" to the element using the appropriate snap select (in this case, keypoint). There is no special key you enter, nor do you have to switch to a special mode. Click on the Snap or Tentative button and accept or reject its location and move ahead with your tool.

As a point of interest, *tentative point* is an older term for *snap*. It is actually a good description of how a snap works. The snap is a temporary holding location, a tentative value as it were, which you are considering but have not yet accepted as the coordinate location to feed to the current tool. You can click on the snap button as many times as you need to, to finally locate the precise spot or element you need for your current design step. At the same time, the snap is also an element locator that, depending on the particular snap you have selected, will influence the outcome of the current tool operation (more on this later).

The tentative point snap provides additional information on the screen. Returning to the illustration of the Tentative Point in action (see following illustration), note how there are two sets of coordinates given for your review. The AccuDraw window always displays the difference or delta distance from the last data point to the snap location.

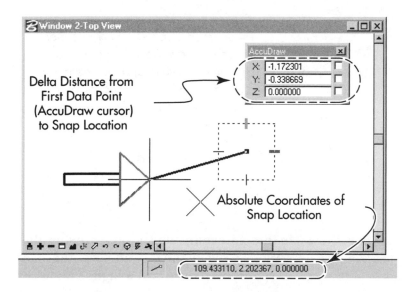

The absolute coordinates shown in the status bar are set by the TPmode value.

By default, the coordinates that appear in the status bar show the absolute coordinates of the snap location. However, you can change this readout using the Set Tpmode <option> key-in. Options include *distance* (gives distance and angle from one snap location to another), *vdelta* (gives XYZ coordinate "delta" distance from one snap location to another), and *locate* (gives absolute XYZ distance from design file's origin). For now, however, leave it at the default (Locate, or absolute, mode).

The Common Snaps: Keypoint and Nearest

Prior to version 5 of MicroStation, there were really only two snap modes: *Project* (now called Nearest) and *Keypoint.* The Nearest snap forces the tentative point to snap to the closest location on the element nearest your cursor at the time you clicked on the Tentative button. Think of it as the intersecting point of your cursor and that element.

Keypoint, by contrast, forces the tentative point to jump to specific locations on the element. In the case of a line, this would be the closer of the two endpoints. A circle keypoint would be the closest quadrant or its center point, a line string the nearest vertex, and so on. The following illustration shows the most common elements and their keypoints.

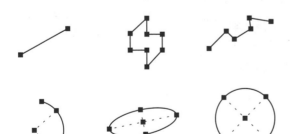

The most common elements and their keypoints. Note how the arc has a keypoint at both endpoints and one at a quadrant distance along its length. Also note how the circle's key snap locations are rotated from the normal XYZ axis, which indicates its placement orientation.

The keypoints shown in the foregoing illustration are for the default snap divisor of 1. You can divide the default keypoint snap locations by setting the keypoint snap divisor to a higher value, as shown in the following illustration. This is done one of the following ways.

- From the Key-in Window, enter *ky=<divisor>*
- From AccuDraw, press the shortcut key K and select or type in the divisor

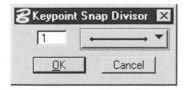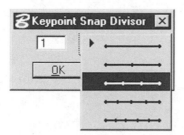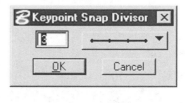

AccuDraw's snap window provides a point-and-click selection for the keypoint divisor. In this example, the divisor is being set from 1 to 3.

By setting your snap divisor to a higher value (for instance, *KY=4* typed in via the Key-in window), you can snap to additional points equidistant between the default element keypoints. With all of the various snap modes available, Keypoint and Nearest are still used the most. In fact, most users select Keypoint as their default. In exercise 4-2, which follows, you will practice using the Keypoint and Nearest snaps.

NOTE: *To identify what snap mode you are currently in, just look at the status bar. In the field just to the right of the message field, you see an icon representing the active snap mode. You should get into the habit of checking this status field as you work.*

EXERCISE 4-2: USING KEYPOINT AND NEAREST SNAPS TO BUILD A PICTURE FRAME

Are you ready to try out the most basic of snaps? Good! The object of this simple exercise is the creation of a picture frame consisting of two blocks, one inside the other, as shown in the following illustration. You will use the snap lock to build the beveled corners.

The dashed lines represent the elements you will be creating as part of this exercise.

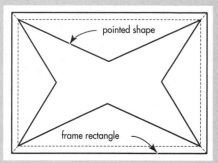

1   From the companion CD-ROM, open *PIXFRAME.DGN*. Note the rectangle and "pointed" shape. You will be using the pointed shape to guide placement of the elements in this exercise.

2   Set the Keypoint snap by clicking on Keypoint on the status bar snap icon (see the following illustration) while holding down the Shift key. This sets Keypoint snap as your default snap mode.

3   Select the Place Block tool. Move the drawing cursor over one of the pointed shape's outer points and click the tentative point button (on a two-button mouse, press both buttons simultaneously; on three-button+ mouses, click the middle button).

4   With the Tentative Point cursor displayed, click the data point to accept the location. If you miss the corner (the tentative point does not align with the pointed shape), simply move the drawing cursor closer to the vertex of the pointed shape and try again. With one corner of your rectangle tied down, you will need to identify the opposite corner.

5   Using the keypoint snap again, identify the opposite corner of the pointed shape. At this point, you have placed the inside edge of your picture frame. Next, you will miter the corners of the frame.

6   Select the Place SmartLine tool. Using the keypoint snap, select the corner of the outer rectangle, and data point.

7   Next, use the keypoint snap/data point to snap to the inner rectangle's corner.

8   Click the Reset mouse button to terminate the new line.

9   Repeat for the other three corners.

10 With the Place Line tool selected, verify that Tentative Point is set to Key-point (hint: check the status bar's snap status field). If it is not set to Key-point, click on the snap status field to display the Snaps pop-up menu and select Keypoint with the Shift key pressed on your keyboard. This sets your default snap to Keypoint.

11 With Keypoint your active snap mode, now move your cursor over the upper left corner of your picture frame. Press your tentative point button. A large cross cursor should snap to this corner, and the element should then be highlighted. Accept this location with a data point.

12 Now, move your cursor over the upper left corner of the inner box and place a tentative point, followed by a data point. The result should be a line that goes from the inner box corner to the outer box corner.

NOTE: *Clicking a snap mode on without holding down the Shift key makes the selected snap mode active for only a single tentative point. To make it the default snap mode, you must press the Shift key while selecting the snap mode.*

Setting Your Snaps Couldn't Be Easier

In the exercise just presented, you were instructed to go to the status bar and set the snap to Keypoint. While there, you saw a number of additional snap options available. Let's explore these various snaps. Setting your snap mode is easy. MicroStation provides no fewer than the following four places at which you can set your active snap.

- From the Settings pull-down menu (Settings > Snaps)

- Via the Snap Mode button bar (Settings > Snaps > Button Bar)

- Via the Locks settings box (Settings > Locks > Full)

- Via the Snaps pop-up menu invoked by either clicking on the Snap Mode field on the status bar or by clicking on the Tentative button with the Shift key pressed

Snaps Pull-down Submenu

Let's look at the Snaps submenu under the Settings menu. This submenu is shown in the following illustration.

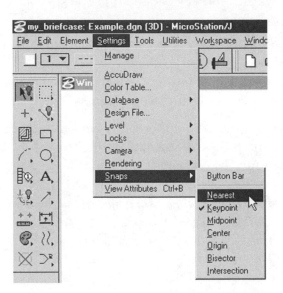

The Snaps submenu is located under the Settings menu.

With the Element Selection tool selected (the arrow icon in the upper left corner of the Main tool palette), the list of snaps available under the Snaps submenu represents what could be considered the generic snaps. Table 4-1, which follows, describes the generic snap options.

 NOTE: *These snap actions always operate on the element closest to the cursor.*

Table 4-1: The Generic Snap

Snap Name	Action
Nearest	Jump to the nearest location
Keypoint	Jump to the nearest keypoint (endpoints, center points, and so on)
Midpoint	Jump to the midpoint of the nearest element segment

Table 4-1: The Generic Snap

Snap Name	Action
Center	Jump to the center point or centroid of the element
Origin	Jump to the origin point of a cell
Bisector	Jump to the midpoint of the entire element (not just the closest segment)
Intersection	Jump to the intersection point of two elements

✔ <u>N</u>earest

*Snap option
check mark.*

When you pull down the Snaps menu, one thing you should see is a check mark next to one of the snap options, as shown at left This check mark identifies the active snap. When you select a different snap, this check mark will move to that snap option. Snap modes selected from this menu are active for only a single snap operation.

Snap Mode Button Bar

The Snap Mode button bar provides an on-screen toolbox. This is most handy when you are constantly changing your snap mode. In addition, it helps stimulate the design process by reminding the user what options are available.

To activate the Snap Mode button bar, select Settings > Snaps > Button Bar. The Snap Mode button bar, shown in the following illustration, will appear on your screen ready for you to choose the appropriate snap mode. Instead of names for each of the snap modes, you are presented with icons representing each snap.

The default shape of the Snap Mode button bar. Like any toolbox, you can change its shape by dragging its border.

Notice how one of the snap mode buttons looks like it has been depressed. This is the default snap mode. If you click another snap mode, you will see that the default button comes up but still remains shaded. This takes a little explaining.

Default Snap Versus Snap Override

To enhance productivity, MicroStation allows you to temporarily change your default snap to another snap option. Called "snap override," this snap option remains active only for the duration of the present operation.

For example, if you are placing a line with the Nearest snap as your default but decide that for this one time you want to use the Center option, you select Center and snap to the arc or circle's center. The dynamic line appears from this center location ready to be completed. With a data point, you not only accept this new endpoint but cancel the Center snap option. This returns the snap to the Nearest option, your default snap. In this example, if you looked at your Snap Mode button bar while Center was active, you would have seen the Center button depress and the Nearest button turn gray (see the following illustration).

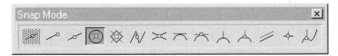

Here, the Nearest option is the default; the Center option is an override for one tentative snap operation.

To actually change your default snap, you must double click on the desired snap button. The reason for this double click has to do with predictability during the design process. If you continually changed the snap option, you could never predict the outcome the next time you clicked on the tentative point button. Instead, you would find yourself constantly checking the Snaps pull-down menu or reading the Command window.

This can be very confusing and downright irritating (remember, you will be doing this day-in and day-out). With the concept of setting your snap to a default selection and temporarily using a different snap, you always know how your tentative point is going to react. When working in a drawing, you find yourself constantly placing elements at the ends of others; thus, Keypoint is the most common default snap.

Your First Visit to the Locks Settings Box

As mentioned earlier, there is more than one way to select a snap option. The Locks settings box provides yet another way to do this. However, it does more than give you access to snaps. It gives you additional control over the snap process.

To access the Locks settings box, shown in the following illustration, select Settings > Locks > Full. You will be presented with a rather large dialog box containing much more than snap options (why do you think they call it the Full Locks settings box?).

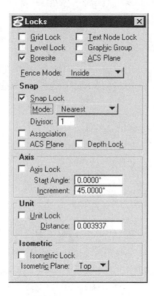

The Full Locks settings box provides access to many options. Note the Snap section.

If you look at the Snap section of this menu, you will see a Mode selection field that lists all available snap modes. Unlike the Snaps pull-down submenu, one click will make your selection the default.

More important, though, are the other options found in the Snap section. This is where you can turn the entire snap operation on or off. This is done by selecting the Snap Lock checkbox.

Another important option located here is the Divisor field. When used in conjunction with the Keypoint snap, this value tells MicroStation how many additional keypoints to use along an element segment. For instance, if you set the Divisor value to 5 and tentative point along a line, you will have six equidistant tentative points along its length. The reason there are six and not five is that the divisor represents the number of element segments, not the number of tentative points to create. By default, the divisor is set to 2, allowing the Keypoint snap to emulate the Midpoint option (after all, when you divide an element in half you have found its midpoint).

TIP 1: *If you find yourself working with a lot of small line segments, be sure to set the divisor to 1. This avoids the hazard of accidentally snapping to something other than the vertex or endpoint you really wanted. In addition, the Keypoint snap will work faster in a congested area.*

TIP 2: *A shortcut for changing the divisor value is the use of the KY= key-in. Entered in the Key-in window, this key-in will set the value of the Divisor field without having to call up the Full Locks settings box.*

For now, the other options found under the Snap section will go undefined. These will come into play later, as you learn more about MicroStation's other features.

Status Bar Snaps Menu

Clicking on the Snaps icon displayed on the status bar will also bring up a snaps menu for immediate selection. This is probably one of the most popular methods for accessing your snaps, as you normally look at this icon to ascertain your current snap mode. An example of a snaps menu is shown in the following illustration.

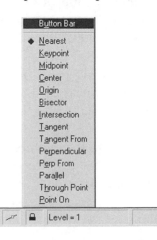

Clicking on the snaps icon located on the status bar brings up the snaps option menu. The precise list of snap options will change depending on the active tool (context sensitivity).

Pop-up Snaps Menu

The last method for selecting a snap mode is the pop-up menu. There are two ways to invoke it. As shown in the following illustration, hold the Shift key and press the tentative point button, and the Snaps pop-up menu will appear at your cursor location. Click on the Snap Mode field in the status bar, and the Snaps pop-up menu will appear next to it, just above the status bar.

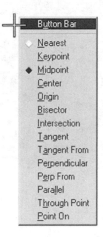

The Snaps pop-up menu is activated at the cursor location by clicking the tentative button in conjunction with the Shift key.

The open diamond identifies the default snap and the filled diamond designates the snap override. If you select an option with the Shift key pressed, that option becomes the default. Otherwise, the snap chosen is only good for the next snap operation.

Using Grids

Analogous to grid paper, the grid lock function allows you to control how MicroStation treats your data points when entering, modifying, or otherwise manipulating elements on the design plane. If you have spent any time exploring the sample drawings Bentley supplies with MicroStation, you have probably seen this grid system. In its simplest form the grid, shown in the following illustration, is nothing more than a regular spacing of dots that can be turned on or off in any active view.

The small dots represent the grid units. The larger tics are the grid reference markers.

As a visual aid, these "dots" are very helpful in sizing up a design project. But if all the grid system did was display a "pretty picture" of dots, it would be of questionable value. Grid lock has the ability to force all data points to the nearest grid point. Any application that requires drawing straight lines at regular intervals can benefit from using grid lock. You can place lines more accurately using the grid system than by trying to "eyeball" the linework.

Grid Controls

Just as you can buy many different types of grid paper, you can adjust MicroStation's grid to suit your needs. The size and type of grid is controlled under the Grid category (shown in the following illustration) of the Design File Settings dialog box, accessed via the Settings pull-down menu. Here you set the size of your grid (Grid Master option) and the number of grid dots between grid reference points (Grid Reference option).

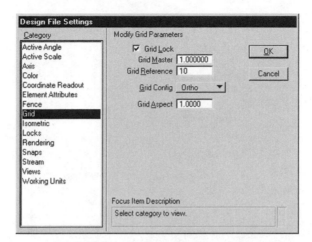

The Grid category of the Design File Settings dialog box controls the grid settings.

Alternatively, you can set these two major values of the grid system via key-ins: the Grid Unit and the Grid Reference. The Active Grid Unit (*GU=*) key-in sets the distance between each individual grid "dot." The second key-in, the Active Gridref (*GR=*), sets the number of grid units or dots between each grid reference "tic."

This is sometimes confusing for the first-time MicroStation user. In the previous grid illustration, if you count the number of dots between each tic, you would find 11 (or 12 segments). In this case, the grid unit has been set to :1 (GU=:1) and the reference grid has been set to :12 (GR=:12).

The result, in this case, is one dot per inch, with a reference tic every 12 inches. If you were to set the grid unit to 0.5 master units (same as 6 inches), the result would be an enlargement of the grid six times. However, the reference grid would still be set to 12. Thus, the distance between each GR tic would be :6 times 12.

To use the grid for precision placement, you must first use the Grid Lock command. Usually selected from the Locks dialog box (available via the Settings pull-down menu) or the Lock icon in the status bar, you can also key in *LOCK GRID ON* in the Key-in window. (See the following illustration.)

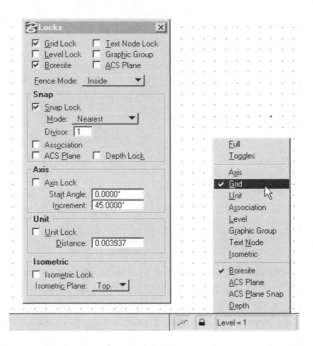

Grid Lock can be turned on and off either from the Locks settings box or by clicking on the Locks icon located on the status bar.

MicroStation lets you know the condition of the grid lock with a "Locks=" message in the status bar, as in the following illustration.

Grid lock status message.

You cannot verify Grid Lock's current status simply by observing what is on a view because grid display is totally separate from grid lock. You can display (or not display) the grid in any active view via the View Attributes settings box (Settings > View Attributes), shown in the following illustration.

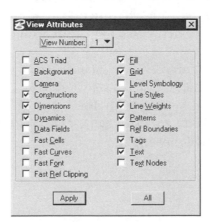

View Attributes settings box.

In addition, as you zoom out, MicroStation will turn off the dots when the density of the grid is such that it obscures the design. This will also occur with the grid reference tics, but at a greater distance out.

Practice with Grids

Now that you have some background on grids, you can get to know them a little better with some practice. Exercise 4-3, which follows, shows you how to place lines with the grid feature.

EXERCISE 4-3: UNDERSTANDING GRIDS

In this exercise you are going to use the grid feature of MicroStation to accurately place some lines.

1   Open the file *GRID.DGN*. View window 1 contains the image shown in the following illustration.

Design file to be worked on.

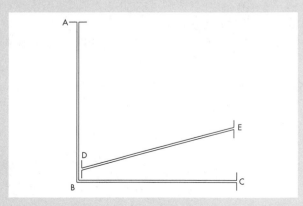

2   Select the Place Line tool. As accurately as you can, place a line between the vertical parallel lines from point A to point B. Try to place it perfectly straight and evenly between these two lines. How did you do? With the grid turned off, you probably did not do very well. Trying to place a line by eye is not easy. Observe the Locks status message, or click on the Lock icon in the status bar. Is Grid Lock on or off? Off! Let's start over with your linework.

3   Use the Undo command to remove the line you previously placed (Edit > Undo Place Line).

4 Turn on the Grid lock. Open the Locks pop-up menu by clicking on the Lock icon on the status bar. To activate Grid Lock, select the Grid checkbox.

5 Place the same line between points A and B. This time the line should fall nicely between these points, as shown at right. Next, click the data point button at point C. All of your lines should be parallel to the existing ones.

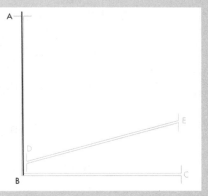

Line between points A and B.

6 Now display the actual grid points on the screen. Activate the View Attributes settings box (Settings > View Attributes). Select the Grid checkbox and click on the All button. What happened?

The large tic marks you see are your reference grid. To see the actual grid points, you need to zoom in a little closer. Click on the Zoom In icon on the View Control bar located along the lower left border of your view window and place a data point to enclose point B. You should now see dots all over your view. Note how the line you placed is on one of these dots.

7 Now fit the drawing in your view by selecting the Fit View icon on the View Control bar. You should see your entire drawing again. Click on the Reset button once.

NOTE: *The view manipulation commands are covered in the next chapter.*

8 Next, try placing a line parallel and evenly spaced between points D and E, as shown at right. What happened? Most likely, no matter how hard you tried, the line just would not stay between the two lines. The reason is simple. MicroStation's grid system is orthogonal in nature, and the two lines you are trying to split with your new line are obviously angular (15

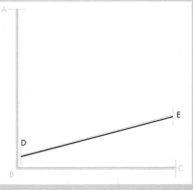

Line between points D and E.

degrees, to be precise). This is a good example of when to use AccuDraw's Polar compass.

9   Place a data point at D. If the rect-angular compass is displayed, press Space to toggle to Polar display. With input focus in the AccuDraw window, press Alt-A (locks the line's angle), Tab, and enter *15* followed by another Tab (or Enter). Now, move the 15-degree locked cursor up to E and data point. The result is shown at right.

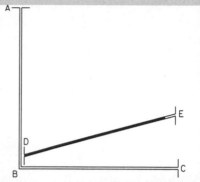

What you should see after step 9.

This exercise showed you the basics of turning on your grid lock and displaying your grid. In addition, it showed in an example where not to rely on the grid for element placement.

Summary

You have been introduced to some of the basic element types used in MicroStation. In addition, you have learned how to place these elements with a certain degree of reliability. By now you may have some idea of the power of MicroStation's tools and how they interact with one another.

In the next chapter you will explore this interrelationship and how you can use MicroStation's various tools to construct your designs. In addition, you will continue your exploration of AccuDraw, as well as learn to navigate within the drawing (the view tools) and to make all of those elements take on characteristics beyond their appearance as black or white lines.

2D BASICS: PART 2

Mastering Precision Input, Screen Control, and Element Manipulation

IN THE PREVIOUS CHAPTER YOU WERE INTRODUCED to the primitive elements that constitute the bulk of all CAD drawings. A key feature of any CAD package is its ability to modify objects once they have been placed in a drawing. In this chapter, you will learn how to navigate within MicroStation's design file environment and how to manipulate previously created elements. First, however, you need to know a bit about AccuDraw and the art of precision input.

AccuDraw Revisited

As mentioned in the previous chapter, AccuDraw is the main method for entering precise coordinate data using a combination of your mouse (or graphics tablet) and keyboard. Definitely an improvement over the previous generation of input techniques (i.e., precision key-ins), AccuDraw brings an intuitive feel to the CAD drawing process.

So, how do you get started with drawing something with MicroStation and AccuDraw? More specifically, how do you lay down that first line? This question often gets overlooked by technical texts. The answer is not always cut and dry, and requires a bit more forethought.

Your Design Approach

As someone who has spent his entire career in computer graphics (note I did not say CAD), the author has developed a lot of different drawing methods and techniques suited for specific types of designs and projects. For instance, did you know that every illustration in this book was generated using MicroStation? Not a trivial thing when you inspect a few of the line art illustrations that look more like they were drawn with Adobe Illustrator.

Over the years I have observed common techniques and ideas that successful users employ in their daily design process. The following are the basic tenets of a fundamental approach—a CAD design philosophy, as it were.

  • Start with a plan.

It happens more often than anyone wants to admit, but there are many a legacy project that began life without any preplanning. Projects begin life as *seed2d.dgn* and go downhill from there. This refers to, of course, starting out a design without any forethought on the goal of that design. True, you can simply open a new design file and begin drawing your design, but the real productive value of MicroStation comes from its consistency and logical approach. This means starting with an idea of what the end results are.

If you are tasked with generating a set of contract documents, you should apply the standard design or drafting practices associated with those documents. This sounds like common sense, but you would be surprised how often we try to "fix it up" after the fact, with the usual outcome: garbage in, garbage out. Fortunately, most companies these days have established standards and practices for CAD design. If your company has not, there are several well-established industry or de facto standards available via the Web (for instance: *http://cadlib.wes.army.mil/*).

  • Use construction elements.

When confronted with a design challenge, many first-time users look for the "magic bullet," that single command or tool they can select that single-handedly resolves the design problem. In most cases, *there is no single tool that can solve the problem*. Instead, you need to work through the design issues to arrive at the correct solution. This involves drawing a lot of what I call construction elements. These are lines, arcs, circles, shapes, and so on that are not the final design but in some way aid in the design's development. This is a nice aspect of MicroStation: it is cheap and easy to create elements, and just as cheap and easy to delete them.

- When it doubt, draw it out!

Do not be stingy. If you need to "sketch out" some part of your design to work it out, do it! You can always clean it up later, the subject of the next chapter. MicroStation is very efficient at storing design elements, so you can have literally thousands of elements in a drawing without making a measurable impact on the storage requirements for your project. As you will learn soon, modifying an element is just as easy as creating it, so if you need to work out a dimensional problem, draw out what you know, and then modify the results as you get more information or the problem resolves itself.

- Trust MicroStation!

As you "sketch out" your design, keep in mind that MicroStation maintains an incredible degree of accuracy for you. Even if you have relatively coarse working units (i.e., a low count of positional units per sub unit), MicroStation SE+ incorporated an additional 15 bits of accuracy hidden away in enhanced precision. What this really means is that you can rely on MicroStation to record your every edit without concern for computation errors or round-offs.

Returning to the discussion of AccuDraw and precision input, in the last chapter you learned that AccuDraw reacts to shortcut keys. By default, there are quite a few shortcuts, shown in the following illustration.

Enter	Smart Lock
Space	Change Mode
O	Set Origin
V	View Rotation
T	Top Rotation
F	Front Rotation
S	Side Rotation
B	Base Rotation
E	Cycle Rotation
X	Lock X
Y	Lock Y
Z	Lock Z
D	Lock Distance
A	Lock Angle
L	Lock Index
RQ	Rotate Quick
RA	Rotate ACS
RX	Rotate about X
RY	Rotate about Y
RZ	Rotate about Z
?	Show Shortcuts
~	Bump Tool Setting
GT	Go to Tool Settings
GK	Go to Keyin
GS	Go to Settings
GM	Go to More Settings
GA	Get ACS
WA	Write to ACS
P	Point Keyin (single)
M	Point Keyin (multi)
I	Intersect Snap
N	Nearest Snap
C	Center Snap
K	Snap Divisor
Q	Quit AccuDraw

You can display the entire list of AccuDraw shortcuts by resizing the shortcuts window.

TIP: *When first exploring AccuDraw's capabilities, it is a good idea to dock the AccuDraw shortcuts window along the left or right (preferred) edge of the MicroStation application window. In this way, you can quickly review the various shortcuts and their purpose.*

AccuDraw's Smart Lock

When you begin placing elements using AccuDraw, you will rely on its axis locking feature (as you type a number, the axis with input focus auto-locks). An amazing amount of most design activity involves orthographic projections from the major X and Y axes of the drawing. Acknowledging this fact and taking full advantage of it, AccuDraw provides a dynamic locking mechanism called *axis indexing.*

After you have placed your first data point for almost any element type, AccuDraw responds with its compass located at that data point. If you move the dynamic cursor along either of the two main AccuDraw axes (X and Y), a curious thing happens. As long as you stay within a given distance of the axis, the resulting dynamic graphic (the "rubber band") stays locked to the axis, as indicated by a heavy highlight (see following illustration).

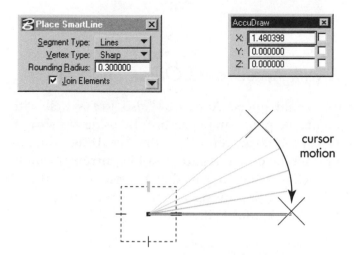

The Place SmartLine tool is demonstrated with AccuDraw active. Note the Y field is zeroed out, although the vertex cursor is not quite on the X axis. This is a result of AccuDraw's index "snap" feature.

If you data point while this axis is indexed, the resulting coordinate will take on the highlighted axis value of 0 (in other words, it stays on the axis). This is accomplished without the need for a shortcut key-in; just move the cursor along the axis and, voilà!, the element is drawn indexed to that axis. This is a major productivity enhancement you will definitely come to rely on.

Smart Lock

To complement the indexing feature of AccuDraw, you can press the Enter key to lock the indexed axis so that you can move the cursor along the other axis for further definition (see following illustration). The Enter key actually invokes an AccuDraw command called Smart Lock. Smart Lock locks the currently selected axis so that no further changes can be made to that axis until it is unlocked. This works very similarly to the X, Y, and Z shortcuts that directly lock that respective axis. Although this sounds complicated in print, it seems so natural when you try it that you will wonder how other CAD products manage to draw anything!

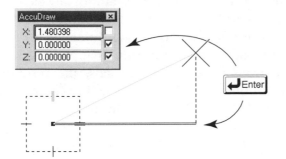

Pressing the Enter key locks the non-indexed axes (in this case, Y and Z, see previous illustration) but still allows you to reposition the cursor for the X value. The Z field in the AccuDraw window indicates this is a 3D design.

Using Snaps with AccuDraw's Origin Shortcut

As you probably could guess, AccuDraw also works well with MicroStation's tentative point snaps feature. By using the Origin shortcut (the O key), you can reposition the AccuDraw compass to any location within the design regardless of its current location. This is a dynamic action, which means that as soon as you press O on the keyboard, the compass moves to your current cursor location, as indicated in the following illustration.

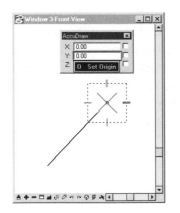
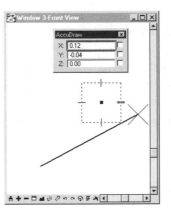

As soon as you press O, AccuDraw moves the compass to the current cursor location. You can see that it does not impede the operation of the element placement tool; the dynamic line continues to track from the first data point location.

It will not take you long to realize that arbitrarily moving the AccuDraw compass does not improve your drawing accuracy unless you perform this move in conjunction with the snap operation. When you click the tentative point snap button on your mouse, AccuDraw's compass does not immediately react—that is, until you press O. At that point, it relocates the compass to the tentative point location and begins to track the distances from that location.

This provides you with the ability work with any MicroStation tool relative to practically any point within your design.

For instance, say you are placing a pair of 350-cm columns flanking the entrance to a building. Furthermore, you need to place the columns 1 meter from the front facade (see following illustration).

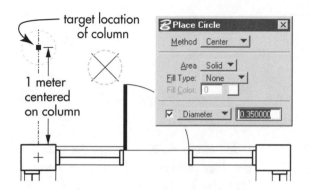

To set the column's diameter, use the Place Circle by Center tool with the Diameter option set to 350 cm (0.35 meters).

Using the Place Circle by Diameter tool, you can set the size of the column and use AccuDraw's Origin shortcut to input the offset from a center snap. This is shown in the following illustration.

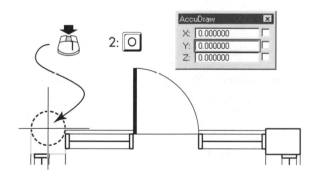

Set the reference (datum) point for the column placement by snapping to the face of the exterior corner of the wall. Use the AccuDraw Origin shortcut to reposition AccuDraw's compass.

As with most MicroStation drawing solutions, there is more than one method for generating the required geometry. The illustrations that follow present one possible "solution" to the column placement exercise. To optimize the number of illustrations presented, some of the serially executed operations involved have been consolidated into one picture.

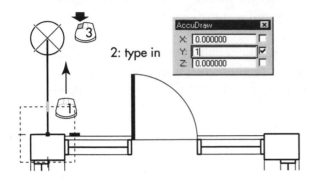

2: type in

Use AccuDraw's axis index lock to move along the Y axis and enter the distance offset from the exterior of the building (1 meter). A data point accepts this offset and places the column (350-cm circle).

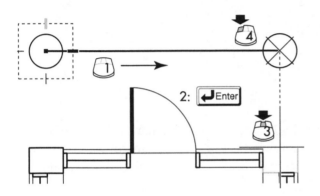

2: Enter

Index along the X axis from the previous data point and use the Smart Lock shortcut to lock the Y axis. A keypoint snap provides the X distance and generates the second column.

Setting the Keypoint Divisor Value via AccuDraw

In two of the steps of the previous example, a snap was used to find the midpoint of an element. This was possible because MicroStation's keypoint snap lets you select the divisor applied to the element to locate intermediate snap locations (remember, open elements such as lines and arcs always have keypoint snaps at their endpoints).

You can set this value, called the *keypoint divisor*, as part of the Full Locks settings window (Settings menu > Locks > Full). However, AccuDraw provides a shortcut that makes setting this value very quick and painless. Whenever input focus is on AccuDraw, pressing K will bring up AccuDraw's own Keypoint Snap Divisor window, shown in the following illustration.

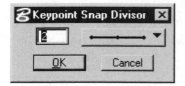

Pressing K in AccuDraw opens the Keypoint Divisor window. Although it appears to be a modal dialog box (has OK and Cancel buttons), this window can be left open, and changes made to the keypoint value will be immediate. Pressing Cancel returns the divisor value back to its pre-K value, a handy trick when you need to change the divisor for only one operation.

Other AccuDraw Snap Shortcuts

It would stand to reason that if there is a shortcut for setting a keypoint value there should be other snap-related shortcuts. A quick study of the AccuDraw shortcuts (remember the "?" shortcut?) shows that there are the following three.

- *i*: Set the snap to intersection of two elements (Intersect mode)
- *n*: Set the snap to nearest point on element (Nearest mode)
- *c*: Set the snap to center of element (Center mode)

Each of these modes is temporary and only affects the next snap operation. Experienced MicroStation users memorize and use these snap shortcuts to great advantage.

Showcasing Other Useful AccuDraw Shortcuts

Using AccuDraw shortcuts provides a very efficient way of cutting down on the number of actions you must execute in order to complete a given design operation. The sections that follow discuss more AccuDraw shortcuts you will find very useful.

Rotate Quick Shortcut (rq)

When using AccuDraw's normal compass orientation, you are restricted to locking along your view window's X-Y and Z axes. However, there are times when you need to lock along another axis. Without going into a full-blown discussion on auxiliary coordinate systems and how to configure, save, and restore them (a subject more appropriate to an intermediate or advanced level book), AccuDraw does provide a "down and dirty" method of quickly realigning its compass axis along a user-defined orientation.

Called Rotate Quick (shortcut: rq), this feature allows you to temporarily rotate AccuDraw's compass to any angle you need before proceeding with the current AccuDraw or drawing tool action. The illustrations that follow show the use of this feature.

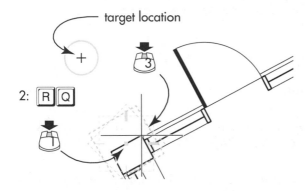

After identifying the center point of the exterior wall segment and invoking AccuDraw's origin option, the next step is to rotate the compass axis using the Rotate Quick (rq) AccuDraw shortcut.

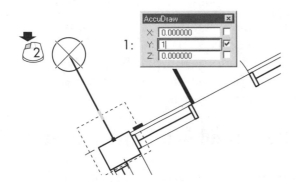

With the AccuDraw compass temporarily rotated to the plane of the exterior wall, all you have to do is index to the Y axis and enter the distance desired (1 meter). Clicking a final data point places the column.

Point Key-in

There are times you need to place drawing elements at absolute 7locations within your drawing. As you may have noticed, AccuDraw always works relative to some point within your view. This point does not necessarily correspond to the absolute coordinate location you desire, so you need an alternative method for entering such a coordinate. Enter the Data Point Key-in shortcut. With AccuDraw as the input focus, pressing P (for point) brings up the Data Point Key-in window. At this point, you enter your absolute coordinate value, in the following format, and press Enter.

```
xcoord,ycoord,zcoord(in 3D only)
```

The Data Point Key-in window will disappear, with the coordinate you typed in sent directly to the current tool. When the Data Point Key-in window is open, you can also enter other coordinate data formats by clicking on the option menu located on the left side of the window, as shown in the following illustration. In most instances, you will use this window to enter the absolute coordinates, but once in a while you might need one of these alternate data entry options.

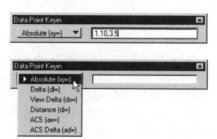

Entering an absolute coordinate in 2D is shown on the left. On the right, the various coordinate options are displayed when you click on the Data Point Key-in option menu.

TIP: *There is a separate keyboard buffer associated with the Data Point Key-in window that stores each coordinate you enter. This allows you to use the up arrow to scroll through previously entered coordinates. Once you have identified one of these coordinates, pressing Enter sends the highlighted coordinates to the active tool.*

Multi Key-in

Sometimes when entering multiple absolute coordinates it is helpful to have the Data Point Key-in window stay open for more than one coordinate set entry. Pressing M will bring up a stay-open version of the Data Point Key-in window. As you enter coordinates and press Enter, this window stays open. Once you have completed entering your coordinate, you must click on the Close button in the window's upper right corner.

This concludes this quick look at some of the additional capabilities of AccuDraw and how it can work for you. As you progress through this book, other AccuDraw features will be introduced on a regular basis. In the sections that follows, you will learn how to navigate within your design using MicroStation's rich set of view controls.

Screen Control at Your Fingertips

Even in these days of telecommuting, most of us do not have the luxury of working where we live. Most times we drive, walk, or even fly to work. At the end of the day, we return home, where we have additional duties to perform (cooking dinner, changing diapers, taking out the trash, painting the house…).

Of course, we cannot perform these home-related duties while at work, and vice versa. Working within a design file is very similar; you perform the various design functions in specific parts of the drawing. You may need to fill out a title block, and then modify a preliminary plan layout or create a scaled-up detail of a complex part. Unless working on a very small drawing, these functions are performed at locations in your drawing that are at a considerable distance from one another. Fortunately, MicroStation provides a set of tools and capabilities for navigating—or commuting, as it were—within or about a design file. Collectively, these are called *view controls*.

Overview of MicroStation's Video Frontier

In previous chapters you were directed to create drawings using MicroStation's many drawing tools. In all cases, you worked within a view window labeled Window 1. The following is a pop quiz question.

Pop Quiz Question: Why do you think this view window is labeled Window 1?

Answer: MicroStation supports more than one view window!

In fact, MicroStation supports a total of eight independent view windows, as shown in the following illustration.

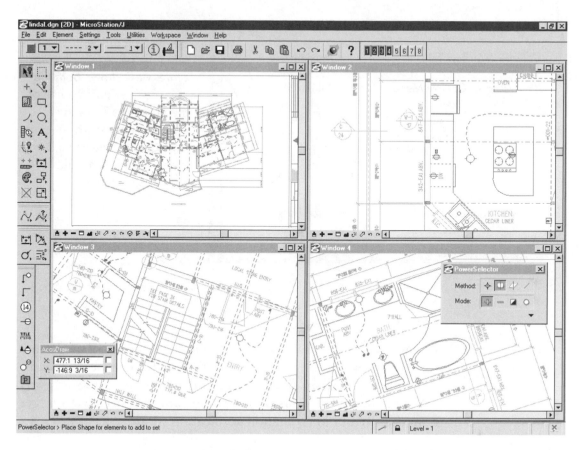

An example of MicroStation's views in use. Note how each view is labeled with its corresponding window number. Also note how all the views are from the same design file (identified by name in the title bar).

Anatomy of a View Window

Each View window contains its own set of view-specific features, as shown in the following illustration. Although many of these features are familiar to most users, there are a few that are unique to MicroStation. To review, let's take a closer look at these features.

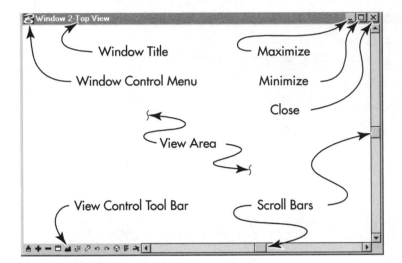

Window Title

Window Control Menu

Maximize

Minimize

Close

View Area

View Control Tool Bar

Scroll Bars

The components that make up the typical View window. Note how the title of the view shows both the view number and the title of the view. The Top view is one of the standard orthographic views built into MicroStation's 3D environment.

Border

Framing each view window is a rectangular border that defines the extent of the individual view. This border can be resized by dragging a point on the border. This works the same as every other Microsoft Windows product. You cannot resize a view larger than the current MicroStation application window.

Title Bar

Located at the top of the view, the primary use of the title bar is to move the entire view. This works the same as with other windows and palettes. To move the entire view, you click and drag the title bar, and the view follows. You can move views off the edge of the MicroStation application window so that only a part of the view is visible. Again, this is very much in keeping with the Windows look and feel.

Clicking on a view's title bar or any part of its border will bring the view to the top, obscuring any portion of other view windows it may overlap with. You can also select a specific view from the Window menu. An arrangement of view windows is shown in the following illustration.

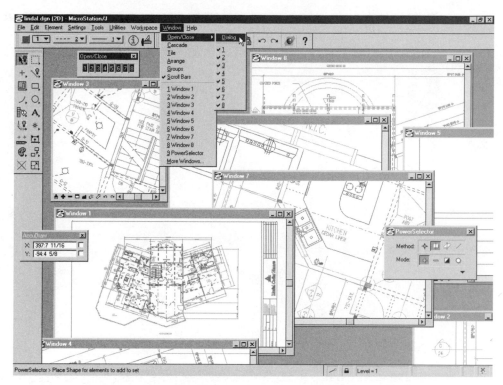

MicroStation with eight views open, as noted by check marks on the Windows > Views submenu, the Window menu itself, and the open Windows Open/Close dialog box. The layout of the views has been purposely jumbled for illustrative purposes. In a real design environment, you would normally apply some organization to your view layout.

Minimize and Maximize Buttons

In the upper right corner of each view are three important view window buttons. Selecting the Minimize button causes the view to shrink to its minimum size. On a minimized window, selecting this button (now labeled Restore) will restore the view to its previous size and location within the MicroStation application window.

The Maximize button is located to the right of the Minimize/ Restore button. Clicking it expands the view to fill MicroStation's application window. When the view is maximized, the Maximize button also becomes the Restore button, so that clicking it again restores the view to its previous size and on-screen location. (See the following illustration.)

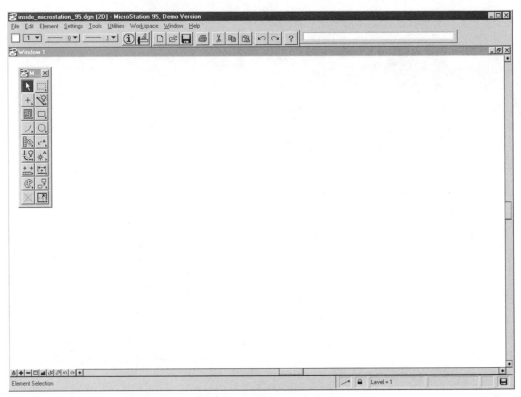

The result of a maximize view operation when the Main toolbox is left floating on the screen . . .

. . . and the result when the Main toolbox is docked along the left edge.

The third view control located in the upper right corner of the View window is the Close button, which, as its name would imply, closes the view window completely. To open a closed view, use the Open/Close command (under the Window option on the MicroStation menu bar).

Window Control Menu

Located in the upper left corner of each view is the Window Control menu, shown at left. Common to all windows in MicroStation, the commands found here are used to manipulate the entire window. Table 5-1, which follows, describes the various Window Control menu commands.

Window Control menu.

Table 5-1: Window Control Menu Commands

Command	Action
Restore	Restores the view window to its former state after a maximize or minimize command.
Move	Provides a method for moving a view window using the keyboard arrow keys.
Size	Provides a method for resizing a view window using the keyboard arrow keys.
Minimize	Minimizes the view window to its minimum size. Also available from each View window (Minimize icon).
Maximize	Enlarges the view window to its maximum displayable size. Also available from each view window (Maximize icon).
Close	Closes the view window. This is the same operation as invoked by double clicking on the Window Control menu icon, or clicking on the Close button on the title bar.
View Attributes	An additional method for opening the View Attributes settings box. Also available from the Settings menu (Settings > View Attributes).
Level Display	Opens the View Levels settings box. Also available from the Settings menu (Settings > Level > Display).
View Save/ Recall	Opens the Saved Views settings box. Also available from the Utilities menu (Utilities > Saved Views).

Views: Windows on Your Design

One of the cornerstones of MicroStation has been its excellent view control commands. MicroStation gives you complete visual control over all of its views. In fact, support for dual graphics screens and the display of multiple views have been major features of MicroStation (or its predecessor, Intergraph's IGDS product) since the PDP11/VAX days (circa 1981). Giving you the ability to "zoom in" for detailed work while maintaining an overview of the design goes a long way toward eliminating the dreaded CAD "tunnel vision" syndrome.

 ## TIMEOUT: An Experiment in Tunnel Vision

The following is a lighthearted experiment you can perform to experience first-hand what is meant by "tunnel vision." It is cheap, quick, and easy to evaluate. Go ahead, give it a try.

Step 1: Procuring Supplies

Go find a discarded cardboard tube, the type found in the middle of paper towel rolls or (heaven forbid!) toilet paper. Found it? Good. Now find a quiet place where your fellow workers will not see you performing the next part of this experiment (you will see why in a moment).

Step 2: The Experiment

Look around the room. Find a picture or another feature on a wall of interest. Turn your back to the feature identified and move to the right, left, or forward (or any combination). Now, place the cardboard tube over one of your eyes, closing the other eye. Now, without prior consideration, turn around and peer through the cardboard tube, trying to locate the upper right corner of the feature or picture. Did you find it immediately or did you have to hunt around a bit.

Now, imagine trying to build that object while peering down the cardboard tube. This is an example of tunnel vision, defined as the "restriction of your primary visual frame of reference" during the design process.

Back in the days when people performed design work at a drafting table, peripheral vision was used to maintain a strong frame of reference with respect to the entire content of a given drawing. By remaining subliminally aware of the entire drawing, one did not have to think about the frame of reference.

In designing with a CAD system, this is not the case. On most systems, when you zoom in to work on a specific area of the design, the rest of the design lies outside your peripheral vision cone, not dissimilar to what you experienced in the previous cardboard tube "experiment." The only way to keep track of where you are within the design is by keeping a mental picture of the entire drawing.

Sometimes, probably more often than we want to admit, we get lost inside our own designs, at which point the only way to reorient ourselves is to back out and reestablish the overview. This constant resetting of the view can be a real time-waster.

The creators of MicroStation recognized this problem very early in its development and devised the multiple-view concept to compensate for tunnel vision. By allowing several concurrently active views within the design file, MicroStation provides a way to get up close and personal with the design data while maintaining that crucial overview (frame of reference).

In the current product, up to eight of these concurrent views can be opened. Although all of these views can be accommodated on one monitor, most users open about four views on each monitor. If you have dual monitors, you can display all of these views (four per monitor), as well as many of the toolboxes and several non-modal dialog boxes.

The "Ship at Sea" View Analogy

A good analogy of "views" are portholes on a ship (see following illustration). When far out to sea, several (if not all) portholes on the side of a ship show the same scene: the ocean, an island, another ship or other objects. All of these portholes have the same virtual point of view to distant objects. Even as the ship were to approach the dock, the portholes would still show much the same view (features of the harbor, docks, warehouses, other ships). At this point, you could call someone in another cabin and point out an object that they would be able to see through their own porthole.

When far out at sea, the view out any given porthole is essentially the same.

But once the ship is docked, this common point of view changes (see following illustration). Whereas you might have a view of a mooring cleat through your porthole, your neighbor may see only a dock pier or a plank. There is no doubt you are both looking at the same dock, just different parts of it. This is yet another example of tunnel vision in action.

Once docked, the view looks very different from each porthole.

The portholes would have the same "zoom factor" because portholes are bolted securely to the same side of the ship. In other words, they approach the dock at essentially the same time. MicroStation does not have this restriction with its "portholes" (view windows). Its views can individually be set to view the entire drawing, or individual details up close, or anywhere in between.

MicroStation's View Controls

MicroStation provides an entire array of view controls. For starters, there are two tools for controlling the distance perspective displayed in a view: Zoom In and Zoom Out. The scale factor applied to the zooming process is user selectable via the Zoom control's tool settings window. By default, the zoom factor is 2:1 per invocation (one data point).

MicroStation' incorporates the view controls into the lower left corner of each view frame. It is next to the familiar horizontal scroll bar of each view window. Additionally, these controls, along with the Copy View tool, can be invoked from the View Control toolbox (Tools menu > View Control).

There are further, view-specific settings found in each View window option menu. This is the menu that appears when you click on the icon in the upper left corner of the view (the Bentley B icon). Finally, there are some view operations you can access by entering commands in the Key-in Browser window (Utilities menu > Key-in), or by using a special mouse operation.

To start the process of understanding how the view controls work, let's do some view navigation using that porthole metaphor from the earlier view discussion. Exercise 5-1, which follows, provides you with practice using a view manipulation command.

EXERCISE 5-1: USING A VIEW MANIPULATION COMMAND

1 Open the file *PORT-HOLE.DGN.* You are presented with what is apparently an empty drawing. However, if you look closely you will discover two tiny dots near the center of View window 1, shown at right. These two dots are the main elements of this drawing. Let's get in closer to these elements.

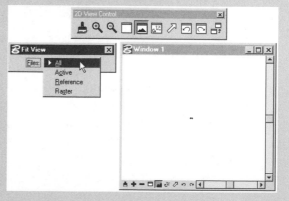

Two small dots at center.

2   Select the Fit View control. You
can select this control from the
View Control bar on the horizontal scroll bar of View window
1 (referred to from this point on
as View 1), or from the View
Control toolbox (see the illustration at right). If you selected
it from the View Control bar,
the extents of the drawing
immediately fit in the view window. If you selected it from the
View Control toolbox, and you
have more than one view win-

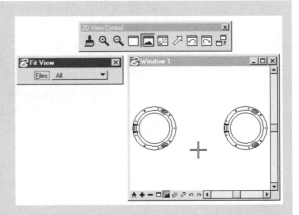

Two portholes visible.

dow open, you are prompted to "Select view to fit," in which case you will
need to place a single data point within View 1.

Two portholes are now present in View 1, as shown at right. Let's zoom in
on one of them.

3   Select the Window Area view control. In response to MicroStation's prompt
"Define first corner point," enter a data point to the left and below the
latch on the rightmost porthole.

4   In response to "Define opposite
corner point," place the second
data point to the right and
above the latch. This defines an
imaginary rectangle enclosing
only the latch of the porthole.
Notice the dynamic rectangular
shape that matches the shape of
View 1. This assists you in getting just the right content of the
overall drawing in your view.
The selected "window" of the
design now fills View 1, as shown
at right.

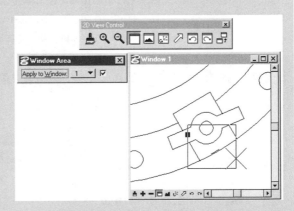

Window filling View 1.

5   Note the small "imperfection" just below the latch. Let's investigate it. You will need to zoom in closer to investigate this item. You will do this by continuing to use the Window Area view control. As an alternative, you could have used the Zoom In control with an appropriate zoom factor specified in the Tool Settings window (the default is 2:1). Chances are, you still cannot make out the "imperfection" in the drawing. No problem, simply continue to use the Window Area control to magnify the area around the "splat."

6   "Window-in" on the imperfection, shown in the following illustration.

As you can see, the "imperfection" is actually the title of this book. Isn't it amazing how much detail you can cram into a drawing?

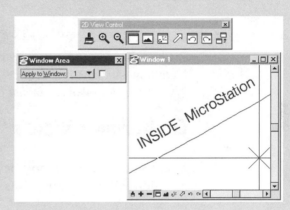

When Can You Use View Controls?

In most instances, you can invoke any view control during any design operation. This means you can pause your line placement operation in mid–data point (so to speak), perform a view manipulation, and then return to the line placement where you left off, as indicated in the following illustration. Most view controls support multiple iterations (for instance, zooming in multiple times), so you have to let MicroStation know you are done with the view control by clicking on the Reset button.

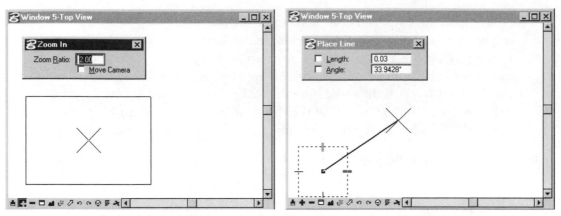

Pressing Reset in Zoom In *returns you to the paused Place Line operation. Note how even the first data point has been retained, which allows you to continue right where you left off.*

Controlling a View's "Point of View"

MicroStation offers many view manipulation commands. Table 5-2, which follows, summarizes these commands.

Table 5-2: View Manipulation Commands

Command	Action
Window Open/ Close	Opens and/or closes a view window
Update View	Redraws the content of a view window
Zoom In/Out	Changes the magnification of a view window
Window Area	Explicitly defines the viewing area in a view window
Fit View All/Active/ Reference/Raster	Redefines a view window's magnification by the extents of the design plane and/or reference files
Rotate View	Redefines a view window's orientation by rotating it based on two specified points
Pan View	Moves a view window's point of view by a distance and direction specified by two points
View Previous	Displays the content of a view window as it existed before you changed it with a view control operation

Table 5-2: View Manipulation Commands

Command	Action
View Next	The reverse of the View Previous command
Copy View	Copies a view window's content and attributes to another view

*Window Open/
Close tool.*

Window Open/Close

The Window Open/Close command is accessed from the Window menu located in MicroStation's application menu bar. You open or close a view window by selecting the number corresponding to it in the Open/Close submenu, as shown at left.

Alternatively, you can activate the Open/Close window (see following illustration), which lets you activate or close multiple view windows without resorting to the pull-down menu. This window can also be docked on the top or bottom edge of the MicroStation application window.

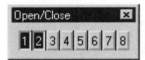

Activated by selecting the Window Open/Close Dialog command, the Open/Close window provides a convenient method for manipulating your views.

Organizing View Windows

When you have several views opened, it quickly becomes obvious you need some tool to help you organize the view windows themselves. MicroStation provides three helpful commands for this purpose. These commands are described in table 5-3, which follows.

Table 5-3: View Window Manipulation Commands

Command	Action
Cascade	Lays out any open views as a cascading set of windows that fill the MicroStation application window from upper left to lower right.
Tile	Lays out any open views in a tile format in which all views are visible and the same relative size. The tiled layout depends greatly on how many views you have open at the time you invoked the command.

Table 5-3: View Window Manipulation Commands

Command	Action
Arrange	This option is a relatively recent addition to the view layout commands. Instead of drastically changing the layout of the views, such as the previous two commands do, Arrange View attempts to maintain the current layout of any open views while optimizing their layout within the MicroStation application window. The cascading and tiling of views are shown in the following illustrations.

Note the order of these views. Cascade always positions the lowest view window number first, in sequential order. Tile starts with window 1 in the upper left, window 2 upper right, window 3 lower left, and window 4 lower right. If you have fewer than four view windows active, the positioning will vary.

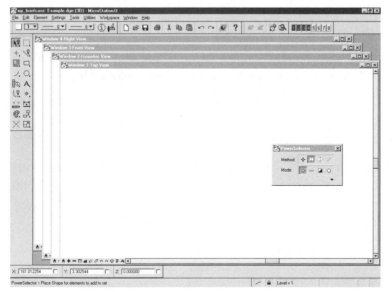

The Window Cascade command stacks all open View Windows like this . . .

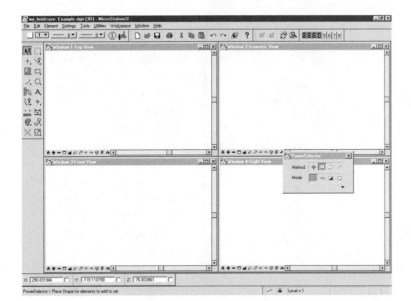

. . . and Tile positions the open view windows like this.

Common MicroStation View Layouts

In day-to-day usage of MicroStation, you will find that two particular layouts are common. The first is the single view maximized to fit the application window. When working on details, a single view is often all you need. Having it fill the application window ensures you have the maximum viewing capability available to MicroStation. You will periodically open additional views on top to perform specific operations, but immediately close them upon completion of those operations.

The other typical view layout is four views tiled. This provides four views laid out "two over two," with all views the same size. In 3D work, this is especially popular, as it closely emulates the standard three orthographic projections (top, front, right) and an isometric view of your work. The following illustration shows an example of view usage in a 3D design.

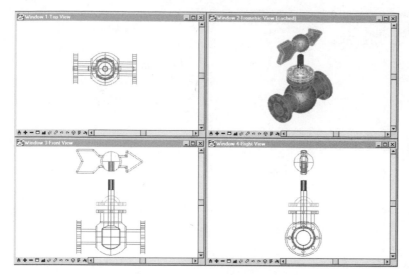

Here you can see a typical example of view usage with a 3D design. Note the names of the views (Top, Front, Right, Iso), which identify the standard orthographic projections commonly used in 3D layouts. Also note the use of a rendered image for the isometric view. In addition, the (cached) label in the Isometric view identifies the use of MicroStation's QuickvisionGL rendering technology.

View Control Tools

Each of the most commonly used view controls is described in the following section. Where appropriate, additional notes and usage details are provided. For the most part, the view controls are selected from the view control tool bar located in the lower left corner of every view window, shown in the following illustration.

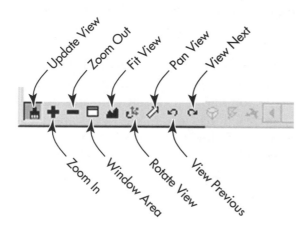

The most-used view controls can be found on the lower left corner of every view window.

Update View

During the course of a design session, your opened views will begin to gather small display "anomalies" such as tiny line fragments not completely erased during a delete element operation, or grid dots that were jumbled by moving a line. This is not an uncommon problem with CAD applications. In MicroStation, this is especially true after performing a dynamic pan (or the horizontal/vertical scroll bars) or an element manipulation command.

The Update View view control tool is used to clean up the appearance of a view. Selecting it from the View Control bar along the bottom edge of a specific view refreshes the content of that window. Alternatively, when Update is selected from the View Control toolbox, you are prompted to select the view window to update. In either case, an Update All Views button appears in the Tool Settings window, which when pressed redraws the content of all open view windows.

Zoom In and Zoom Out

Zoom In and Zoom Out affect the magnification scale of the view window. Once you have selected the control for a specific view, MicroStation prompts you to identify a point about which to zoom. The data point you place becomes the center of the new, zoomed view. The default zoom ratio in the Tool Settings window is 2; however, you can change it to whatever value you choose. Once you have selected the Zoom command, it stays active until you click on Reset. This means you can perform multiple zooms in various open view windows.

Window Area

Used to define a view window's exact content, the Window Area view control lets you identify a specific area of your design for display in a selected view window. You enter two data points to define the rectangular area around your area of interest, and the view zooms as necessary. If you want to display the rectangular area in a view window other than the one from where you selected the view control, identify that view in the Apply to Window field of the Tool Settings window prior to selecting the area.

This ability to let you select a separate target view different from the source means that if you have view windows 1 and 2 active, you can define the area you want to view in view window 1 and have the outcome placed in view window 2. This feature can be very convenient. It lets you display the entire drawing in one view window, on perhaps a secondary screen, while conveniently displaying "window areas" for detail work in another view window, thus minimizing the need for constant zoom and pan operations. (See the following illustration.)

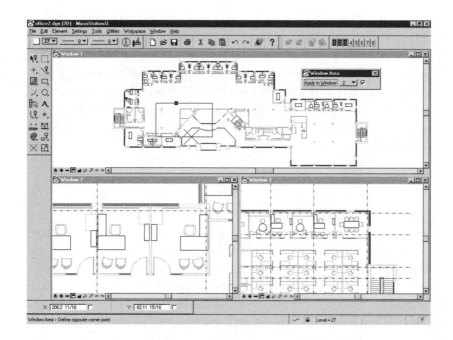

An example of displaying a window area from window 1 to window 2.

Fit View

Sometimes after you have zoomed, windowed, and gotten in close to your work, you need to view the overall perspective of your design. To do this requires either a number of Zoom Outs or the use of the Fit View control. When selected from the View Control bar on the horizontal scroll bar of a view window, the content of your design file, along with its reference files, zooms to fit your view. When selected from the View Control toolbox, MicroStation prompts you to select the view to fit. When you key in *FIT*, Micro-Station prompts you to select the view to fit, but only fits the active design file content, not the reference files.

When the command is invoked from an icon, the four options on the Tool Settings window are All, Active, Reference, and Raster. These fit the design file, along with reference files, in various combinations.

Rotate View

The Rotate View control allows you to rotate the content of a view window for display purposes. The actual elements in your design file are not rotated or moved in any way; only the display axes of the selected view window are rotated. Why would one want to do this?

Consider the layout of a highway that runs in a northeasterly direction. Rather than drawing it along a diagonal direction on a standard rectangular border sheet, it is conventional to draw it running left to right, parallel to the border sheet edge. Now, for the sake of maintaining its precise north and east coordinates, you do not want to rotate the lines and arcs that constitute the highway; you only want to rotate the view to plot the highway along an orthogonal direction.

When you invoke the command, you are asked to define the first point, followed by another point to identify the X axis of the view. If you want to negate the view rotation operation, you can select the Unrotated option on the Tool Settings window.

Pan View

The Pan View control allows you to relocate the view's point of view within the design file. MicroStation simply moves the view along the vector (length and direction) specified by the two data points it requires. A dynamic arrow attaches to your cursor as you click the first data point. This arrow indicates the location and direction the initially clicked data point will move to.

 NOTE: *Do not overlook the scroll bars located on every view. Clicking on an arrowhead on the scroll bar moves or scrolls the view in that direction by a distance one-tenth of the view. To pan a screenful at a time, click on the scroll bar between the slider box and the arrowhead in the direction desired.*

Dynamic Panning

Pressing the Shift key down while dragging your mouse (hold down the mouse button) activates MicroStation's dynamic panning function. As you move your mouse farther from the initial point at which you started the drag operation, MicroStation responds by increasing the panning speed. As the image moves across the screen, MicroStation updates the portion of the screen in the direction of your pan. This results in a continuous update process along the edge of your view. In exercise 5-2, which follows, you will practice panning within a view.

NOTE: *An excellent tool for fine-tuning your view's location, dynamic panning may be a little slow if you are moving from one end of your drawing to the other. In such cases, you might want to use the Window Area command in conjunction with a view in which your entire design is displayed.*

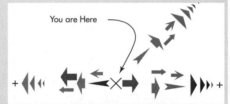

You are Here

EXERCISE 5-2: MOVING ABOUT IN THE DRAWING

1   Open the file *Arrows.dgn,* shown at right.

2   Using the View Open/Close selections (Window menu), ensure that views 1, 2, and 3 are open. Move and resize the views so that they look like those shown in the following illustration (hint: use the view borders to do this).

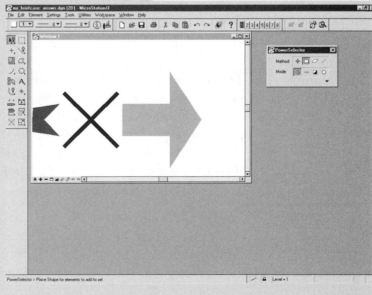

Views as they are to appear after step 2.

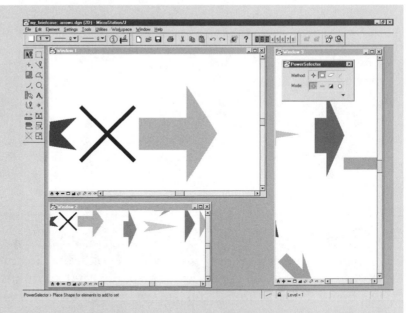

3   In Window (View) 2, use the Fit View control to see the extents of the design. Note the variety and orientation of the shapes shown in the following illustration.

Extents of the design.

4   In Window (View) 1, move to the right using the horizontal scroll bar (click on the arrow, not on the horizontal bar). Keep doing this until you see the red target (+), shown in the following illustration. If you see a green target, you have "moved" in the wrong direction.

The target (red +) you
should see.

5   In Window (View) 1, use the horizontal scroll bar to move to the left until
    you see the green target, shown in the following illustration. This time,
    however, click on the horizontal bar, not on the arrow. Was this faster?
    Using the horizontal scroll bar, navigate back to the starting point (the X
    symbol).

The target (green +) you
should now see.

Extra Credit Assignments:

*   Use the Window Area to navigate to different parts of the design in View 1,
    but use View 2 as your selection window (hint: set the Tool Settings to View
    1).

*   Investigate View 3 and try to figure out why it does not look right when
    compared to views 1 and 2. Fix the view so that it looks the same as views 1
    and 2 (hint: the key is Unrotated).

View Previous/Next

Similar to the Undo and Redo commands for drawing and editing operations found under the Edit menu, this set of commands lets you undo and redo view operations. View Previous negates a view operation. View Next negates a previously negated view operation. Because you can manipulate each view window independently, MicroStation lets you perform this command separately on each. Up to six previous view operations are maintained in each view buffer.

Copy View

MicroStation supports the ability to copy one view's parameters into another view. Click on the Copy View icon in the View Control toolbox (Tools > View Control > 2D). This tool is not found in the view control bar. You are prompted to select the "source" view, and the "destination" view. All of the view parameters of the first view are "copied" into the second view. Do not confuse Copy View with the Copy Element tool.

This is a very helpful tool when you have certain view-dependent features set up the way you like. Rather than going through the tedious process of setting each of these parameters for each view, just copy the established view to other views and then perform your additional view navigation operations.

NOTE: *View controls are also available as a pop-up menu at your cursor location. Press the Shift key and click on the Reset button in a view window.*

Basic Element Manipulation Tools

Element manipulation is one of the major cornerstones of every CAD application. If you could not modify or manipulate objects placed in your design file, CAD would just be an expensive "electric pencil." Fortunately, MicroStation has an enormous array of manipulation tools, one of which you have already encountered, the Delete Element tool. Table 5-4, which follows, describes the basic element manipulation commands available in MicroStation.

Table 5-4: Basic Element Manipulation Commands

Command	Action
Move	Moves one or more elements a user-defined distance*
Copy	Copies one or more elements a user-defined distance
Delete	Deletes one or more elements
Scale	Changes the size of one or more elements by a set scale or user-defined distance
Rotate	Reorients one or more elements with respect to a view's Z axis
Mirror	"Flips" one or more elements' orientation around a defined axis (X, Y, or user-defined)

* You will learn more about identifying more than one element later in this chapter.

All Elements Are Created Equal

All MicroStation elements can be manipulated with these tools. There are no exceptions. This simple statement is an extremely important concept related to CAD. By treating all of the various elements as equals subject to these manipulation commands, MicroStation gives you the opportunity to create your design without worrying about how the pieces are going to be affected by the design process. When it comes to modifying an element (changing only a single aspect of its geometry, such as only one endpoint of a line), this is not always true, as you will learn later.

Manipulate Toolbox

So, where do these element manipulation commands reside? Appropriately enough, you can find them in the Manipulate toolbox, accessible from both the Main toolbox and the Tools menu (Tools > Main > Manipulate). Many users open and dock this often-used set of tools. The Manipulate toolbox is shown in the following illustration.

Manipulate
Toolbox

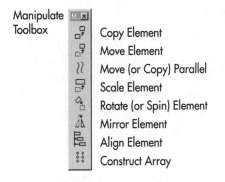

Copy Element
Move Element
Move (or Copy) Parallel
Scale Element
Rotate (or Spin) Element
Mirror Element
Align Element
Construct Array

*Vertical orientation of
MicroStation's
Manipulate toolbox.*

Copy Element

Copy Element, shown at left, is a fundamental element manipulation tool. You select the target element with a data point, whereupon a dynamic copy of the selected element appears on your cursor. A second data point establishes the new location for the element.

The location on the target element where you select it is very important. It becomes the "from" point. The second data point selects the "to" location for the new element. By snapping to the end of an element and entering an absolute coordinate (via AccuDraw's O, for Origin shortcut), you can control the absolute location of the new element. If you data point a third time, an additional copy of the selected element appears. To release the element being copied, click on Reset. An example of the use of the Copy Element tool is shown in the following illustration.

Copy Element tool used to copy a general aircraft outline. Note how the first data point (placed mid-wing on the aircraft) is referenced for the new element.

Move Element

The Move Element tool is similar to Copy Element, except that it moves the original element instead of generating a copy of it. Copy and Move are so closely related that selecting Make Copy in Move Element's tool settings results in the Copy Element tool. The converse is also true (deselecting Make Copy in Copy Element results in Move Element tool activation).

NOTE: *Although you may have no element selected, the copy command can still be active. If you are not careful you may end up with extraneous copies of elements in your design. Always check your status bar for the active command.*

Scale Element

Scale Element, shown at left, is the first of the manipulation tools that requires additional tool settings data to operate properly; in this case, a parameter set collectively called the *Active Scale*. Consisting of two values for scaling along the X-Y and Z axes of the view. The scale is entered as a ratio of the new size to the original size of the element. To scale an element to 50% its original size, you enter .5. To double the size of the element, you enter 2, and so forth. Other tools within MicroStation use the scale parameters, henceforth collectively known as the *Active Scale*. (See the following illustration.)

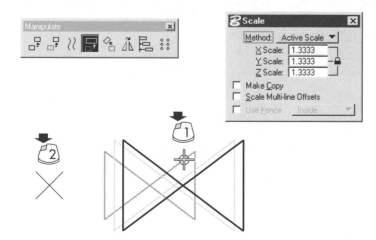

Scale Element uses the Active Scale (set to 1.3333 or 133% of original size). Note the small lock symbol to the right of the X Scale and Y Scale fields, which indicates the fields are locked together. Clicking on this lock will "unlock" the two scale fields, allowing you to enter a different scalar value for each axis.

Although you can specify each of the axes' scalars, in most instances you work with all the scales set to the same value. Recognizing this as a common usage, MicroStation provides you with a locking option (the option appears as a padlock either "locked" or "unlocked") that automatically sets all axes' scales to the same value should you change any one of them via the tool settings window.

It does not matter where you select the element with your first data point. This second data point determines the location of the newly scaled element (more on this in material to follow).

Make Copy Option

Rather than presenting two separate tools, as is the case with Move and Copy Element, MicroStation modifies the behavior of the single Scale Element tool by providing the Make Copy checkbox.

Understanding Scale Element's Seemingly Eccentric Behavior

Once you have selected the element to scale, you will see this element move in relation to the cursor on the screen. However, you will notice that it does not necessarily behave the way you would expect. Sometimes it seems like the dynamic element is moving away from your cursor! Not to worry, this is due to a complicated bit of mathematics being applied to this operation.

Many times when you scale an element, you want it to maintain some sort of relationship to the original element being scaled. For instance, a gear train may need to be designed with an inside gear (known as a planetary gear), nested at a convenient tangent point on the original circle.

In order to guarantee that the newly scaled circle will lie on this tangent point, Scale Element applies the active scale to the distance from the original object. In this way, if you select a one-third scale and tentative point on a circle, the resulting circle will touch the original circle on its circumference no matter where you move the cursor. In exercise 5-3, which follows, you will practice scaling a circle.

EXERCISE 5-3: SCALING A CIRCLE

This is a very simple exercise designed to illustrate the behavior of Micro-Station's Scale Element tool. Take a moment to try this exercise out. Use the Undo command to reset to the beginning of the exercise should you want to try it again and again.

1   Open the example design file *SCALEXER. DGN*. A large circle, the subject of this exercise, is displayed in View 1, as shown at right.

2   Select the Scale Element tool from the Manipulate toolbox.

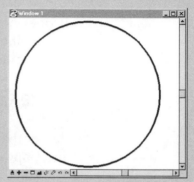

3   Enter *.25* in the X Scale data field. Note how the Y scale changed to match the X scale value, as shown in the following illustration. If it did not, the lock to the right of these fields would appear unlocked. Click on the lock to close it, and reenter the *.25* value.

Beginning design file.

Change in the Y scale.

4   Before continuing, you need to set one more option: Make Copy. Click on this option in the Tool Settings window. Note the tool name change in the status bar as a result of selecting the Make Copy option.

5   Data point once on the circle. A circle 25% the diameter of the master circle should now appear attached to your cursor.

6   Now move the cursor around the circumference of the master circle. Notice anything strange? No matter where you move the cursor along the circle's perimeter, the smaller circle is tangent to the larger one, as shown in the following illustration. MicroStation always performs the scaling operation with respect to the center of the selected object, no matter where you data pointed to select it.

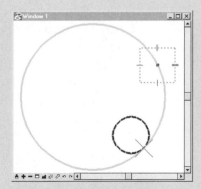

Circles tangent to each other.

7 Tentative point/data point anywhere along the source circle. A circle 25% the size of the previous circle appears. Again, it remains tangent to its parent circle.

The secret to the Scale Element tool is how it applies the active scale to both the object and the distance from the selected object's center point. This applies to all elements, not just circles.

Go ahead and try it with a block, shape, or polygon. Try it with other scales as well. Once you see how the distance scalar works, it will no longer be a mystery why the scale command acts as it does. Many long-time MicroStation users do not understand how this distance scaling works, so you are one up on them.

Active Scale Setting

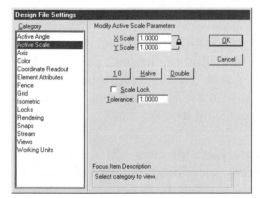

Design File Settings dialog.

Another method for entering the scale parameters is via the Active Scale category in the Design File Settings dialog box, shown at left. Accessed from the Settings pull-down menu (Settings > Design File), this dialog box allows you to set both the X and Y scale values. As with the Scale Element tool, you can lock the two values together via the lock symbol located to the right of the scale fields. Clicking on the lock symbol toggles its effect.

Also found on the Active Scale category of the dialog box are three buttons for performing some common scale manipulations. The 1.0 scale button resets the active scale to 1 (equivalent to entering an *AS = 1* key-in). The Halve and Double buttons change the present active scale values. If the active scale was set to 3.0, and you were to click on the Double button, the result would be an active scale of 6.0.

The final feature of the Active Scale settings box is the Scale Lock. By activating this feature and selecting an appropriate tolerance value (normally set to 1.0), you can force the scale value to be a multiple of the tolerance. Leaving the tolerance set to 1 essentially provides you with a method of forcing the active scale to an integer value.

The active scale can be any number, large or small. For instance, if you wanted to reduce an object by 50%, you would enter .5. Typing in *20* would multiply an element's size twenty times. The default value is 1, meaning no scaling is done to the element.

TIP: *Via the Key-in window, you can also change the active scale by entering* AS=<scale value> *via the Key-in Browser window (Utilities menu > Key-in). One advantage to using AS= is its support of simple arithmetic expressions. For instance, to double the current active scale, you can enter* AS=*2. *Note, however, that any such change is subject to the current scale lock.*

Rotate Element

Now that you have mastered the Scale Element tool, let's look at the tool most closely related to it, Rotate Element, shown at left. Instead of applying a scale to a selected element, Rotate Element applies an angular rotation. As with the Place Arc tool, Rotate Element needs a sweep angle to perform its task.

In its most basic form, Rotate Element gives you the ability to rotate an element by graphically specifying the angle of rotation. Called the 2 Points method (set in the Tools Settings window of course!), you provide three data points: one to select the target element, one that specifies a pivot point, and one that defines the angle of rotation.

The actual angle of rotation is calculated by the angle between the view's X axis passing through the pivot point and the third rotation termination point. You will find your first encounter with this tool to be fun. Do not fight it! Go ahead and spin that element a few times before clicking the data point button.

Using the Active Angle Method

Although the "spin cycle" of the Rotate Element tool is an important method for certain design circumstances, most users operate this tool with the Active Angle method, an example of which is shown in the following illustration. By keying in a specific angle value and selecting the element to be rotated, the user maintains precise control over how MicroStation rotates this element.

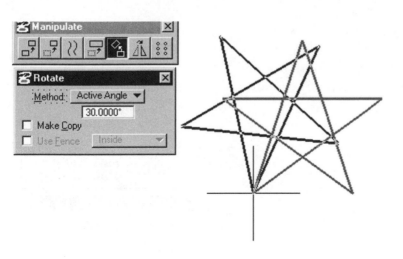

The Rotate Element tool in use with the Active Angle method.

By entering an angle of 45 degrees, you can select any element and MicroStation will rotate it for you. The first data point you supply will select the element; the second data point selects the point of rotation. If you tentative point to the actual element, the result will be a rotation about this point. This results in the figure rotating about that point. As is the case with the scale command, there is the Make Copy option. In exercise 5-4, which follows, you will practice rotating an object.

TIP: *A relatively new user enhancement to the Rotate Angle is the predefined angle increment feature. Shown as a set of up/down arrows to the right of the angle data field, this productivity feature allows you to adjust the current angle up or down by a predefined increment (by default, 90 degrees). You can change this increment in the Design Settings dialog box (Settings > Design File > Angle category). The author normally sets this value to 15, 30, or 45 degrees, depending on the current design project (2D isometric schematics are almost fun with this setting!).*

EXERCISE 5-4:   ROTATING AN OBJECT

1   Open *ROTEXER.DGN*. This design consists of a single diamond shape, as shown at right. In this exercise, you will be creating rotated copies of this shape.

2   Select Rotate Element from the Manipulate toolbox. Set the method to Active Angle in the Tool Settings window and enter an angle of 30 degrees in the appropriate field. Enable the Make Copy option. (See following illustration.)

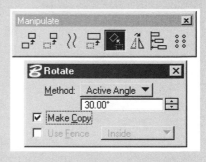

Diamond shape.

Settings for step 2.

3   Place a data point anywhere on the diamond shape. A dynamic and rotated copy of the diamond will appear on your cursor, as shown in the following illustration. Notice how it follows your cursor if you move it along the perimeter of the original element.

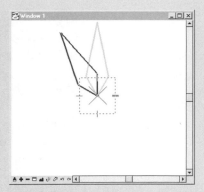

*Altered diamond shape on
the cursor.*

4 Place a tentative point, followed by a data point, at the bottom vertex of the diamond.

TIP: *This part of the exercise is much easier to perform if you use AccuDraw. Snap to the bottom vertex of the diamond shape, whereupon the AccuDraw compass will appear. Further data points are guaranteed to be at this same location, as long as you keep the cursor "indexed" to the compass' center point.*

Another dynamic shape appears, rotated an additional 30 degrees, as shown at right.

5 Place a tentative point on the second data point at the same bottom vertex of the diamond (if you are using AccuDraw, you do not need to tentative point first, just data point while indexed to the AccuDraw compass).

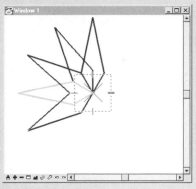

Additional dynamic shape.

Another copy of the diamond appears rotated a further 30 degrees. You can continue the tentative point/data point sequence several more times to create an interesting flower-like pattern. Because each successive data point was in the same location, the result is a circle of shapes that returns to the original element's initial location.

Try placing additional shapes and apply other angles of rotation or alternating data points. In most instances, the results will be a pleasant rosette type pattern, as seen in this exercise.

Active Angle Parameter

As with the Active Scale setting, MicroStation's Active Angle is a global parameter used by a number of commands and tools; and, as with the Active Scale parameter, there are a number of ways to enter active angle values. Changing the active angle value with any tool that uses it will change it for all tools you subsequently select.

Many first-time users of MicroStation miss this aspect and ask the question "Why is my geometry rotated?" This key-in of the active angle automatically updates any active angle data fields presently displayed on your screen. In addition, you can enter the active angle via the Design File Settings dialog box.

Active Angle Setting

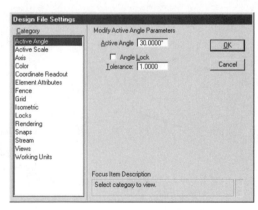

Active Angle category.

The Active Angle category on the Design File Settings dialog box (Settings > Design File), shown at left, gives you another way to set the active angle. You simply key in the angle in the data field provided. Note the presence of the Angle Lock checkbox and the Tolerance field. If you enable Angle Lock, you can only set the active angle to a multiple of the tolerance value specified. Thus, if the tolerance is 1.0000 and the Angle Lock checkbox is enabled, any active angle you key in is rounded to the nearest integer.

MicroStation always measures angles from the horizontal in the counterclockwise direction. However, there are two other methods for specifying angles used in design work: azimuth and bearing. An example of an azimuth angular reading is shown in the following illustration.

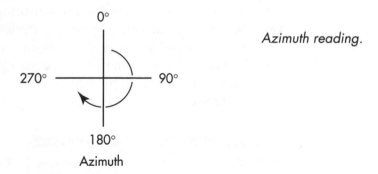

Azimuth reading.

Azimuth refers to angular readout from the vertical, as in a north direction, and following a positive angular degree clockwise through 360 degrees. Bearing angle measurement follows the four points of a compass and is expressed in a primary direction followed by a degree of rotation and a secondary direction. For example, N30E refers to North 30 degrees, toward the East. An example of a bearing angular reading is shown in the following illustration.

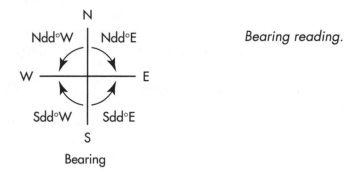

Bearing reading.

Recognizing the need to support these systems, MicroStation is equipped with the ability to accept input and display angular information in both azimuth and bearing methods, as well as the "conventional" counterclockwise from horizontal readout shown. MicroStation can also use and display the degrees, minutes, and seconds necessary for most surveying and mapping work.

Coordinate Readout Setting

The various display formats for angular information are set in the Coordinate Readout category of the Design File Settings dialog box (Settings menu > Design File), shown in the following illustration. In the Angles section, with the Mode options menu item, you select the appropriate angular measurement system, and the format and accuracy of the readout.

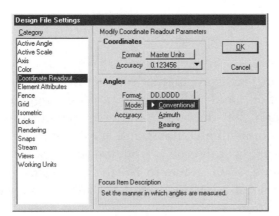

Coordinate Readout category.

In a fashion closely related to the angular readout, using the options presented in the Coordinates section of this dialog box, you can also set how MicroStation displays its XYZ coordinates. You can select whether MicroStation displays such information in working units, subunits, or master units.

Mirror Element

Another fundamental function of MicroStation is its ability to mirror one or more graphic elements around an arbitrary axis. This is accomplished using the Mirror Element tool, shown at left.

To illustrate how the Mirror Element tool relates to your design, you can perform a very simple experiment using a small pocket mirror (sorry, you will have to provide your own mirror). If you take a mirror and stand it vertically on any drawing, the resulting mirrored image shown in the mirror is controlled by how and where you place it on the drawing (see the following illustration). The bottom edge of the mirror is the axis about which the image is mirrored.

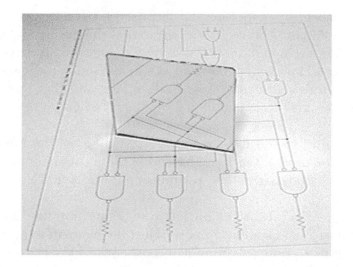

The Mirror command uses the bottom "edge" of the mirror to define the axis.

When using the Mirror Element tool in MicroStation, you must define this axis. To help you identify two of the most obvious axes (X and Y), the Mirror Element tool provides the Mirror About option in its tool settings. You can select to mirror about the X axis or the Y axis, or define an arbitrary line to mirror about, as indicated in the following illustration. As with most of the element manipulation tools, the Make Copy option creates a new element rather than mirroring the original one selected.

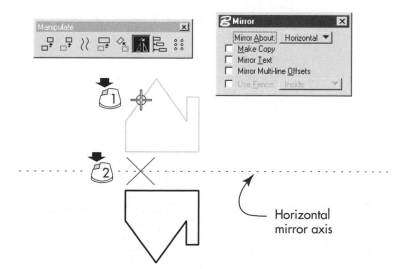

When using the Mirror Element tool, you define the axis rather than the direction of the mirror operation.

Horizontal mirror axis

First-time users of this tool may be surprised by its initial results. Because the Mirror About specifies the axis of the mirror and not the direction, the operation seems to be the opposite of what you would expect. However, defining an axis makes much more sense when you think about your design as a model and that you are really specifying a reorientation of its geometry based on this axis.

In other words, to flip the element horizontally, use the Mirror About/*Vertical* option. Once you have used these commands a few times you will understand their relationship to the X and Y axes.

The Mirror About/Line option lets you define any line about which to mirror an element. You first select the element to mirror, specify the first point on the mirror axis, and click a third data point to identify the direction of the axis. The following illustrations should help you visualize the mirror axis. In exercise 5-5, which follows, you will practice mirroring an object.

EXERCISE 5-5: MIRRORING AN OBJECT

1   Open *MSHAPE.DGN.* An obviously asymmetrical shape is displayed in View 1, as shown at right. In this exercise, you will mirror this shape along various axes to get a feel for how the Mirror Element behaves.

2   Select the Mirror Element tool from the Manipulate toolbox, shown in the following illustration. Set Mirror About to Vertical and enable the Make Copy option.

Shape shown in View 1.

Mirror Element tool of the Manipulate toolbox.

3   Data point somewhere on the shape. Once you
have, watch the dynamic copy of the element
(shown at right) as you move your cursor
around the view. Note how the distance from
your first data point to the current cursor location is also mirrored.

4   Data point to the left of the original object.
The Mirror tool will continue to prompt you
for a new location of the copy. To release the
shape, click on Reset. Before proceeding,
select the Undo command from the Edit menu
to remove the mirrored copy of the shape.

Dynamic copy of the shape.

5   Next, try the Mirror About/Line option. With
the Line option set, data point anywhere on the
original shape. Next, click a tentative point followed by a data point at the location labeled A in
the diagram, as shown at right. A dynamic mirrored image of the shape will appear to rotate
from this location. Do not be fooled, however;
this rotation is actually the mirror process shown
as the axis of mirror passes through your last data
point and the current position of your cursor.

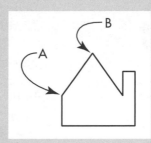

*Tentative point and data
point at location A*

6   Click a tentative point at point B in the previous illustration. The sloping
line from point A through point B becomes the axis of mirror, as shown in
the following illustration.

Axis of mirror.

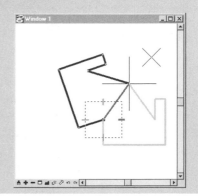

Remember that all mirror operations occur about the selected axis. Vertical and Horizontal refer to the axis, not the mirrored results.

Much Ado About Undo

Now that you have learned a few of the more powerful element manipulation tools, it is time to revisit for a moment that safety net feature called Undo. Say you are about to take on a particularly nasty design session that will markedly alter the design file. You are not sure that what you are about to do will create a solution to your design problem; however, you must try. You also believe it may involve so many changes that you are not sure the Undo buffer will hold them all.

Not to worry! MicroStation provides for such design process considerations with the Set Mark and Undo Other commands, shown in the following illustration. By "setting" a mark (Edit > Set Mark) prior to starting the design process, you can return the design to the premarked condition by using the Undo Marked (Edit > Undo Other > To Mark) command. This provides the equivalent of a "What if" function.

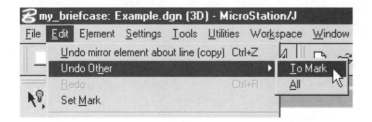

The Undo To Mark command provides the ability to return your design to a predetermined state identified by previous use of the Set Mark command.

An additional option under the Undo Other pull-down menu is the Undo All command (Edit > Undo Other > All). This command essentially reverses all of the work performed on a design file since it was called up in the current MicroStation session. Because of its destructive nature, and the fact that there is no Redo All equivalent, you are presented with an Alert in which you have to OK the completion of this operation.

Manipulating Multiple Elements

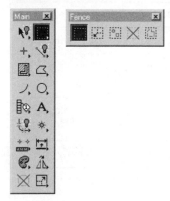

Use Fence option.

With all of the element manipulation tools just described, you may have noticed an option associated with them called Use Fence, shown at left. This option refers to MicroStation's ability to act on more than one element at a time. Collectively called the Fence capability, this feature is an important part of how you work with MicroStation.

Fence Manipulations

To have MicroStation operate on the content of a "fence" with the element manipulation commands discussed earlier, you first need to place a fence around those elements.

This fence element you create is not really a true element but rather a temporary construct that allows you to identify a part of the drawing you wish to modify. When the Use Fence checkbox is enabled for a tool, MicroStation acts on the elements enclosed in the fence.

NOTE: *Fence manipulations can operate on a large number of elements. If the Undo command does not restore your manipulations, you may need to increase the size of the Undo buffer.*

Place Fence

Place Fence, shown in the following illustration, provides six methods of placing a fence. Table 5-5, which follows, describes these methods.

The Place Fence tool implements six ways to invoke it, as seen in the Fence Type pop-down field in the Tool Settings window.

Table 5-5: Fence Placements Methods

Fence Type	Effect
Block	Places a rectangular fence when its diagonal points are specified
Shape	Places an irregular-shaped fence with data points supplying its vertices
Circle	Places a circular fence upon specifying its center and a point on its circumference
Element	Uses an existing closed shape in a design file and places a fence around it
From View	Places a rectangular fence around the perimeter of a view window
From Design File	Places a rectangular fence that surrounds all elements in a design file (works only with 2D design files)

The most commonly used type of fence is the Block, a rectangular shape. You place it by clicking two diagonal points. The Shape fence is also commonly used when trying to operate on a specific group of elements in a congested area. The best way to learn the different fence placement options is to try them out. You are sure to find good uses for them as you get more acquainted with MicroStation.

Using the Fence with Your Tools

By placing a fence and then selecting the Use Fence option, the tool will act on those elements enclosed by the fence. Once you have surrounded the elements with your fence and have activated the appropriate tool, additional data points are needed to direct the tool's operation. Unlike the single-element operation, wherein your data point typically identifies the element and the focus point of the tool, fence operations require you to provide this from/to coordinate information with two data points. In all cases, the element manipulation tools function the same as their single-element cousins.

It is easy to see how a fence manipulation command can modify an element totally enclosed within the fence. What is less obvious is what happens when an element crosses the fence boundary.

Associated with all of the element manipulation tools and the Fence tools is something called the Fence Mode. Seen as a pop-down field next to the Use Fence option, these modes control what happens when a fence-enhanced operation is executed. Table 5-6, which follows, describes the Fence Modes.

Table 5-6: Fence Modes

Mode	Action
Inside	Acts on all elements totally enclosed within the fence
Overlap	Acts on all elements within the fence and those that overlap the fence boundary
Clip	Acts on all elements within the fence and only that portion of any element that crosses the fence
Void	Acts on all elements completely outside the fence
Void-Overlap	Acts on elements outside the fence and those that overlap the fence
Void-Clip	Acts on all elements outside the fence and only that portion of any element lying outside the fence

The last three fence modes—Void, Void-Overlap, and Void-Clip—complement the first three modes. Think of the void modes as the "inside out" modes. In most cases, you will want to avoid the default use of these three modes; instead, use them only when needed and return to one of the three normal fence modes.

WARNING: *In fact, it is safest to return to the Inside Fence mode at the conclusion of most operations, as the accidental use of the Clip operation could be disastrous. When you invoke a fence manipulation command you are affecting a large number of elements at one time. If you are indiscriminate with the placement of your fence, this manipulation may lead to changes you did not anticipate or desire. For this reason it is a good idea to understand the fence modes thoroughly.*

Clip Fence Mode

Clip is by far the most "aggressive" (read: dangerous!) option of all. In this case, dangerous is not too strong a word to use. This fence mode chops any elements that cross the fence into separate pieces lying inside and outside the fence.

The ones lying inside the fence are then affected by the fence command chosen. Those lying outside the fence are "left alone," but the act of chopping has, of course, changed those elements as well.

The Clip option also points out another powerful feature of MicroStation: constructing one type of element out of another. If the overlapping element were a circle, the result—after, say, a Move Element operation—would be two arcs. A block would become two line strings.

The indiscriminate use of the Clip option can have nasty consequences. How do you avoid potential problems? Develop the habit of always selecting the Fence Mode/Inside option when you have completed the use of a fence clip manipulation. This also applies to the use of the Overlap option. This does not mean that you should avoid using the clip function. It is a very important and powerful function and has its place. Just remember to turn it off when you are finished with it.

Modify Fence

The Modify Fence command, shown in the following illustration, allows you to relocate a specific vertex of a previously placed fence, or to move the location of the entire fence. You select the tool, then the option (Vertex or Position), from the Modify Mode pop-down list in the Tool Settings window, and click to identify the vertex or fence. Finally, another data point identifies the new location for the vertex or the fence.

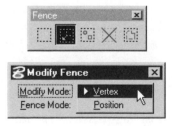

The Modify Fence icon implements two commands: Modify Fence Vertex and Modify Fence Position (or Move Fence).

Manipulate Fence Contents

If you enable the Use Fence checkbox while using tools in the Manipulate toolbox, MicroStation switches to operate on the content of the fence. However, this is not the only method, or even all the commands available, for manipulating the content of a fence. There is the Manipulate Fence Contents tool in the Fence toolbox. In addition to providing another way to copy, move, rotate, scale, or mirror fence content, it implements the Fence Stretch command, shown in the following illustration.

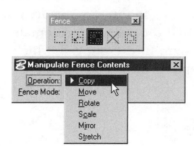

The Operation pop-down list for the Manipulate Fence Contents tool includes the Fence Stretch command, not found on the Manipulate toolbox.

Table 5-7, which follows, summarizes the function of each fence operation.

Table 5-7: Fence Manipulation Operations

Operation	Action
Copy	Copies the content of a fence after you specify the "from" and "to" points
Move	Moves the content of a fence after you specify the "from" and "to" points
Rotate	Rotates the content of a fence, but does not offer the 3-point rotation method otherwise available if the Rotate Element tool with the Use Fence option is used
Scale	Scales the content of a fence
Mirror	Mirrors the content of a fence about the vertical or horizontal axis
Stretch	Only available from the Fence toolbox, it allows relocation of the vertices of elements that fall within a fence

In general the tools found on the Manipulate toolbox, with Use Fence enabled, offer greater flexibility than similar tools in the Fence toolbox.

Delete Fence Contents

Remember our old friend the Delete Element tool? You may have noticed that this important tool has no options. There are times, however, when you may want to delete entire portions of your design. As a safety precaution, MicroStation provides this Delete Fence Contents tool in the Fence toolbox, which is shown in the following illustration. In this way, you are reminded of the current fence mode so that you will not accidentally delete the wrong portion of your design. In exercise 5-6, which follows, you will practice fence manipulations.

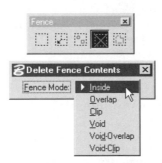

The Delete Fence Contents tool on the Fence toolbox.

 NOTE: *Whenever you select the Use Fence option, the status bar displays the current fence mode. You should always check this information prior to performing your tool's operation.*

EXERCISE 5-6: FENCE MANIPULATIONS

1 Open the *FENCEXER.DGN* design file. You should see the elements shown at right.

2 Place a fence block around the highlighted elements in the center of this design.

3 Next, move the elements inside the fence using the Move Element tool. Select the Use Fence option and the Fence Mode/Inside. Click a data point once at the center of the elements (the big

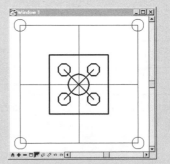

Beginning design file.

X) inside the fenced area and about a quarter of the way up from the bottom of the screen, as shown in the following illustration.

The elements enclosed in the fence moved to the new location based on your data points. Note how the fence shape travels with your cursor. Only those elements totally enclosed within the fence block were moved. Those elements crossing over the fence were not affected. This was due to the Fence Mode setting of Inside.

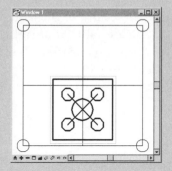

Moving the elements.

4   Now, set the Fence Mode to Overlap. Click a tentative point, followed by a data point on the center point of the objects you just moved, and click a data point once at the center of the large square. (Hint: Use the Center snap mode.) Now click a data point again. The result should be that elements that overlapped were moved as well, as shown at right.

What happened? The elements moved back to the center of the square but the intersecting lines have now moved down. Because they crossed the fence boundary, they were moved as well.

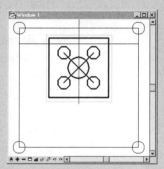

Overlapping elements moved.

5   Next, try the Clip fence mode. Use Place Fence Shape and create an irregular fence that overlaps various elements in the middle of the design. Cut through the circles and the square, shown at right.

6   Select the Move Element tool once again and select Use Fence with the Clip fence mode. Click a data point once inside the fence and a second time down and to the right of the fence. Instead of circles and blocks you have arcs and line strings. The

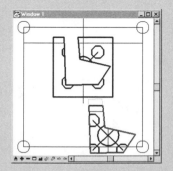

Using the Clip fence mode.

clip mode radically changes the appearance of your design file. This is why the clip lock is so dangerous. In fact, if you had a large number of elements modified in this way, and your Undo buffer was not large enough, you might not be able to recover from such an inadvertent command.

7   Make sure to set your fence lock to Inside (shown at right) before proceeding—an excellent habit to develop, starting now.

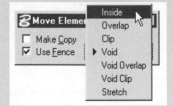

Fence Lock set to Inside.

The Selection Set

An alternative method for working with more than one element is the Selection tool. *Selection set* is the name given to the collection of elements you have either identified with the Element Selection tool, selected with the Select All command, or established with the Select by Attributes facility (Edit menu > Select by Attributes). This last feature is discussed in a later chapter, but the first two are easy to understand.

Select All

This command is very simple. Once activated (Edit > Select All), it selects every element in your current design file as your selection set. Any command or tool selected after this will act on all of the elements in your design file.

WARNING: *Be careful! This command is similar to using one of the more dangerous fence modes, and you may not be able to undo it.*

Element Selection Tool

Located on the Main tool frame, this arrow-like tool is the main method for selectively identifying elements for incorporation into the selection set. The Element Selection tool, shown at left, is simplicity itself. You identify elements by clicking on them one by one while holding down the Ctrl key. You can select more than one element by clicking and holding the data point button and dragging the cursor diagonally across the elements desired. In either case, the selected elements will be highlighted with small filled boxes called handles, as shown in the following illustration.

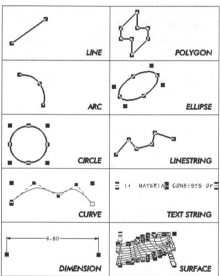

These handles serve two functions. First, they identify elements that are already part of the selection set. Second, by clicking on a high-lighted element's handle you can change the location of the endpoints of a line, the diameter of a circle, or the corner of a shape. If you click and drag somewhere along an element's length, the result is similar to the Move tool. This is as close as you can get to a modeless operation in MicroStation.

Handles associated with selected elements.

NOTE: *With the release of MicroStation/J, the display of the element handles has been replaced with an element highlight color. Most users prefer the highlight color; however, you can reinstate the display of the handles (Workspace menu > Preferences > Input > uncheck Highlight Selected Elements).*

Selection Set and the Element Manipulation Tools

One of the main features of the selection set is its effect on many of the element manipulation tools. When you have a selection set active, all you have to do to act on these elements is select the appropriate tool. The act of selecting the tool actually executes its operation. No "Accept/Reject." No "Data point to accept." Even tools such as Delete Element offer no prompting when invoked with a selection set active. In fact, with a selection set active, all you have to do to delete these elements is press the Delete key on the keyboard.

In the case of a tool that may require a data point to proceed, the selection set is treated like a fence. For instance, the Move Element tool requires two data points to perform its task. One minor difference in the operation of Manipulate Element tools with selection sets is how they terminate. With the "normal" tool, like Move Element, you are prompted to keep moving the element (or fence) until you hit Reset. With a selection set active, the Move tool prompts you for the obligatory from-to data points, drops the element handles, and then returns active control to the Element Selection tool. The elements in the selection set remain highlighted, waiting for your next command.

Deactivating the Current Selection Set

To discontinue the use of the selection set feature, you click in an empty area on your design plane without holding down the Ctrl key. This assumes the Element Selection tool is active. You can deselect an individual element of the selection set by selecting it with the Ctrl key held down.

PowerSelector!

PowerSelector.

Without a doubt, one of the most innovative tools introduced with MicroStation SE was PowerSelector, shown at left. This is a tool that allows you to add, remove, and invert elements in a selection set. Previously, you could only remove or add elements by pressing the Ctrl key while clicking or dragging across elements. PowerSelector introduces a much more user-friendly way of setting up your selection set.

PowerSelector's Methods

PowerSelector provides a wealth of methods for selecting elements in your design for further manipulation. Instead of the traditional

Method option menu you have seen in several tools to this point, PowerSelector provides an optimized iconic set of options in its Tool Settings window. This user interface change emphasizes the almost continual usage this tool gets by a majority of MicroStation users (it is second only to AccuDraw in popularity).

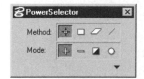

Each icon associated with PowerSelector's tool settings adjusts both the method by which you identify elements within the design file and how PowerSelector will treat those elements (the Mode settings).

The best way to learn PowerSelector is simply to use it. There are some additional capabilities within PowerSelector, which are covered later in the book. The only one you should really be aware of now is the toggle between overlap and inside operation of the selection method. Double clicking on any of the method icons changes the icon graphic and the way PowerSelector responds to your selection action, as indicated at left.

A broken line appears across the icons, which indicates that the selection process will highlight elements both inside the dynamic selection box and any elements that cross the selection box. This is similar to the Fence Overlap operation demonstrated earlier. Double clicking on a method icon will toggle this operation back to inside, as indicated in the following illustration.

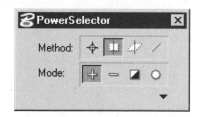

A double click on a method icon within AccuDraw's tools settings toggles between inside and overlap operation. Note the dashed appearance of the block and shape icons, which indicates overlap mode is active.

More on PowerSelector's special features is presented later in the book, but for now, just practice with it and soon you will learn to love it! In fact, you can set PowerSelector to be your default tool (highly recommended!). You do this via the Preferences dialog box (Workspace menu > Preferences > Tools category > Default Tool/PowerSelector).

Summary

You have now been introduced to most of MicroStation's primary element manipulation tools, as well as how to navigate around within your design. There is much more to MicroStation but, by now, you are beginning to grasp the very powerful nature of MicroStation itself. Time will show that it is, indeed, a very capable CAD program. More importantly, it truly enhances the design process.

A couple of parting thoughts before the next chapter. First-time users of MicroStation tend to underutilize the views and overuse some of the element manipulation tools. This is natural. The ability to copy, move, and rotate elements gives you a freedom you have never before experienced with drawing.

On the other hand, navigating about the design takes getting used to. Do not worry, as you get comfortable with MicroStation and its viewing ability you will begin to use more and more views. Before long, you will think nothing of having all eight views active and displaying totally different parts of your design. At that point, you will be an old hand at MicroStation.

The next chapter is a very important one. There you will learn about element attributes, additional element types, and how to modify your elements (as opposed to manipulate). The next chapter also revisits snaps and how you use them with the tools to construct the objects of your design.

2D BASICS, PART 3

More about MicroStation's Elements and Tools

IT SEEMS THE MORE YOU LEARN ABOUT MICROSTATION, the more there is to discover. As you move into the final areas of MicroStation's basic functions, the tools discussed should seem somewhat familiar. This is due to the uniform operation of most MicroStation tools. Once you have mastered a certain number of these tools, the rest will be easier to comprehend, even if you have never used them.

Element Attributes

So far you have been dealing with elements that look, well, rather plain. When drawing on paper with traditional drafting techniques, you use various line widths and styles to represent the various features of a drawing. In MicroStation you perform the same function by setting your elements' display characteristics or attributes. Elements have several such attributes, including weight (thickness), color, style, and level.

Setting Line Weights

When you use a technical pen to create a drawing, you expect uniform and predictable line widths. You put up with the maintenance that comes with all technical pens for the excellent results such drawing instruments provide. When you use a CAD system, you expect the same uniform results.

MicroStation gives you 32 different line weights, which are comparable to 32 different technical pens. The term *line weight* is used as opposed to line *width* or *thickness* because MicroStation does not actually assign a specific line thickness to each element. Instead, an element's weight is considered a logical (or virtual) characteristic. You assign the weight to true thickness at the time you generate the plot or web output. In most instances, this weight-to-line-width assignment is handled by the plotter or printer driver specification file.

This all sounds very complicated, but in practice is rather simple. Seldom, if ever, will you find a need to use all 32 weight values. In fact, most of the time you will find that the first eight or so will suffice. When generating plots of your designs you will find that there are some nice benefits to using the various weights.

When you select a line weight from either the Primary Tools toolbox, shown in the following illustration, or from pop-down fields in any other tool, it becomes the *active line weight*. When you create new elements in your design file, they will automatically take on this weight attribute. Because these weights are an abstract, logical characteristic of an element, how they are displayed on the screen can be somewhat disconcerting the first time you see them. MicroStation uses weights something like line thicknesses. This, however, can be very misleading when trying to associate a weight with a specific line thickness. In exercise 6-1, which follows, you will gain some experience in working with line weights.

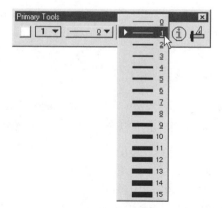

The first 16 line weights are directly accessible from the Primary Tools toolbox.

EXERCISE 6-1: WORKING OUT WITH WEIGHTS

1   Activate the *LINEWT.DGN* design file. Note the two circles, shown at right. The bottom circle is apparently "thicker" than the top one. This is due to its use of a different line weight.

2   To verify this, use the Element Selection tool (the arrow icon) to select the top circle. Then select Analyze Element from the Primary Tools toolbox. The weight field displays the number 0, as shown in the following illustration.

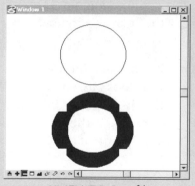

LINEWT.DGN *file.*

The Analyze Element tool is useful for assessing the details of a specific element.

3 Analyze the lower circle. This time, the circle has a weight of 31, the maximum value allowed. This explains the difference in appearances between the two identical circles.

4 Close the Information window.

5 Select the Zoom Out tool from the View Control menu (the magnifying glass with the minus sign). Data point once between the two circles.

6 Surprised by the result? The heavy-weight circle appeared to get thicker, as shown at right. This is actually an optical illusion. In reality it remained the same as before, the difference being that the circle's diameter shrank in response to the zoom action.

7 Zoom out again. It now appears as though the heavier circle is overwhelming its lesser-weight cousin.

8 Zoom out a few more times. The "fat" circle now looks like a cross. This again is due to the display characteristic of line weights. Even when the circle is too small to really make out, the weight routine will still "paint" it at the weight given.

Effect of the zooming action.

As you can see, the effect of the heavy weights is dependent on the scale of the view. MicroStation paints all weights with the same number of screen pixels no matter what the zoom scale of the view.

Working with Line Styles

The other major attribute most closely associated with manual drafting techniques is the use of various line styles. Dashed, dotted, and even phantom lines are immediately recognized in a drawing and do not require further description. In MicroStation there are two distinct categories of line styles. The first category is the set of eight "line codes" that are hard-coded into the product.

The second category is the user-definable (custom) line styles. These were introduced in 1995 and allow you to create your own line styles defining stroke (dash and gap) patterns, point symbols (embedded and repeated drawing symbols), and complex styles (a combination of stroke patterns and point symbols).

To see what MicroStation offers for line styles, simply select the line style options menu from the Primary Tools toolbox, shown in the following illustration. The resulting pop-down menu shows the standard hard-coded line styles as defined since before there was a MicroStation. (They were introduced with Intergraph's IGDS minicomputer based CAD product in the early 1980s.) The pop-down menu also contains the selection Custom.

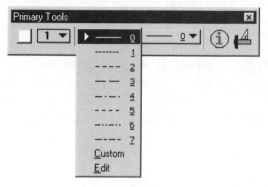

Pop-down line style options menu.

Selecting Custom brings up the Custom Line Styles dialog box. Here you will find a couple of dozen additional line styles, ranging from Border to Wide Dash. All of these are available for use in your design file. To see the actual elements of the line style selected, select the Show Details option. However, before you go totally nuts and start drawing timberlines and railroads, you need to take a look at the basic line styles that are an integral part of MicroStation, such as those shown in the following illustration.

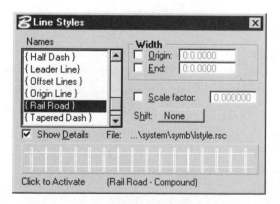

MicroStation is delivered with dozens of predefined custom line styles. The ones shown here are part of MicroStation's default workspace.

The actual list of available line styles you see will depend on a number of factors, including the workspace you have active, any corporate styles defined by your system administrator, and those that are unique to your current project. Suffice it to say, custom line styles are a very powerful feature of MicroStation.

Instead of descriptive names, as appear in the Line Styles settings box for custom line styles, the built-in line styles (sometimes referred to as line codes) are identified with a single-digit number. When you select a line code from the Primary toolbox, it becomes the active line style. Any new elements you place take on that line style. The illustration that follows shows the eight basic MicroStation line style types.

NOTE: *For the sake of compatibility with older versions of MicroStation, you can select line codes by typing in* LC=<linecode #> *via the Key-in Browser (Utilities menu > Keyin). Thus, if you enter* LC=6, *the line style field in the Primary toolbox updates to display the line code 6 as your active line style; in this case, a large-small dash pattern.*

MicroStation's Line Styles (LC=)

0 ——————————————————
1 ·
2 -
3 — — — — — — — — — — — — — ·
4 -·-·-·-·-·-·-·-·-·-·-·-·-·-
5 -
6 —··—··—··—··—··—··—··—
7 -

The eight basic line style types supported by MicroStation.

As with weights, the on-screen appearance of the built-in line styles may differ markedly from that on the final plotted output, where the plot driver defines the ultimate appearance of these styles.

NOTE: *Line styles are discussed again later in the book. For now, you should stick with the eight default line styles. Your system administrator may restrict the use of the custom line styles to those developed for specific projects or company standards. There are few things worse in a project's final output than to see a mix of line styles representing the same category of drawing objects.*

Using Color

Color is another major tool used to convey distinctions between various components of a design within MicroStation. In the past, color was primarily relegated to its use as a drawing aid and was seldom used in final output. However, with the deployment of more color output technologies (the HP DesignJet series of color plotters for instance), the use of color is now on the increase.

With that said, color is still used as a primary method of differentiating between various components of your drawing. For instance, in a "typical" drawing, the design subject could be colored yellow, the dimensions and other drawing annotation shown in red, and the drawing sheet border colored green. At a glance you can identify precisely which drawing features belong to which "classes" of drawing components.

Another application of color in a CAD design is the differentiation between similar but separate design elements. For instance, in a roadway redesign project, colors can be used to differentiate between an existing road alignment and a proposed alignment.

MicroStation includes excellent color support. Depending on your video board's capability, MicroStation is capable of displaying up to 16.7 million colors. However, MicroStation provides direct use of only 256 colors via a color table associated with each design file.

That is not to say MicroStation cannot use more than 256 colors. In fact, MicroStation provides a very powerful image-rendering capability for generating photographic-quality images from your design file using all available colors.

MicroStation allows you to assign any of the 256 colors in the color palette to any element in your design file. If you click on the color picker in the Primary Tools toolbox, shown in the following illustration, the first field in the toolbox, you will see a palette of 256 colors. Selecting one of these colors sets the active color. The next element you place will be drawn in this color.

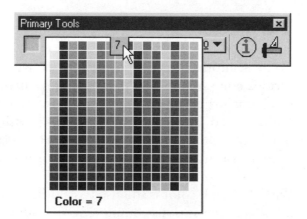

The color palette is most readily accessed from the color picker in the Primary Tools toolbox.

As you select a color, MicroStation displays that color's numeric identifier along the bottom of the palette. The numbering of the colors in the palette goes from 0 to 15 along the top row, 16 to 31 on the second row, and so on.

Color 254 is used by MicroStation as the background color of the design. Most companies use black as the drawing's background color, but more and more people are opting for white or a lighter color to better match the final output appearance (ink on white paper). There is even an option within MicroStation (Workspace menu > Preferences > View Windows category > Black Background > White) that will automatically invert the color 0 and background color to black on white. All of the screen images throughout this book were generated using this option.

TIP: *Using the Key-in window, you can directly enter the color's number using the* CO= *key-in, followed by a number matching the color desired. In addition, if you are using the default color table, you can also enter the color name for the first seven colors in the table, which follow.*

Size	Standard Colors
0	White
1	Blue
2	Green
3	Red
4	Yellow
5	Violet
6	Orange

NOTE: *Unlike AutoCAD, MicroStation assigns element attributes to every element in your design. Later on, you will learn about level symbology, an override feature that allows you to redefine the appearance of your design based on the level attribute.*

MicroStation provides a separate dialog box for adjusting the color values for each color number (Settings menu > Color Table). Double clicking on a color brings up the Modify Color dialog box. Here you can define the color makeup one of several ways (as indicated in the following illustration), including selection from a color palette, from a list of color names, or by directly adjusting the RGB or CMY color values. You can save custom color tables using the Save command (Color Table File menu > Save As) or open an existing one (Color Table File menu > Open). Once you have adjusted the table, simply click on the Attach button to update the color table on your active design.

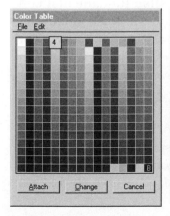

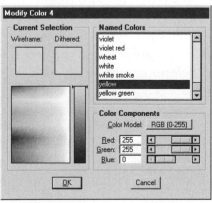

The values of each color in the color table can be adjusted to suit your needs.

Drawing on Different Levels

All of the element attribute settings discussed so far have been easy to relate to their manual drafting counterpart. Now, however, comes a new type of attribute that although not unique to MicroStation is one of the most exploited features found in today's engineering CAD operations.

Overlay Drafting

Overlay drafting has been around quite some time, and is closely related to the levels or layer concepts associated with today's CAD systems, an example of which is shown in the following illustration. Overlay refers to the drawing of a design on multiple sheets of drafting media, usually a semi-transparent or transparent Mylar. To create a final print of a specific drawing, the appropriate sheets are merged using the photographic or lithographic process. The fundamental idea behind overlay drafting is "don't draw the same linework twice."

When you design a house, you show the walls of the building on different sheets in the plans. When using non-overlay manual drafting techniques, this requires the redrawing of these same walls on several different drawing sheets. Using overlay drafting, these walls are drawn only once, usually on their own drawing sheet. Each major feature of the building is then drawn on a separate overlay sheet. By combining the various sheets, you construct the complete design (an electrical plan, for instance).

MicroStation's level attribute follows this same fundamental concept. By drawing your walls on one level and windows on another, you can control how your final drawing will look. MicroStation supports 63 separate levels in the active design (but you can have up to 255 reference files, each with its own set of 63 levels, discussed in a later chapter).

There are several commands used to control the levels. Using a variety of tools, you control which levels are displayed in each of Micro-Station's views. Chief among these level control tools is the Level Manager dialog box (Settings > Level > Manager) which, in turn,

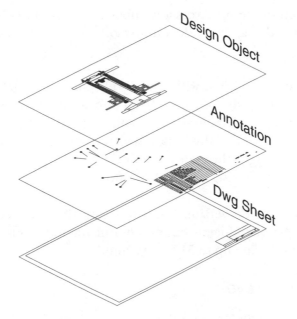

A visual example of levels.

provides you with full access to all levels of your active design file. The Level Manager dialog box is shown in the following illustration.

The Level Manager dialog box is used to select the levels to display in each view. There is a lot more to this dialog box than meets the eye.

In the example of the View Levels settings box shown, levels 1 through 44 are currently turned on in view 1, whereas levels 45 through 63 are not displayed (turned off). Using your mouse, you can select levels to display or not display. Holding down your

mouse and dragging across a number of levels will either turn on
or turn off those levels, depending on the state of the first level
you touched.

When you are satisfied with your level selections, clicking on the
Apply button applies this level selection to the view shown in the
View Number selection box. Using the View Number option
menu, you select the MicroStation view to which you want to apply
the level settings. Alternatively, you can apply your level selection
to all views by clicking on the All button.

V8: MicroStation version 8 will dramatically enhance the Level
Manager by introducing unlimited and named levels. Each level
will accommodate up to 512 characters.

The Active Level

With regard to the previous illustration, you may be asking yourself,
"What's with the round symbol on level 14?" The round symbol designates a very important value, your active level. When you create
elements in your design file, they must be placed on one of the 63
levels associated with every design file, thus the active level. To
change your active level, either double click on the level in the View
Levels settings box or select it from the level selection palette on
the Primary Tools toolbox, as shown in the following illustration.

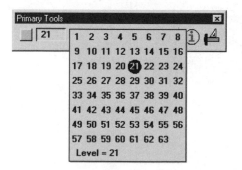

*You can always set
your active level
directly from the
Primary tool bar.*

NOTE: *MicroStation will not let you turn off the display of your
active level in any view, which is why it appears in the same highlighted
color as the other displayed levels in the View Levels settings box.*

You should be aware that when you select a different level as your active level, the previous active level is not removed from the list of levels displayed. To turn off that level, you must use the View Levels settings box.

TIP: *Try holding down the Shift key while dragging a data point diagonally across levels 1 to 63 in the Level Manager dialog box.*

Using the Level Control Key-ins

Using the View Levels settings box is fine, but you may want to try a "quick and dirty" way of manipulating the display of levels and your active level. You can do this by using MicroStation's level key-ins in the Key-in window (Utilities menu > Keyin). The following are three shortcut key-ins.

- *LV=*   Sets the active level

- *ON=*   Turns on the specified levels and prompts for identification of the view(s)

- *OF=*   Turns off the specified levels and prompts for identification of the view(s)

If you had to turn on or off each level individually it would be a tedious operation if a drawing consisted of many levels. Fortunately, you can turn on or off individual levels and groups of levels using a combination of commas and dashes. Examples follow.

- *ON=1,3,5*   Turn on levels 1, 3, and 5 only

- *ON=1-20*   Turn on levels 1 through 20

- *OF=3,6,11,22-40*   Turn off levels 3, 6, 11, and 22 through 40

- *ON=6,1,11-42*   Turn on levels 6, 1, and 11 through 42

In the last example, notice how you can specify your levels in any order and even give a range of levels. Once you have keyed in your levels, MicroStation prompts you to select a view. Data point on each view you want this level change to affect.

NOTE: *If you are going to be using levels a great deal, it is a good idea to generate a standard-level description. There is no greater time waster than trying to find a specific bit of information in a drawing by turning levels on and off looking for it. Most design departments establish standard level usage for various types of drawings or projects. If this does not exist where you work, it would be wise to develop one.*

Naming a Level

An alternate level-definition facility introduced in MicroStation in the early 1990s is the ability to assign a textual name and description to each level number, as indicated in the following illustration. The Level Names dialog box (Settings > Level > Names) is used to define and review the level names. When selecting an active level or just displaying a level in a particular view window, you can type in this name. For instance, if you have already set level 10 to represent exterior walls, you could set this as your active level by keying in *LV=EXTWALL*.

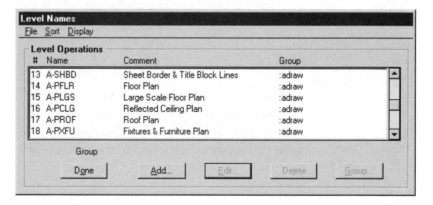

By selecting a level and a name you create a cross-reference that can be used interchangeably in the various level commands. The example shows a portion of the AIA layer (level) naming standards provided as a sample with MicroStation.

Saved with the design file itself, these level names are transferable by saving them from the File menu of the Level Names dialog box (shown in the previous illustration).

V8: Version 8 will enhance MicroStation/J's level names by replacing numeric assignments with the alphanumeric names described here.

Changing Attributes

This discussion of color, weight, and style would not be complete without giving you some method for changing an element's current attribute values. If you select the color red, for instance, and place a line, but realize the color should be green, you could reinsert the line with the correct color. However, a more efficient method would be to simply modify the errant colored line using the Change Element Attributes tool.

Change Element Attributes

Located in the Change Attributes toolbox (Tools menu > Main > Change Attributes), this tool allows you to selectively modify an element's attributes via options boxes and pull-downs that closely match those of the Element pull-down menu options. Once you have chosen the attribute you want to change and have set its value, simply select the element and accept the change. In addition, note that this tool provides the same Use Fence option as found in the various Manipulate Element tools discussed in the previous chapter. Change Element Attributes is shown in the following illustration.

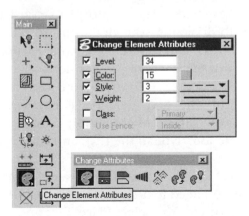

Change Element Attributes allows you to adjust one or more existing element's display attributes via its extensive tool settings options.

Change Element Attributes and Selection Sets

You can also change the attributes of a selection set by using the Primary toolbox. The act of selecting the weight, line style, or color from this toolbox immediately updates the affected attribute for all elements in the selection set.

More Element Types

Previous chapters covered some of MicroStation's most basic drawing elements. These included such favorites as line, circle, and shape. However, one very important element type was overlooked: the text string. Before moving on to other aspects of MicroStation, spend a few minutes looking at this fundamental building block.

Working with Text

"One picture is worth more than a thousand words."

—Chinese proverb

"A picture shows me at a glance what it takes dozens of pages of a book to expound."

—*From* Fathers and Sons, *Ivan Sergeyevich Turgenev, 1818–1883*

The expressiveness of a picture is undeniable. However, it is the written word that rules the world of construction contracts. In addition to depicting your design in graphics, you need to label them with text, and describe the construction sequence in notes on your drawings, create component schedules, even describe properties in sentence form on your drawings. All of this is only possible with the text element. Fortunately, text creation and manipulation within MicroStation is very easy. You select the appropriate tool, type in the text, and place it.

Well, almost. Let's say you have created a section detail for a new design; all you need to do is give it a name by placing a string of text in the design file. You select Place Text and type in *Wall Section* via the Text Edit window, which automatically pops up. You place the text with a data point, as shown in the following illustration.

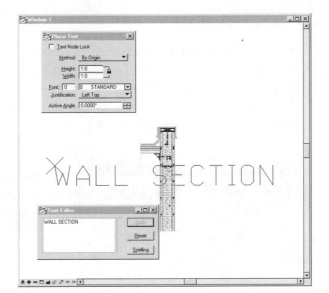

Placing text.

You might ask if the resulting text would look better if the text were drawn in some style other than that atrocious "computer stick-figure style." It would also be nice if the text were a little larger (or smaller), and maybe even centered below the design. You probably realize by now that there are a number of additional factors that affect the appearance of your text. You need to consider several aspects about your soon-to-be placed text. These include the following.

- Size (height and width)
- Text font
- Angle (not to be confused with italics or slant)
- Justification or orientation about the data point location

All of these parameters are unique to text, with the exception of angle (this is the same as Active Angle). The standard element attributes also apply to text (weight, line style, color, and level).

NOTE: *Setting the optimum text parameters requires you to consider the final output scale of your design. It just would not do to create a fabulous drawing and end up with unreadable text.*

MicroStation's text attributes are controlled by various text-specific parameters set via the Text settings window (Element menu > Text), shown in the following illustration. In addition, the most-often-modified text parameters are also available in each text-related tool's Tool Settings window.

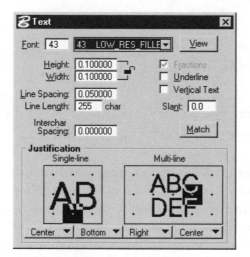

The Text settings box provides access to all of the text-specific attributes.

Setting Text Attributes

Most users today are familiar with text fonts and the major text attributes associated with them. This is due in no small part to the popularity of the graphically intensive operating system in use today (Windows, MacOS, Solaris). As part of the push to make documents more readable, fonts have improved to the point that these days you can find literally tens of thousands of fonts.

The good news is that you can use most of these same fonts within MicroStation. However, unless your final output device supports such fonts, your final plotted results (and the time it takes to generate them) can be less than optimal. For this reason, some prudence should be used in font selection and usage within MicroStation.

Text Font

This does not mean you cannot generate great-looking text. On the contrary, MicroStation comes with an assortment of very clean and precise text fonts. Font 1 (name: Working), for instance, is a match to the standard Leroy text associated with technical pen lettering templates still in use today. To review the fonts available to you within MicroStation, select the View button from the Text settings box. This activates MicroStation's Fonts settings box, shown in the following illustration.

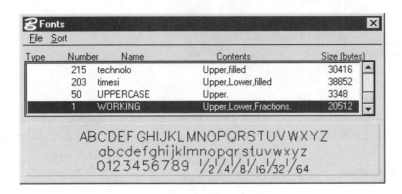

The Fonts settings box displays examples of MicroStation's installed fonts.

Just scroll through the list of fonts and select one of interest. Doing this displays the font in the lower half of the settings box. To set this font as the active text font, data point on the example text.

You can also set your active font from the Text settings box. This is done either by entering the font number in the Font field or by selecting it from the pop-up option field, as indicated in the following illustration.

So what is a font number? This is the shortcut number you can enter in the Key-in window using the *FT=* key-in (or the longer version command name *Active Font* #). This directs MicroStation to use this font for your next text placement. For instance, *FT=1* will give you the Working font.

MicroStation supports a variety of font formats as part of its Font Installer utility (Utilities menu > Install Fonts). You can install any commercial fonts using Adobe Postscript, AutoCAD .SHX, or Apple/Microsoft TrueType data format. The font installer will

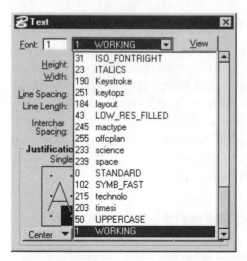

Selecting the Font pop-up field presents you with a list of all fonts installed in MicroStation.

convert these formats into MicroStation's own internal font resource format (stored in *.rsc* files). However, before you indiscriminately install various fonts into MicroStation, keep in mind that font usage is normally the responsibility of your system administrator or project manager.

There are many issues involved with the use of commercial fonts. These include font ownership (fonts are licensed from font foundry companies such as Adobe), distribution rights (you may not be allowed to freely distribute a font to your client at the conclusion of the project), and many others. Leave it to your administrator to worry about these minor details; just use the fonts installed in MicroStation for now.

V8: MicroStation version 8 will support direct reading of Windows TrueType fonts and AutoCAD SHX fonts. You will no longer have to import these fonts!

Text Size

The most crucial decision you must make when setting your text attributes is the text size. You set your text's size using the Height and Width fields in the Text settings box. Note the padlock icon to the right of these fields. By "locking" the two fields, any changes made to the text height will result in the same value for the text width and vice versa. As with the previous font setting, there are key-in shortcuts associated with the text size parameters. These are spelled out in table 6-1, which follows.

Table 6-1: Text Size Key-ins

Text Height Key-in	Text Attribute Affected
TH=	Height
TW=	Width
TX=	Height and width together

Text is one of the element types affected by the final plot scale of your drawing. To ascertain the proper text height to use in MicroStation, you need to do a very small bit of math. The formula for setting your target text height is as follows.

Formula for Calculating Text Size

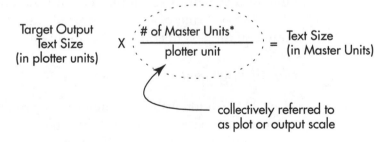

* from Working Units (Settings menu > Design File > Working Units).

Suppose you are constructing a floor plan at 1/4 scale, a common U.S. architectural drawing scale (1 inch of plotter paper = 4 feet of design space). If your design standards require a uniform text size of 1/8 inch, you would plug the various values into the following text scale formula. Targeting 1/8 inch as your final output text size at 1/4 scale results in a text size setting in MicroStation of 1/2 foot (or 6 inches).

$$1/8 \ (0.125) \, ^{''}X \ \frac{4\,'}{1\,''} \ = \ 1/2 \ (0.5)\,'$$

The following is a simple metric example.

```
Target text height: 2.5mm, drawing scale: 1:50
2.5mm text height × (50mm design units ÷ 1mm plotter
  units) = 125mm design (TX=125)
```

 NOTE: *If you change the scale of the plotted drawing, you will also need to change the size of any text you place accordingly. Text already placed can be updated using the Change Text Attributes tool.*

Line Spacing

Line spacing refers to the amount of space used between multiple lines of text. When you place notes and other multiple text lines (such as this paragraph), you need to set the amount of space between the bottom of one line of text and the top of the next line of text. This is accomplished with the Line Spacing attribute. As a general rule of thumb, you will want to use line spacing at one-half the text height. The related key-in is *LS=*.

Justifying Text

Text Justification is the parameter that tells MicroStation where to place your text with respect to the data point. MicroStation supports the following nine justifications.

- Left Top
- Center Top
- Right Top
- Left Center
- Center Center
- Right Center
- Left Bottom

- Center Bottom

- Right Bottom

Even if you are familiar with using text justification in other programs, MicroStation probably includes several you have never seen before. The most recognizable is Left Bottom. This means that the text's origin is at the lower left corner of the text. Center Center orients the text symmetrically around the data point in both axes. This justification is often used with callout symbols and other annotation features where the text is placed within a circle or polygon. Left Top is often used when placing several lines of text (as in general drawing notes) as one element. Right Center and Left Center are often used on the right and left sides of drawing details to annotate features.

As with the other text attributes, there are key-ins for setting text justification (see following illustration). The key-in is *ACTIVE TXJ xx*. In the preceding list, the first letter of each word is used with this key-in in place of the *xx*. For instance, to set your justification to Right Center, you would key in *ACTIVE TXJ RC <Enter>*. For Center Center, you would key in *ACTIVE TXJ CC <Enter>*.

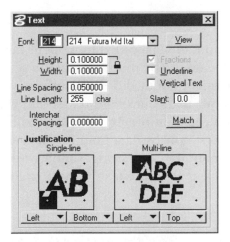

The Text settings window is used to set all parameters associated with the text element type. Note that the Justification font matches the currently selected font.

The Text settings window graphically displays the current justification as well as the current font. Simply click on the justification you want to use to highlight it. The justification can be selected from the pull-down menus located underneath the justification preview. With all this talk about text parameters, it is easy to forget there is a set of text-related tools you can use to place and modify text. These are shown in the following illustration.

V8: MicroStation version 8 will incorporate a greatly enhanced set of text capabilities, including several new text attributes and a new text editor window.

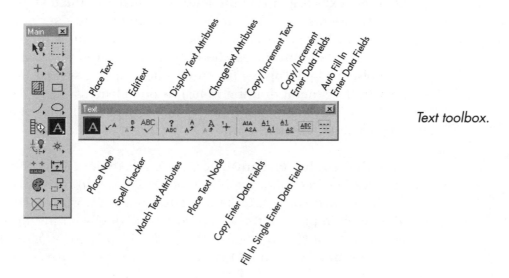

Text toolbox.

Place Text

Upon selection, the Place Text tool opens the Text Editor window. You enter the text to be inserted into the active design file. As you move your cursor about the view windows, you will see a dynamic version of your text. Clicking a data point will set the location of the text. You can continue to place the same text string elsewhere in the design simply by data pointing at each location. When you are ready for another text string, simply revise it in the Text Editor window. To clear all of the text in the Text Editor window, click on Reset.

NOTE: *You can "edit" the text currently displayed in the Text Editor window and place the modified text. If you click on Reset, the text in this window is cleared.*

Note how even after you have placed a data point in your design, the dynamic cursor and text are still active. This allows you to place more than one copy of the text. The following illustration shows the Place Text tool in action.

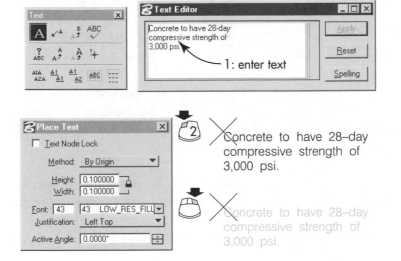

Place Text tool in action.

Text Editor Window

The primary method for entering text in MicroStation is the Text Editor window. Beyond just allowing you to enter text, the Text Editor window is a simple word processor in its own right. By using the mouse and various Control key combinations, you can cut and paste text within this window. The following are the primary functions performed with the Text Editor.

- Entering text for new text placement
- Editing previously placed text
- Entering text into *enter_data* fields (described later)

When you select any tools associated with these functions, the Text Editor window is automatically activated. To enter your text, just type it in. Note that when you type in more text than can appear on the same line in the Text Editor, the text scrolls horizontally. Pressing the Enter key results in a multi-line text entry with a small dot placed at the end of each line of text, as shown in the following illustration.

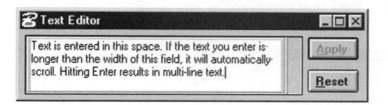

An example of text being entered in the Text Editor window. Note the use of small dots at the end of some lines to indicate that this is a multi-line text entry.

Also note the grayed Apply button in the previous illustration. This button does not "apply" to text you are placing for the first time. Hitting a data point places or "applies" the text into the design file.

The Text Editor window can be resized by dragging its borders. There will be times when you will want to increase the size of this window (as when you are entering general notes), and other times when the window just takes up too much space (as when you are entering callouts or labels).

Whether entering text for the first time or editing previously placed text, there are certain functions you may want to perform on text in the Text Editor window. To facilitate these functions, the Text Editor provides limited text manipulation. Table 6-2, which follows, spells out the Text Editor functions.

Table 6-2: Text Editor Functions

Command Name	Description
Insertion Point	The vertical cursor indicating where keyed-in text will be inserted.
Insert/Overwrite Toggle	Normally the Insert key, this toggle controls whether text you enter over-writes the existing text or is inserted at the current text insertion point. The default condition is Insert.
Text Selection	By selecting and dragging the cursor over text in the Text window you select text for further operation. Double-clicking text will highlight the word.

Table 6-2: Text Editor Functions

Command Name	Description
Cut Text	Normally Ctrl-X or Shift-Delete, this command "cuts" the selected text and places it in the cut-and-paste buffer. The selected text is deleted.
Cut Text command, Text Editor Copy Text	Normally Ctrl-C or Ctrl-Insert, this command "copies" the selected text to the cut-and-paste buffer. The selected text is left intact.
Paste Text	Normally Ctrl-V or Shift-Insert, this command "pastes" any text previously cut or copied into the Text window at the current insertion point cursor.

Text Placement Methods

When you select the Place Text tool from the Text toolbox, shown in the following illustration, you are presented with the most commonly changed text attribute fields but also one special parameter. The Method parameter specifies how the Place Text tool will place the text you entered into your design file. The simplest method (and the default), By Origin, will place the text wherever you click a data point using the current text parameters. There are several other text placement methods. These are described in table 6-3, which follows.

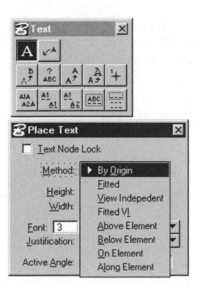

The Place Text method setting controls how the Place Text tool responds to your data points.

Table 6-3: Text Placement Methods

Method	Description
By Origin	Places text at the data point (default).
Fitted	Places text fitted between two data points. Text size and rotation angle are adjusted so that text fills the distance between the two points.
View Independent	Places text at the data point with View Independent attribute turned on. Text placed thus will always read normal to the view. Most effective in 3D design files.
Fitted VI	Same as Fitted with View Independent attribute.
Above Element	Places text the line spacing (LS=) distance above a selected element at the point of selection. Left/Center/Right justification applies to text orientation.
Below Element	Places text the line spacing (LS=) distance below a selected element at the point of selection. Left/Center/Right justification applies to text orientation.
On Element*	Places text on the element, cutting an opening in the element and orienting the text along the axis of the element.
Along Element*	Places text above or below (defined by second data point) and tangent to the element selected. The text "flows" to follow the contour of the element.

* See the text that follows for further information on these methods.

 TIP: *Double clicking at the beginning of a line of text in the Text Editor window highlights the entire text line. Dragging a data point just inside the left side of the Text Editor window highlights multiple lines of text.*

In the case of the On Element method, the element you select for its insertion *will be modified.* The selected element is literally cut and the text is placed in the opening. Keep this in mind because from this point on the target element becomes two separate elements. The text will be placed by the Left/Center/Right current justification.

The Along Element method (shown selected in the following illustration) is one of the more interesting text placement methods found in MicroStation. To place text that follows the target element precisely (especially curved elements such as arcs or b-spline curves), each character in your text string must be placed tangent to that element at each character's individual location. To accomplish this, each character is placed as a separate string of text. This makes the text much more difficult to modify later, so use this command only when you are dealing with curving elements, and use the Above or Below method when you are working with lines and line strings.

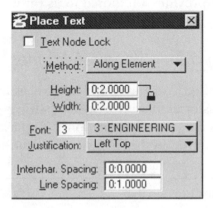

The Place Text Along Element method relies on the Intercharacter Spacing value to keep characters from overlapping.

Because Along Element can place text along very convoluted elements, an additional parameter is required: intercharacter spacing. As a general rule, if the element along which you are placing text is more than a little curved, use a spacing of 50 percent of the text width. Be sure to set this back to zero when returning to the "normal" text placement methods; otherwise, subsequent text w i l l a p p e a r v e r y s t r a n g e i n d e e d !

Setting the Text Angle

In the last chapter you learned how the Rotate Element tool is used to establish the active angle parameter to define the degree of rotation during its operation (*AA=degrees* of rotation). Text placement is also affected by the active angle. For instance, setting your active angle to 45 degrees (*AA=45*) will cause all of your subsequent text to be rotated to 45 degrees.

Unfortunately, there is no indicator of the active angle in the Place Text tool settings window. However, you will know right away if the active angle is set to something you do not want, because your text will appear rotated prior to actually placing it. Should this happen and you have the Key-in window open, press the Escape key, which returns temporary control (i.e., "focus") of MicroStation to the Key-in window. Once here, key in *AA=0.* This will return your active angle to 0 degrees.

Additional Text Tools

MicroStation provides additional tools for working with text. The following sections briefly describe each tool.

Displaying a Text String's Attributes

The Display Text Attributes tool, shown at left, displays the text parameters associated with a specific string of text you identify with a data point. This information is displayed in MicroStation's status bar, as shown in the following illustration.

Display Text Attributes > NN=54, LL=255, LS=0.050000, LV=1, FT=43

Selecting a text string with the Display Text Attributes tool results in the information displayed in the left side of the status bar.

Match Text

Like the preceding tool, the Match Text Attributes tool allows you to identify previously placed text. Upon accepting the text, this text element's attributes become the active text attributes. These include the text's height, width, font, line spacing, and justification. The Match Text Attributes tool is shown at left.

Change Text to Active Attributes

The complement to Match Text Attributes, the Change Text to Active Attributes tool, shown at left, will set any text you select to the current text attributes.

Copy and Increment Text

How often have you drawn an assembly or created drawing notes that use consecutive numbers? "All the time" is probably your answer. MicroStation provides a fast method of copying and incrementing such text. The Copy and Increment Text tool, shown at left, is used to select a previously placed text string.

As the tool creates copies of the text, the last number of the string will be incremented by the tag increment value (default: 1). This means that if you have a text string such as *Note 1*, the Copy and Increment Text tool will create a new copy of the text string: *Note 2*, *Note 3*, and so on. The following illustration shows the Copy and Increment Text tool used for a particular application.

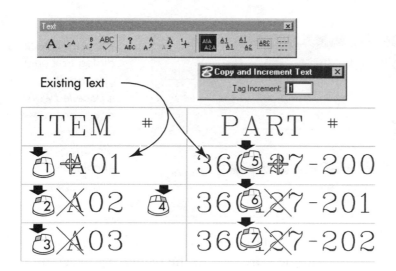

The Copy and Increment Text tool is used here to increment two separate text strings. Note the Reset between the two sets of operations.

Edit Text

Not to be confused with the Text Editor window, this text element tool, shown at left, allows you to change any text strings you have already placed in your design file. Once you have selected a string of text to be edited, the Edit Text tool brings up that text in the Text Editor window. You can edit and change any text in the selected string. When satisfied with your changes to the text string, you select the Apply button, or press Ctrl-Enter or Alt-A on the keyboard, to update the text in your design file. If for any reason you decide the changes should not be made, click on Reset. The original string will reappear in the Text Editor window.

Spell Checker

Introduced in 1999, the Spell Checker is a valuable tool that does as its name implies, checks the spelling of all text strings in your active design file. The Spell Checker includes a fairly comprehensive dictionary of English words. It works with the current selection set, so you will need to use the Select Element or PowerSelector to identify the portion of the design to be checked. When a word is found that is not in the current dictionary, the Spell Check dialog box appears, as shown in the following illustration.

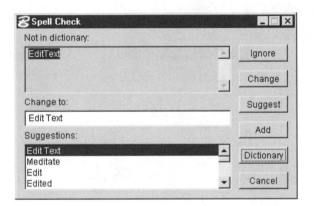

The Spell Check dialog box only appears if a misspelled word or a word not found in the dictionary is encountered.

You can change the suspect word to one of the suggested spellings, manually edit it, add the word as-is to the dictionary, or ignore it. Clicking on any of the buttons on the right side of the dialog box except Cancel will resume the search for misspelled words.

Working with Multi-lines

How many times have you found yourself drawing parallel lines to represent some very basic structure. Obvious examples that come to mind are walls in an architectural plan, pipes in a plant design, and air conditioning duct work. Suffice it to say, parallel lines are used enough to warrant their own tool: Place Multi-line.

Place Multi-line

The Place Multi-line tool, shown in the following illustration, operates in a similar fashion to the Place SmartLine tool introduced in an earlier chapter; you identify the path of your multi-line with data points and terminate it with a final Reset to accept it. In response, the multi-line tool places a series of parallel line strings at a predetermined offset from the data points. The amount of offset, the display attributes, and the number of lines are determined by the multi-line parameters active at the time you placed the multi-line.

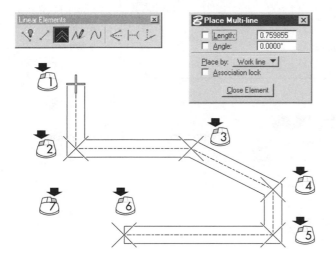

Place Multi-line tool in action.

The definitions of the multi-line are set in the Multi-lines window (Element menu > Multi-lines). You define the number of lines (up to 16) that make up the multi-line, the spacing between the lines, the treatment of the end caps (placed at the beginning and end point of the multi-line), and the display attributes of each line.

Multi-line Settings Box

When activated, the Multi-lines settings box may intimidate you. However, it is relatively easy to use. See the first of the following illustrations.

A multi-line consists of three distinct components: the lines, joints, and caps, as shown in second of the following illustrations. Each component has its own settings in the Multi-line window. The first step is to select the component to define. This is done via the Component option menu. In most instances, you start by defining the line component.

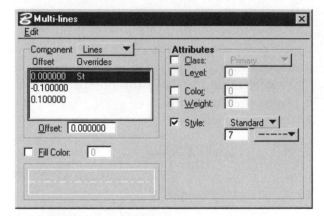

Activated from the Element pull-down menu, the Multi-line Settings window is the "heart" of the multi-line element definition.

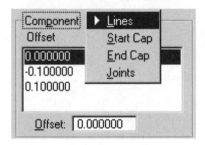

Lines, caps, and joints options.

Each time you exit MicroStation, the multi-line definitions set up in the Multi-lines settings box are lost. There is a way to "save" your multi-line definition by using the Settings Manager facility (an advanced topic reserved for a later time).

The default multi-line consists of three lines. The first line listed shows a zero offset. This means it is drawn from data point to data point. However, it is not just a plain line. If you look to the left side of the following illustration under Attributes, you will find the Style option selected and at the lower right a line style of 7 (long dash/short dash) entered. The word *Standard* refers to the use of MicroStation's eight standard line codes. You can also use any custom line style.

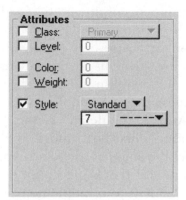

Attributes dialog.

The second line definition is that of the "bottom" edge of the multi-line. In this case, the line does not pass through the data points. Instead, it is offset by one negative master unit. By entering a negative number, the line is drawn below the center line. The offset is entered via the Offset field (shown in the following illustration) in the Component section.

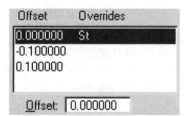

Offset field.

The third line also shows an offset of one master unit, this time a positive number. This pushes the line above the center line.

Insert option.

To incorporate additional lines in your multi-line definition, you must insert a new definition. This is done via the Edit pull-down menu, shown at left, located on the Multi-lines settings window. The new line definition is added above the currently selected line definition. Select the new definition and adjust its values to your needs. In addition to Insert, there are Delete and Duplicate commands on the Edit pull-down menu. The Delete command deletes the currently highlighted line segment. Duplicate creates a new line definition based on the one presently highlighted.

Line

Outer Arc

Inner Arc

Inner &
Outer Arc

Cap styles.

The next two components of the multi-line affect the appearance of the multi-line's beginning and end points. Called caps, this multi-line component has several options: a perpendicular line linking all of the line segments, an arc linking the outermost line segments, an arc linking all inner paired line segments, or any combination of these three. Examples of these are shown at left. Again, you can override the default element attributes by selecting the appropriate checkbox under the Attributes section. Finally, you can set the angle of each end cap. For instance, if you set the Line option and the Angle to 45, the result is a beveled effect on the end point of your multi-line.

The final multi-line component over which you have control is the joints or intermediate vertices. As you place data points to define your multi-line, the joints component definition affects how each vertex is treated. You can display a perpendicular line that accentuates the joint "seam" or let the vertices appear as continuous line strings.

Finally, the entire multi-line can be displayed with a fill. This is a solid color that fills the entire area of the multi-line. You can set the fill color so that the individual line definitions are displayed on top of this fill. Filling an outside wall of a building will make it stand out during the design process; during the plotting process, you can opt to turn it off. Exercise 6-2, which follows, takes you through the process of placing multi-lines.

EXERCISE 6-2: PLACING MULTI-LINES

In this exercise you will be placing a series of multi-line elements with a variety of multi-line definitions.

1   Open the *MULTILIN.DGN* design file. Three boxes labeled A, B, and C are displayed in view 1. Using the Window Area tool, center on box A, as shown in the following illustration.

Centering on box A.

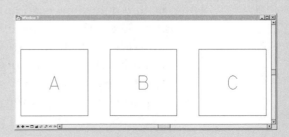

2 Open the Multi-lines settings window (Element menu > Multi-lines), shown in the following illustration. Note the definition for a simple three-line, multi-line representing an 8-inch-wide concrete block.

Multi-lines settings window.

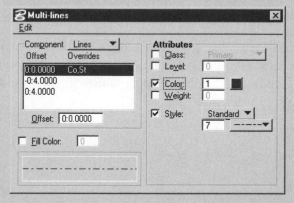

3 Selecting the -0:0.000 definition, change its color attribute by selecting the Color checkbox and selecting the color blue from the color selection menu. Note in the previous illustration the change in the Overrides column (changes from St to Co,St). Also, note the change in the preview image.

4 Using the Multi-line tool (Tools menu > Main > Elements > Place Multi-line), place a tentative point/data point on the lower left corner of box A. Continue this tentative point/data point sequence at all four corners.

5 Click on Reset to complete the placement of the multi-line, shown in the following illustration. Note the "ugly" termination of the multi-line. To avoid this, you could have used the Close Element button, located on the bottom of the Place Multi-line tool settings window. The result would have been a clean corner.

Placement of multi-line.

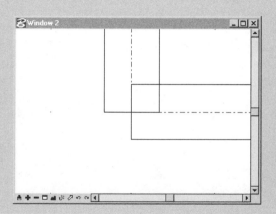

Next, you will change the color of the two outer lines in the Multi-lines settings box.

6   Select the -0:4.0000 definition and set its color to Green. Do the same for the 0:4.0000 definition. Note how the end caps in the sample multi-line still show White.

7   Change the end caps to the same color as your outer lines (green). To do this, change the Component setting to Start Cap and set its color. Do the same for the End Cap, as shown in the following illustration.

Changing the end cap color.

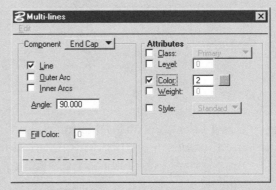

8   Using the view controls, navigate to block B.

9   Return to the Place Multi-line tool. Before placing your multi-line, set the *Place by* field in the Tool Settings window to Minimum.

10   Use the tentative-point/data point sequence on the lower left corner of block B to start building your wall, and continue clockwise around the block. Note how the multi-line is now created to the inside of the block. In the "block A" segment of this exercise, the multi-line was placed by the center line. The wall built around block B is shown in the following illustration.

Wall built around block B.

11 Before selecting the lower left corner to conclude this wall, click on the Close Element button. This time your wall is continuous and clean. Return to the Multi-lines settings box. This time, add a line to the definition.

12 With the Component field set to Lines and the 0:4.0000 line highlighted, select the Insert command (Edit menu > Insert). A "0:0.0000" line definition will appear just above the 0:4.0000 definition. With this definition selected, change its offset field to 0:2, and assign it the color red and the dotted line style.

13 Using the Place Multi-line tool, place another wall, around block C. This time, set *Place by* to Maximum. Tentative point/data point on the three corners and click on the Close Element button one more time.

14 Perform a Fit View on view 1. The results are shown in the following illustration. Note how the three walls all appear to have slightly different sizes, although the three starting blocks were exactly the same.

Result of Fit View.

As you can see from this simple exercise, there is much to the multi-line. You can radically change the appearance of the multi-line by selecting the various options. One other powerful aspect of the multi-line is its ability to be modified for the various intersection requirements found in typical drawings. Before leaving the multi-line element, you need to take a look at one more of its features—its ability to be modified. You may want to try changing your multi-line definitions further. Try setting the Joints to On and see what happens when you place your lines.

Multi-line Joints

When you used the Close Element button in the previous exercise, the Multi-line tool did an amazing thing: it cleaned up after itself! The result was an intersection as you would expect it. No extra work was required to clean it up. You simply obtained a nice, clean intersection.

MicroStation includes other multi-line "maintenance" tools, which are contained in their own toolbox (Tools menu > Multi-line Joints). These tools are used to enhance the appearance of previously placed multi-lines. You can, for instance, cut openings in a multi-line wall for doors, windows, and other components of a floor plan without generating a whole host of extra elements. You can also clean up multi-line intersections using several tools located in the Multi-line Joints toolbox, shown in the following illustration.

Multi-line Joints toolbox.

TIP: *The best way to learn how multi-lines work is to try out each of the tools and see what it does. If you feel you will be using the multi-line in your design work, plan on setting aside some "play time" to try out each of the multi-line tools.*

Element Modification Tools

In the previous discussion, a category of tools specific to the modification of multi-lines was introduced. These tools work exclusively on the multi-line, having no effect on other element types. There are, however, more general-purpose modification tools designed to operate on most types of elements.

In the previous chapter you learned about tools that "manipulate" elements in a global or arms-length manner. You could change an element's overall size using the Scale Element tool, mirror it, and even delete it. What you cannot do is modify a single aspect (for instance, a single vertex of a line, or one endpoint of an arc) of an element. To perform this type of operation you need to use a different set of tools, the Element Modification toolset, shown in the following illustration.

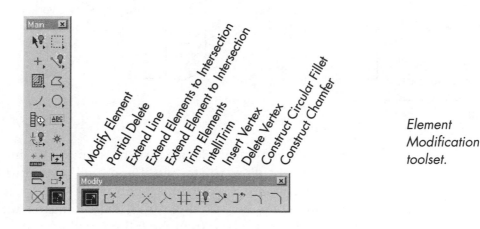

Element Modification toolset.

The Modify Element tools have their own toolbox (Tools > Main > Modify). These tools provide you with the ability to adjust individual aspects of the elements already placed in your design file.

Modify Element

Without a doubt, the most versatile tool in MicroStation, Modify Element, shown at left, is used to change the geometry of most element types. You can modify individual vertices on lines, line strings, and shapes; change the radius of circles; even modify features of dimensions. The following illustration shows the Modify Element tool used on a shape element.

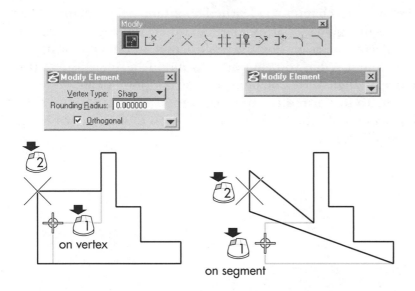

The Modify Element tool during use on a shape element. Observe the tool's behavior change (especially its tool settings options), depending on where you select the shape (left shows vertex, right shows segment selection).

When you select linear elements such as lines and line strings, Modify Element allows you to "push" and "pull" the individual vertices associated with the element. When you select arcs, it lets you modify the endpoints, or the radius, depending on where you select the element, as shown in the following illustration.

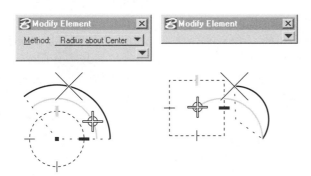

You can modify an arc's radius or endpoint with Modify Element, depending on how you select the arc.

Partial Delete

Just like the venerable eraser shield, the Partial Delete tool is used to cut out segments of elements.

Similar to Fence Delete with the Clip Fence mode selected, the Partial Delete tool, shown at left, appears to remove a portion of your target element. However, appearances can be deceiving. What the tool really does is replace the element you have chosen with one or more new elements consisting of the remainder of the target element. Depending on the target element type, you can even change it from one type of element to another. Exercise 6-3, which follows, takes you through the process of partially deleting a line.

EXERCISE 6-3: PARTIALLY DELETING A LINE

1   Open the *PARTIAL.DGN* file, shown at right.

2   Select the Partial Delete tool from the Modify toolbox.

3   Place a data point at the midpoint of the line, as shown in the following illustration. Note how a portion of the line "disappears" as you move your cursor along the line, as shown in the following illustration.

PARTIAL.DGN *file.*

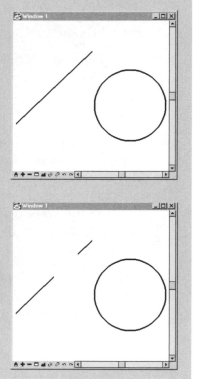

4   Data point toward but not past the line's upper endpoint. The result is two line segments replacing the original line.

A portion of the line has "disappeared."

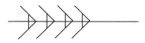

As you can see from this mini-exercise, Partial Delete is very straightforward in its operation. However, its behavior changes a bit when you use it on a closed element such as a circle or shape. By its very nature, Partial Delete creates endpoints. Circles and shapes do not have endpoints, so Partial Delete must replace the original element with its "open" cousin (shape becomes a line string, a circle becomes an arc, and so forth).

In addition, Partial Delete requires an extra data point in its operation because you must identify the portion of the closed element you want to remove. Using the "from-through-to" rule, your data point selection order becomes very important to the outcome of the partial delete operation. In exercise 6-4, which follows, you will practice partially deleting a circle.

Exercise 6-4: Partially Deleting a Circle

1   Returning to *PARTIAL.DGN*, select the Partial Delete tool again.

2   Data point on the bottom of the circle (the "from" location). You are prompted to "Select direction of partial delete," as shown at right.

3   Data point the "through" location on the left side of the circle. As you move your cursor, the gap in the circle will extend from the first data point through the second and toward your cursor, as shown in the illustration below right.

4   Data point the "to" location on the top of the circle. The result is a single arc replacing the original circle.

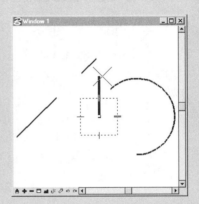

Prompt for partial delete direction.

Gap in the circle

The result of this exercise will be an arc of the same radius as the circle. The endpoints of this new arc will correspond to the first and third data points you entered. The second data point is lost with the deleted segment of the circle. This "from-through-to" selection procedure is also used to erase portions of shape and ellipse elements.

Extending Lines

Quite often you place lines in a design more for orientation than for length. "Striking" a line to temporarily define an edge of a design object is a common design practice. Once the planes of the object have been defined, all that remains is to clean up the various intersections and set the element attributes.

That is precisely what the Extend Line tools do. You "push" or "pull" a line's endpoint along the axis of the line. The difference between this command and the Modify Element tool discussed earlier is the rigid adherence to the orientation of the line.

One other useful function of Extend Line is in finding obscure intersection points. Using some of the tangent and perpendicular snaps with the Place Line tool, you can define key relationships between various elements. The Extend Line tools let you finesse these relationships by providing access to those critical intersection points.

Extend Line

The Extend Line tool, shown at left, does just that. It changes the endpoint of a line in association with your data point while maintaining the direction of the selected line. This means that if you select a line for extension and data point past its endpoint, Extend Line will stretch the line to its closest approach to your second data point. Conversely, if your second data point is somewhere along the length of the existing line, Extend Line will reduce its length to that point.

So how does MicroStation know which end to stretch? It does so by selecting the endpoint closest to the selection data point (the one you used to select the element), as shown in the following illustration.

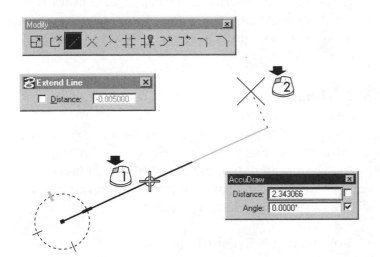

Extend Line is used to modify an endpoint of an existing linear element while maintaining its original planar orientation. Note the appearance of the AccuDraw compass. This provides an easy method for modifying the overall length of the element without choosing a special option.

Extend Line by Key-in

When you select the Distance option, you activate a slight variation of the Extend Line tool, Extend Line by Key-in. This tool moves the endpoint the distance entered. To extend an element, key in a positive value. A negative value results in a trimming operation.

Bringing Two Elements to an Intersection

Extend 2 Elements to Intersection, shown at left, is one of the most useful forms of the Extend tool. By selecting two elements in turn, MicroStation will extend or trim them to their intersection. This applies to lines, line strings, and arcs.

As with most of the Modify Element tools, where you select each element is important. In the case of overlapping elements, you will want to select the element on the side of the intersection you want MicroStation to keep. Selecting the wrong side of the intersection will lead to the element being trimmed opposite to the side you wanted. Undo will let you back out of this situation. The following illustration shows the Extend 2 Elements to Intersection tool in action.

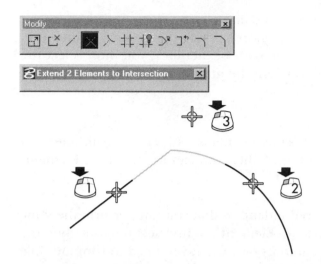

The Extend 2 Elements to Intersection tool in action.

Extend Line to Intersection

Of course, there are times when extending two elements to an intersection is overkill. For that reason, MicroStation also includes Extend Line to Intersection, shown at left.

This tool works like one-half the Extend 2 Elements to Intersection tool. You select the element to be modified and the element to extend to. Again, it is important to select the portion of the first element that corresponds to that you want to keep. Use of this tool is shown in the following illustration.

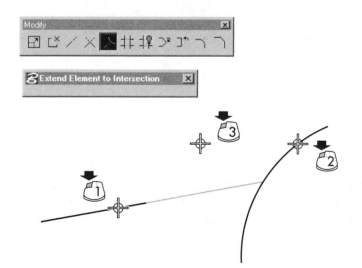

The Extend Line to Intersection tool in action.

One note about selecting the intersecting element: In the case of an element that has two or more possible intersection points (e.g., a circle that lies across the path of the line), you must select the element on the side you want the intersection to be calculated to.

Trim Elements

When you have intersecting elements and want to trim one back to its intersection with the other, you can use the Trim Elements tool, shown at left.

This tool may seem redundant, in that you can perform the same action with the Extend Element to Intersection tool, but the former really comes into its own when you need to trim multiple elements against a single element, as the tool can dramatically cut down the number of clicks needed. This is shown in the following illustration.

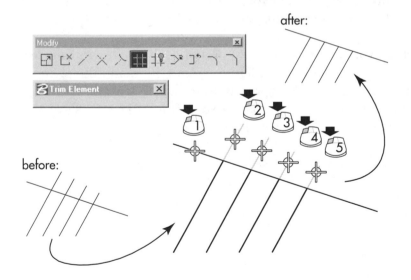

The Trim Elements tool is very efficient in clicks needed when trimming multiple elements against a single element.

When you activate the tool, MicroStation prompts you to select the cutting edge. This is the element you want to use as an edge to cut intersecting elements to. It next asks you to identify the element to trim. The side of the element you identify gets trimmed.

Another use for this tool is to trim a segment from an element that intersects and lies between two other elements. The trick to using the tool this way is to first highlight the cutting edges with the Element Selection tool, and then select the Trim Elements tool, as indicated in the following illustration. Remember, you need to hold the Control key down when making multiple elements a part of the selection set.

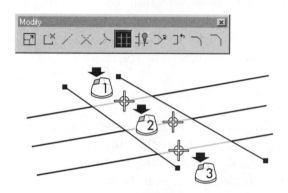

The two cutting edges were first selected with the Element Selection tool prior to invoking the Trim Elements tool.

IntelliTrim, MicroStation's Intelligent Element Trimming Tool

Introduced with MicroStation SE, IntelliTrim, shown at left, enhances the trim tool just described by optimizing the number of data points you enter and making better use of the selection set. IntelliTrim allows you to trim or extend one or more elements to the intersection of one or more identified elements in a single operation. In addition, you can change what portion of the targeted element gets trimmed before accepting the trim operation.

Cut Operation Option

IntelliTrim provides a more direct option for identifying precisely what you want done to selected elements. You can trim, extend, or cut the targeted elements. Trim and Extend control how IntelliTrim modifies the elements and are self-explanatory. The Cut operation is like taking a pair of scissors and "cutting" the element into two pieces along the cutting element path.

The cutting element is one or more elements that, once selected, become the trim or extend limits of the elements being trimmed.

Both open (lines, arcs) and closed (circles, shapes) elements can be selected as your cutting element.

Quick Mode

IntelliTrim works in one of two modes of operation: Quick and Advanced. In the Quick mode you can quickly identify a single cutting element, followed by a selection of elements to be trimmed. The selection method used is similar to Power Selector's Line method, where you identify elements by dragging a line across some portion of the elements to be trimmed.

Advanced Mode

Instead of a single cutting element, the advanced mode allows you to choose multiple elements for both the cutting elements and the elements to be trimmed. You can choose the order in which elements are identified (cutting versus to be trimmed). In addition, this mode works well with the selection set, so you can preselect the cutting or to-be-trimmed elements using PowerSelector.

Once you have identified the cutting and to-be-trimmed elements, you can fine-tune the intended results by placing strategic data points on the portions of each element to identify the portion you want to keep. An example of application of the Advanced mode is shown in the following illustration.

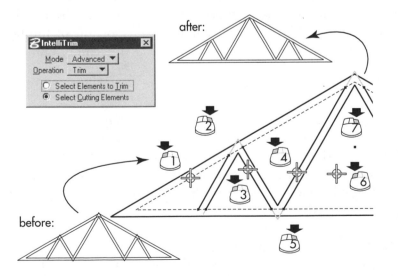

IntelliTrim's Advanced mode provides maximum flexibility in selecting both the cutting elements and the elements to be trimmed. Note how it intelligently cleaned up the truss web elements wherever it crossed the cutting element (the inside edge of the main stringer).

Modifying Vertices

As described earlier in this chapter, the Modify Element tool allows you to change the location of element vertices; however, it does not allow you to add or remove them. That is the job of the Insert Vertex and Remove Vertex tools.

Insert Vertex

The most important point to remember about Insert Vertex, shown at left, is that the segment where you select the line or line string is the portion along which your vertex will be inserted. A second data point sets the actual location for the new vertex.

Another important aspect of this tool is how it modifies a line element. When inserting a new vertex in a line, MicroStation converts the original line into a line string, as indicated in the following illustration. This is because, by definition, a line can only have two vertices: its two endpoints.

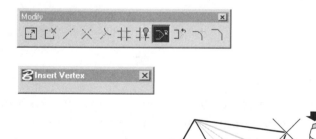

Adding a vertex to a line converts it to a line string that has two segments.

Delete Vertex

The Delete Vertex tool, shown at left, does what its name suggests: it deletes existing vertices. Selecting an element with this command will delete the vertex closest to your selection data point.

This tool works as you would expect, except when you reach the minimum configuration of a particular element type. A shape is defined as having at least three points; otherwise, it would not be a shape. When you try to delete a vertex on a three-sided shape, MicroStation will respond with the message "Minimum Element Size." This is also true when you try to delete a vertex on a line string having just two endpoints. Use of the tool is shown in the following illustration.

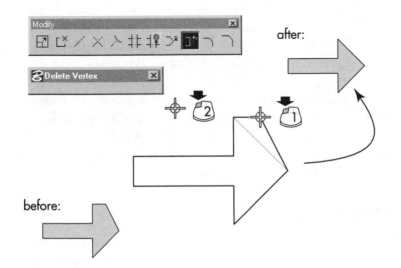

Delete Vertex is used
to remove unwanted
vertices in shapes
and line strings.

So, what happens when you delete a vertex on a line? It turns into a point. A point is nothing more than a line with both endpoints at the same location. This is one ability the line string cannot emulate. If you accidentally reduce a line to a point, just use the Insert Vertex tool to "pull" it back out.

Placing Fillets

First-year drafting classes inevitably include exercises in placing fillets. In mechanical design, this feature eases angled faces, often to relieve potential stress points or to show the true line of a casted part. On a 2D drawing, a fillet is nothing more than an arc of a given radius tangent to two lines.

Construct Circular Fillet

As part of its operation, the Construct Circular Fillet tool, shown at left, can be directed to truncate the elements used to define the arc's location. The following are the three possible truncation scenarios.

- Truncate both elements.
- Do not truncate either element.
- Truncate only the first element.

Which truncate scenario is used depends on the Truncate option in the Construct Circular Fillet tool settings. In addition, a radius is required to set the size of the resulting fillet. Where you select the source element is critical. The fillet will face in the direction of these data points.

If you create a fillet facing the wrong direction, do not worry. Use the Undo command to undo your change and try again. An example of use of the Construct Circular Fillet tool is shown in the following illustration.

The Fillet command can actually be used to place a fillet between arcs and circles, as well as lines. The truncate option will not affect closed elements (circles, shapes), regardless of the Truncate option; however, the fillet will still be placed.

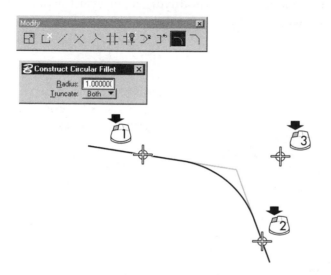

The Construct Circular Fillet tool shown in use with the Truncate Both option.

Construct Chamfer

The chamfer is a close cousin of the fillet. Instead of an arc, a straight "face" (actually a line) is placed between two intersecting lines. The Construct Chamfer tool is shown at left.

The chamfer is defined by providing a point along each selected line where the chamfer originates. If these distances are equal, a 45-degree chamfer is placed. The following illustration shows the distance entry fields within the Chamfer Element tool.

The Chamfer Element tool includes two fields in which you enter the distance along each element where the chamfer begins.

Modifying Arcs

Although the Modify Element tool can be used to modify arcs, the way it behaves depends on where on the arc you click your data point. If there is a specific aspect of the arc you need to modify, you might want to use one of the three arc modification tools found on the Arcs toolbox, shown in the following illustration. Each of these tools specializes in modifying a different aspect of your arc. Table 6-4, which follows, describes the modify arc tools.

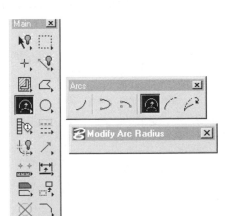

The modify arc tools are found in the Arcs toolbox, not the Modify Element toolbox.

Table 6-4: Arc Modifiers

Tool Name	Description
Modify Arc Angle	Changes the "sweep" angle of the arc
Modify Arc Radius	Changes the radius value of the arc
Modify Arc Axis	Changes the axis of the arc

Modify Arc Angle

The Modify Arc Angle tool, shown at left, is unquestionably the most useful of the three arc modification tools on the Arcs toolbox. By selecting either end of the target arc, you can specify the sweep of the arc. Although the Delete Part of Element tool can perform the same duty when it comes to shortening an existing arc, it cannot lengthen an arc's sweep.

Modify Arc Radius

The Modify Arc Radius tool, shown at left, "stretches" an arc's radius and at the same time moves the center point of the arc. This is necessary because the endpoints of the arc are maintained. The result is sort of like a soap bubble expanding.

Modify Arc Axis

The Modify Arc Axis tool, shown at left, changes an arc's usually equal minimum and maximum radii to create a partial ellipse. Again, the arc's endpoints are maintained.

Manipulate Element Revisited: The Move and Copy Parallel Tools

Now that you have been introduced to the major element modification tools, you can return for a moment to the Manipulate Element tools and visit one of the author's favorite tools, Copy Parallel.

There are many times in the design process when you want to create an element that is parallel to an existing feature. Parallel is defined as a set distance from the selected object at all times. This is quite different from, say, the Copy Element tool, where the distance between an element and a copy is set only for the single location you have selected. Instead, the Parallel tools actually recalculate the offset value of each vertex of the parallel element and construct this new element using the selected element as a template.

Move Parallel by Distance

Although not found on the Modify toolbox (actually it is found on the Manipulate toolbox), the Move Parallel tool, shown at left, resembles an element modification tool more than a manipulate tool.

In its no-options-selected mode, Move Parallel does just that. It modifies the vertices of the selected element so that they will be parallel to their original location. Most users find the two options associated with the Move Parallel tool more useful: the Distance and Make Copy options.

Copy Parallel by Distance

Selecting the Make Copy option in the Tool Settings window invokes the Copy Parallel by Distance tool. As its name implies, the selected element is left intact and a parallel copy is made at the distance specified by a second data point, as indicated in the following illustration.

Copy Parallel by Key-in

By selecting the Distance option in Copy Parallel's tool setting, you can specify the exact distance between the original element and the new copy. This is extremely useful when you are generating a series of parallel elements from a single source element. True, you can accomplish precise distance offsets using AccuDraw, but the Distance option is still more convenient, as you only have to place multiple data points and do not have to worry about AccuDraw's previous distance index, about slightly shifting the mouse, and so on.

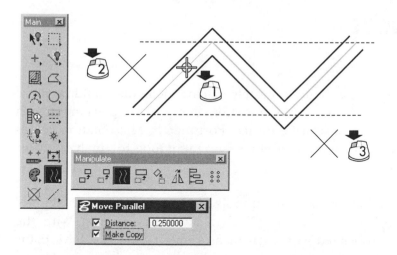

Copy Parallel maintains the parallelism of the new element to its parent by adjusting the length of each element segment. Note the selection of the Make Copy and Distance options in this example.

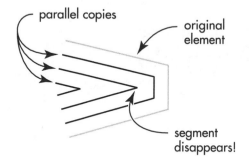

When you use the Copy Parallel tool on a line string that has a relatively short segment, you will find that in the resulting copy you may not have this line segment at all, as indicated in the following illustration. This is due to the parallel distance exceeding the length of the line string segment.

parallel copies

original element

segment disappears!

If you try to copy parallel this line string, you will find that more and more of the middle segment will disappear the farther to the left you go.

TIP: *One of the reasons the author favors the copy parallel tool is its use as a sketching aid. By concentrating on schematic features using line strings, arcs, and other elements, and not worrying about offset values until later, you can sketch out the basic features of many types of designs. Once you have finalized the design, a quick pass using Copy Parallel and IntelliTrim will complete your design in record time.*

Does Your Drawing Measure Up?

All of this talk about accurately placing elements would be more reassuring if you could somehow check their placement with some sort of measuring device. Fortunately, MicroStation comes equipped with a number of measurement tools for analyzing your design.

MicroStation's measurement tools can be found on the Main tool frame's Measure toolbox (Tools > Main > Measure). With the options associated with each measurement tool, you have many more measuring techniques than the six icons shown in the following illustration.

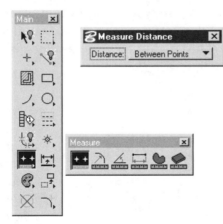

Measurement tool options.

Measure Distance

The simplest of the measurement tools, Measure Distance, shown at left, relies on your setting the distance measuring method through the Distance pop-down field in the Tool Settings window. Each of the four Distance options provides a method for measuring your design file elements. The sections that follow discuss each option.

Between Points

The simplest of the Measure Distance options, Measure Distance Between Points, shown in the following illustration, prompts you for two data points. Upon entry of the second data point, the Measure Distance Between Points tool will return the minimum distance between them as a number in the status bar.

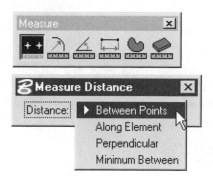

Measure Distance Between Points tool.

The distance is measured as a straight line from the first data point to each subsequent data point, and the results are cumulative (click on Reset to reset the measurement). The results are shown in the status bar, as indicated in the following illustration.

Measure Distance Between Points > Define distance to measure	Dist = 12mu 96:947su

The results of the Measure Distance Between Points command are displayed in the status bar.

The units displayed are in master units (the 12mu) and subunits (the 96:947su). This coordinate readout is set in the Coordinate Readout category of the Design File Settings dialog box (Settings > Design File), shown in the following illustration. You will see the master units, subunits, or working units, with various levels of accuracy, depending on how you set these options.

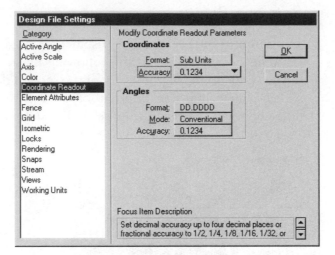

Display of units under Coordinate Readout.

Along an Element

There are times when a straight line distance is not enough. You may need to know the distance along a nonlinear element—an arc, for instance. The Along Element option comes in handy here. By data pointing on an element at the point from which you want to know the length, all subsequent data points on the element will result in a measured distance. This applies to arcs and circles, as well as to lines, line strings, and chained elements.

Perpendicular from an Element

Sort of a right-angle version of the Along an Element, the Perpendicular option prompts you for an element. Upon selecting an element (a data point), a perpendicular dynamic line appears. When you data point again, the distance from the chosen element to that point is measured along this perpendicular line, as indicated in the following illustration.

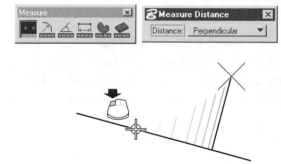

When the Measure Perpendicular Distance from Element tool is invoked, a perpendicular dynamic line is displayed until you place your second data point.

Minimum Between

In many instances you need to know how close one object is to another. That is the purpose for this last option of the Measure Distance tool. When two objects are selected and accepted, the distance is given and a dynamic line is displayed, representing the closest approach between the chosen elements, as indicated in the following illustration.

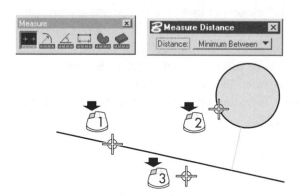

Selecting two elements results in a dynamic line (shown in gray) displaying the minimum distance between the elements. This is helpful for finding both the distance and where it occurs.

Measure Radius

The Measure Radius tool, shown at left, is used to find out the radius of either a circle or an arc. Selecting and accepting such an element results in the display of its radius value.

Measure Angle

The Measure Angle tool, shown at left, prompts you to select two lines (or line string segments). It then returns the positive angle between the chosen elements. The format of the angular data is also set via the Coordinate Readout category in the Design File Settings dialog box.

Measure Area

The Measure Area tool, shown at left, provides a number of advanced measurement options. For now, discussion is limited to only two of the most common area-measuring methods: Element and Points. These are discussed in the sections that follow.

By Closed Elements

The Measure Area tool calculates the area occupied by closed elements. Blocks, circles, and closed complex elements are examples of closed elements. By their very definition, all of these element types enclose a finite portion of the design plane, thus allowing MicroStation to interrogate them for this data. Selecting and accepting an element results in its area and perimeter being reported in the status bar, as indicated in the following illustration.

A=0.3476 SQ " , P=2.0900

The Measure Area tool displays its results on the status bar.

Measure Area Points

Measure Area Points allows you to data point a fence-like area. Upon closure of the fence, the tool reports the area enclosed.

Compressing Your Design

As you add, delete, and modify elements on your video screen, MicroStation is continually updating and changing your design file. Referring back to the shopping list analogy from previous chapters, these changes consist of two distinct operations: those that modify an element in position within the design file, and those that add elements to the design file. The Move tool and all of the Change Element Attributes tools are examples of in-position modification tools. Copy Element and Delete Partial are both examples of commands that add elements.

Notice that nothing was said about the Delete command. This is because Delete is a special type of manipulation command. Whenever you delete an element from the design, you are telling MicroStation to ignore that element. However, the element still resides in your design file, and theoretically can be resurrected through the use of the EDG utility (which is beyond the scope of this book).

The actual elimination of these elements occurs when you invoke a special file maintenance command: the Compress Design command (File > Compress Design). When you invoke Compress Design, MicroStation suspends interactive operation (i.e., will not let you work on the design file) while it goes out and physically deletes the elements marked for deletion.

Of course, as these elements are eliminated, the result is a design file with empty records. MicroStation takes care of this by moving all of the "good" elements up in the design file to eliminate these empty records through compression. The result is a clean design file ready for further work. An overview of this operation is shown in the following illustration.

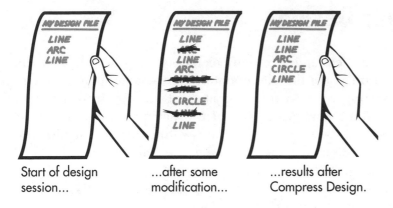

Start of design
session...

...after some
modification...

...results after
Compress Design.

The effect of the
Compress Design
command on a
design file.

Compressing your simple designs will not produce a great deal of
improvement in performance. However, in monster design files
or after large fence deletes, the improvement is substantial. So,
how often should you compress? There is no set rule for when it is
best to compress, but the following notes and tip offer some
guidelines.

NOTE: *Once you compress, you cannot undo any operations per-*
formed prior to the compression.

TIP: *A good habit to develop is compressing your design file just prior*
to exiting the design.

NOTE: *When working with a ProjectBank DGN enabled design, you*
normally do not have to perform the Compress Design operation, ever.
As part of the "Commit changes to the project server" operation,
MicroStation/ProjectBank analyzes your current design and sends
your changes to the server. It then reconstitutes your working file from
the local ProjectBank component store (your briefcase). This results in
a de facto compressed design file ready for additional edits, as indicated
in the following illustration.

V8: MicroStation version 8's new internal file structure will eliminate the need to compress your design file.

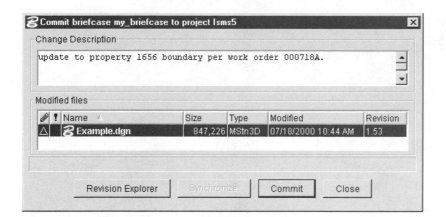

Performing a ProjectBank Commit operation performs a de facto compress of your working file.

Line Construction Tools

Earlier in the book you learned how to use MicroStation to create lines by snapping other elements. There are times, however, when the Place Line command needs a little extra help. For instance, you may need to create a line bisecting the angle between two elements. Unfortunately, there is no snap mode designed to give you this. However, there is a separate tool for performing this very task. In addition, there are other Construction tools for performing a variety of specialized design tasks. Many of these tasks are critical to the design process.

Construct Angle Bisector

A drawing technique commonly used in design work is the division of an angle. In MicroStation, this can be accomplished using the Construct Angle Bisector tool (Tools > Main > Linear Elements > Construct Angle Bisector), shown at left. Although it does not require you to snap to an existing element to provide the source angle, it is not unusual to snap to such lines in your design.

By tentative-point snapping to the endpoints or center points of elements, you provide the angle that this command then divides in half. A line is placed at this angle. The length of the line is set by an imaginary line drawn between the "horns" (the first and third data points) of the angle, as shown in the following illustration.

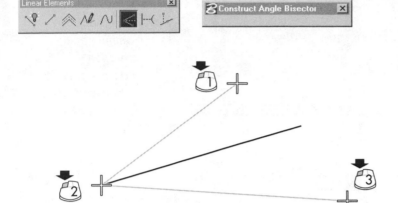

The Construct Angle Bisector tool in action.

Construct Minimum Distance Line

The Construct Minimum Distance Line tool, shown at left, is very useful when you are trying to establish a line perpendicular to an arc and a line. To do this manually would be a tedious job. With MicroStation you need only select the two elements in turn, and the result will be the perpendicular line.

Why is it called "construct minimum distance line" instead of, say, "construct perpendicular to two elements"? The answer is that perpendicularity only results when working with arcs or circles. Use of the tool is shown in the following illustration.

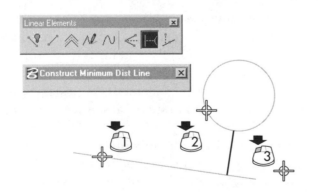

Construct Minimum Distance Line is a handy way to construct a line perpendicular to a line and an arc.

Other times, the resulting line will not be perpendicular. Thinking back to the line perpendicular to an arc, you will recall what happens when the line comes to the end of the arc. A similar thing happens here. The minimum distance between two skewed elements will not have any resemblance to the perpendicular lines seen so far, as shown in the following illustration.

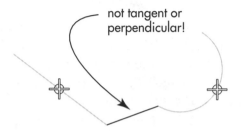

not tangent or
perpendicular!

Note how the resulting line forms nothing that is tangent or perpendicular.

Lines with an Angle on Things

Another important drafting function performed with lines is setting various angles. Quite often you need to strike a given angle off an existing line in order to develop your object. Using the Place Line tool, you could use a combination of tentative points and the *DI*= key-in to perform this function. This, of course, assumes you know the angle of the baseline from which you are drawing your elements. More often, this angle is unknown.

Construct Line at Active Angle

When constructing a line at an angle from another element, you have two distinct methods for calculating the position and length of the constructed line. You can choose a point on the target element and have the tool create the line from this point (the From Point method). Alternatively, you can choose a point in space from which the Construct Line tool will project the line back to the target element (the To Point method). The Construct Line at Active Angle tool, shown at left, provides a Method pull-down field for selecting one of these two methods.

From Point

The From Point method creates the line from the point at which you selected the original element. If your active angle is set to zero, this command will give you an error message such as the following.

```
Unable to construct line at 0 or 180 degrees
```

It will not let you proceed with the command. Entering a value in the Active Angle field will take care of this, as indicated in the following illustration. In this illustration, the data point d2 is perpendicular to the line being constructed. This is how MicroStation calculates the length of the line for this command.

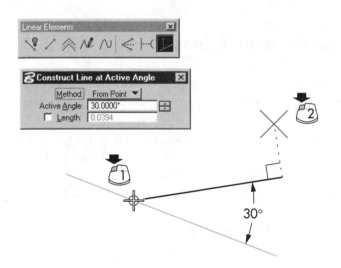

The Construct Line at Active Angle tool used with the From Point option requires the entry of the angular information prior to selecting the target element.

To Point

The To Point method allows you to select the element from which the new angled line will appear dynamically attached to the cursor. Once satisfied with its location, you data point to set it. If you select an active angle greater than 180 degrees, the result may not be what you expect.

Because you set which side of the line the angled line should emanate from, angles greater than 180 degrees can produce strange results. In the case of an active angle greater than 180 degrees,

MicroStation subtracts 180 from the active angle and uses this in subsequent line placements.

This, combined with the fact that you can place such a line in your design file that does not even touch your original line, can lead to some confused results. Try placing a line with an active angle of 185 degrees. The result will usually be an enormously long line set at an angle of 5 degrees.

Length Option

As with other tools, Construct Line at Active Angle can create a construction line with a fixed length. By selecting the Length option and entering a value, you force the tool to do just this.

Summary

This concludes the basic discussion of MicroStation's construction and modification tools. You have now learned enough about MicroStation to attempt to use these tools in a real project. In the next chapter you will learn how to use many of these tools, and various snaps, to construct some real-life designs.

CONSTRUCTION TECHNIQUES AND TOOLS

Using MicroStation to Solve Real Design Problems

UP TO NOW YOU HAVE USED MICROSTATION as you would a drafting table, drawing some lines, and perhaps adding a bit of text. As with any tool, you need to spend time with MicroStation learning to apply it to your engineering or design problems. In this chapter, you will be presented with a series of design challenges, followed by step-by-step solutions that emphasize MicroStation's design features.

A Flower Vase

In the first exercise you are presented with a problem that, on the surface (pardon the pun), looks relatively simple. Given the known dimensions shown in the first illustration of exercise 7-1, your job is to create a 2D drawing of the vase.

Looks can be deceiving. Although you have all of the information needed to draw the vase, it is not necessarily obvious. Review the previous illustration and see if you can spot the missing information. In exercise 7-1, which follows, you will begin creating the flower vase.

EXERCISE 7-1:   CREATING A FLOWER VASE

The key to creating this vase is to start from what you know. In this case, you know the heights and widths of the base and mouth, and the radii of the arcs. In addition, you know that the vase is symmetrical. Therefore, you can develop half the vase and use the Mirror tool to complete the project. However, you need to locate the curves that constitute the sides of the vase, as shown in the following illustration.

The challenge is locating the curves that constitute the sides of the vase. Note the tangency between the curves.

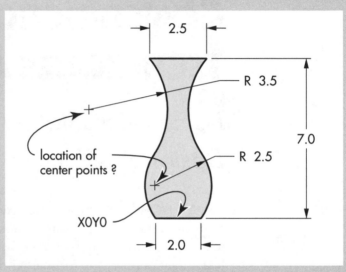

1   Open the design file *VASE.DGN*. This is an empty design file, so you will need to establish a starting point for this project.

2   Draw some lines establishing the base and mouth of the vase. Use the bottom, center of the vase as the origin point for the base (X0Y0).

3   Select the Place Line tool (Main toolbox > Linear Elements > Place Line) and place two lines: one for the base (1 unit long), and one for the mouth of the vase (1.25 units long). Use AccuDraw to accurately locate the two lines, as shown in the following illustration.

NOTE: *For orientation purposes, the completed vase, shown in the following illustration, is shown in light gray. This image is not found in the file* VASE.DGN.

With these two lines established, you can use a geometric construction technique to identify the first of the curves (arcs) that constitute the sides of the vase. This is accomplished by striking two arcs from known points to establish the centerpoint of a third arc. In this case, you know that the lower 2.5-unit arc passes through the base of the vase and is tangent to the upper arc. The radius of each arc and the location of the mouth and base are all you need for locating these arcs.

Base and mouth lines established.

4   Select the Place Arc tool (Main toolbox > Arcs > Place Arc). Select the Center method. Select the Radius option and set the value to 2.5 units. Set the Start Angle to 90 degrees and the sweep angle to 90 degrees. (This will draw an arc 90 degrees counterclockwise from the 12 o'clock position.) Using the Keypoint snap mode, tentative point/data point on the right end of the line that constitutes the base, as shown in the following illustration.

Because we know that the two arcs that constitute the graceful curve of the vase are tangent, we also know that the centerpoint of the bottom arc must lie the combined distance of the upper arc's radius and the bottom arc's radius from the vase's mouth. If you strike an arc (or circle, in this case) of this combined value from the upper right line endpoint through the previously placed arc, the intersection between these two arcs will locate the centerpoint of the bottom arc.

Using the Keypoint snap mode on the base of the vase.

5   Using the Place Arc tool again, set the Radius option value to 6.0 (R2.5 + R3.5), the Start angle to 180 degrees, and the sweep angle to 90 degrees. Using the Keypoint snap, tentative point/data point to the right end of the line that constitutes the vase's mouth, as shown in the following illustration.

At this point, you should have two intersecting arcs that locate the centerpoint of one of the vase's side curves. This is the trickiest part of this exercise.

Using the Keypoint snap mode on the mouth of the vase.

6   Set the Place Arc tool Radius back to 2.5 and deselect the Start Angle, but keep the Sweep option set to 90 degrees. Snap to the right endpoint of the line indicating the base. The arc appears in dynamics, with one endpoint locked to the end of the line indicating the base.

7   With input focus on AccuDraw, press the I key (shortcut for the intersect snap), and snap to the intersection of the crossing arcs. Data point to place the arc.

An arc of the desired radius should now touch the endpoint of the bottom line, with a sweep of 90 degrees. This is further along the curve than is needed, but you will trim it later. At this point, you can delete the two arcs you placed earlier to establish the origin of the third arc. However, it is highly recommended you simply change the line type of these arcs to simply differentiate them from the elements that constitute the vase itself.

Next, you need to locate the upper curved portion of the vase. This is another arc that is tangent to the arc placed in the previous step and that passes through the line representing the mouth of the vase. If you were drawing this by hand, you would locate the centerpoint of this arc in a manner similar to the previous steps (strike two arcs with a compass). This time, however, you will use a slightly different technique, which uses the Tangency Snap mode to help establish this final curve.

8   Select the Place Arc tool again (if it is not already selected). Change the Method to Edge, and set the Radius to 3.5 (the radius of the upper arc). Deselect the sweep angle option.

9   Set the Snap to Tangent (from the Status menu, or Settings menu > Snaps > Tangent). Snap to/data point on the arc that constitutes the lower part of the vase. A dynamic arc with a fixed radius of 3.5 appears tangent to the lower arc.

10 Using the Keypoint snap, tentative point/data point to the upper line's right endpoint. The arc that constitutes the neck of the vase appears. This

arc passes through the endpoint of the vase's mouth, and is tangent to the vase's body arc, as shown in the following illustration.

All that remains in completing the vase is to clean up the first arc and mirror-copy your elements.

Arc of the neck of the vase.

11 Select the Modify Arc Angle tool (Main toolbox > Arcs > Modify Arc Angle). Identify the vase body arc near the end closest to the neck of the vase.

12 With Keypoint snap mode, tentative point snap/data point to the lower end of the arc that constitutes the vase neck (the tangency point established in the previous step), as shown in the following illustration.

At this point, your vase is all but complete. All that remains is to select the two lines and two arcs that constitute the vase and mirror-copy them around the vertical axis of the vase.

Tangency point on the neck of the vase.

13 Select the PowerSelector tool (Main toolbox > Element Selection > PowerSelector), and select the two arcs and two lines.

14 Select the Mirror tool (Main toolbox > Manipulate > Mirror). Set the Mirror About option to Vertical, and select the Make Copy option. Snap to the centerline of the vase (the left end of the line indicating the base is acceptable) and data point, as shown at right.

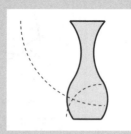

Snapping to the centerline of the vase.

Your vase is now complete! The main purpose of this exercise was to get you thinking about how snaps work and how to use geometric construction to find the missing pieces in your design. All

MicroStation users develop techniques such as those demonstrated in this exercise, and use them on a daily basis. Quite often, the real difference between a new user and an "old-timer" is not the number of tools they know but knowledge of how to apply tools to difficult problems.

Creating a Mechanical Part

The subject of the following series of exercises is an adjustable bracket consisting of a pivot hole and a semicircular adjustment slot with a double mounting flange. The dimensions shown in the following illustration are all you need to complete this layout. The design solution is presented as a series of exercises, each of which focuses on a single aspect of the design, highlighting a particular set of tools and techniques.

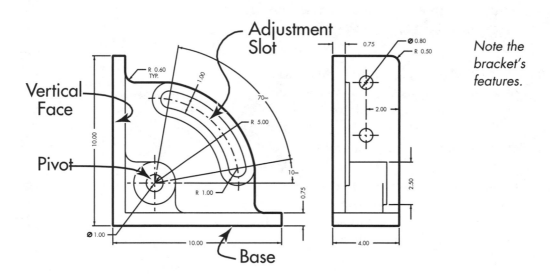

Note the bracket's features.

Although a specific sequence of tool use is followed in these exercises, this does not represent the only solution to the design problem at hand. The intent of this exercise series is to demonstrate how you apply several design tools and techniques to a single design problem. For this reason, this exercise avoids repetitive use of a single drawing technique. In exercise 7-2, which follows, you will establish the drawing origin and place the first few elements of the design.

EXERCISE 7-2: ESTABLISHING THE DRAWING ORIGIN AND PLACING THE FIRST ELEMENTS

To establish the drawing origin and place the first elements of the design, perform the following steps.

1 Open the design file *BRACKET.DGN*. You are starting with an empty mechanical design file, shown in the following illustration. For the record, the settings are shown at left in the illustration.

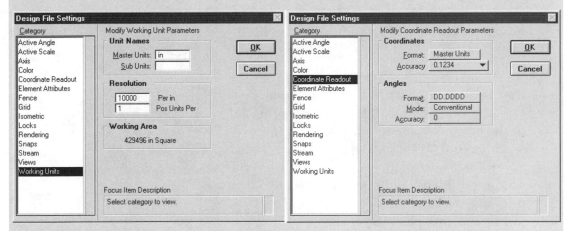

Starting design file.

2 To get started, you need to identify an origin point for the design. A quick review of the design problem shows two obvious candidates: the lower left corner of the bracket and the centerpoint of the pivot hole. For this project, the pivot point has been chosen. This will become the X0Y0, or origin, point of the drawing.

3 Select the Place Circle tool (Main toolbox > Ellipses > Place Circle). Set the Method to Center. Check the Diameter/Radius option and set it to Diameter. Enter the value *1.0.* You could also use the Radius option; but if you do this, remember to halve the pivot point value to 0.5. Next, place the circle.

4 With input focus in the AccuDraw window, press the P key. This opens the Data Point Key-in window, so that you can enter the X0Y0 value.

5 Type in *0,0* in the Data Point Key-in, and press Enter. The first of two circles appears.

6 Continuing with the Place Circle tool, set its Diameter value to 2.5, and place it at the same location, as shown in the following illustration. (Hint: Use AccuDraw's index axis feature to index to the last data point location.) You can also use the tentative point/data point sequence to snap to the center of the first circle.

At this point, you have two concentric circles centered on the design file's X0Y0 origin. Next, you will create the base of the flange using the Place Line tool. You could key in each location of the line, but where is the fun in that? Instead, you will be using a few of the Place Line options to establish the base.

Creating a second circle.

7 Select the Place Line tool (Main toolbox > Linear Elements > Place Line). Select the Length option and set it to 10 (the length of the base). While you are at it, select the Angle option and set it to 0.

8 With input focus on the AccuDraw window, snap to the center of the circles you just placed, and press the O (oh) key. This sets the AccuDraw compass origin to X0Y0.

9 In the AccuDraw X field, type in *-2.5*, press Tab, type in *-2.5* again, and data point. A horizontal line appears, which represents the base of the bracket, as shown at right.

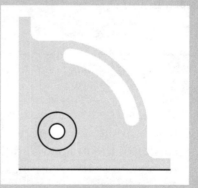

Horizontal line indicating the base of the bracket.

You could place the vertical face by simply changing the angle of the line. However, as stated in the introduction to this exercise, the intent is to demonstrate as many different tools and techniques as possible within this exercise. The vertical face is shown in the following illustration.

10 Select the Rotate tool (Main toolbox > Manipulate > Rotate). Select the By Active Angle option and set it to 90 degrees. Select the Make Copy option. Data point on the line created in the previous exercise. A dynamic line appears. To place it in the right location, snap to the left end of the base line element and data point once. The vertical face of the bracket appears. Press the Reset button to release the dynamic line.

Vertical face of the bracket.

In exercise 7-3, which follows, you will create the bracket's slot. You have successfully completed the first part of the bracket's design. Next, you will define the centerline of the slot, which you will use, in turn, to define the slot itself. You will actually place only one arc, but through the use of the Copy Parallel tool it will become five arcs.

EXERCISE 7-3: CREATING THE SLOT

To properly locate the adjustable slot, you must first establish the slot's centerline. A quick look at the dimensioned drawing shows it to have a 5-unit radius. It starts 10 degrees from horizontal (MicroStation's 0-degree orientation) and sweeps through 70 degrees. You can place this centerline arc in one step.

1 Select the Place Arc tool (Main toolbox > Arcs > Place Arc). Set the Method option to Center. Select the Radius option and set it to 5.0. Select the Start Angle option and set it to 10 degrees. Select the Sweep angle option and set it to 70 degrees. A dynamic arc appears on the cursor. All of the arc's parameters are set, except for its location. These settings are shown in the illustration at right.

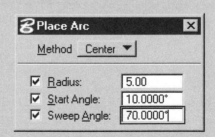

Establishing the slot's centerline.

2   Snap to the center of the pivot circle (X0Y0) and data point once.

You have now established the centerline of the adjustable slot. The next step is to create the edges of the slot and the slot shoulder, shown in the following illustration. This is performed using the Copy Parallel tool.

Slot edges and shoulder.

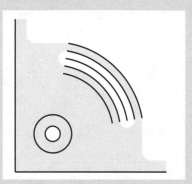

3   Select the Move Parallel tool (Main toolbox > Manipulate > Move Parallel). Select the Distance option and set it to 0.5. Select the Make Copy option. Note how selecting these options changes the tool name displayed in the status bar from Move Parallel to Copy Parallel by Key-in.

4   Data point on the slot's centerline arc. As you move your cursor to either side of this arc, a dynamic arc appears parallel to this arc, exactly a half unit over. Data point twice on each side of the centerline arc to define both the inner and outer edges of the slot. This results in five arcs equally spaced (see following illustration). Press the Reset button to release the arc.

Copy Parallel by Key-in > Identify element

Status bar showing the Copy Parallel by Key-in option.

Before finishing the slot, let's take a break and work on the base and vertical face by adding a flange plate (a thickness). In exercise 7-4, which follows, you will use the Copy Parallel tool again, but this time a little clean-up using an element modification tool is needed.

EXERCISE 7-4: CLOSING THE BRACKET'S BASE

To close the bracket's base, with the flange plate shown in the following illustration, perform the following steps.

Flange plate to be added.

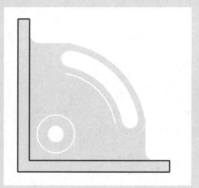

1   With the Copy Parallel by Key-in tool still selected, enter a new distance value of 3/4 (or *0.75*). Data point once on the horizontal line you placed earlier. Move your cursor above the line indicating the base and data point once. A second horizontal line appears above the base line. As before, another dynamic line appears, ready for a data point to set (but do not do that!). Press Reset to release the line.

2   Repeat the Copy Parallel operation on the vertical face line. Note how the intersection of the two new lines is not as it should be, as shown at right. This is where the Extend tool comes into play.

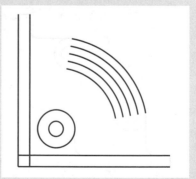

Problem at the intersection of the two new lines.

3   Select the Extend 2 Elements to Intersection tool (Main toolbox > Modify > Extend Elements to Intersection). Data point once on each overlapping line. A third data point initiates the Extend (or trim, in this case) element operation.

NOTE: *You must identify each line on the portion of the line you wish to keep; otherwise, the results will not be what you expect.*

Before leaving the base and face, let's close up the end of the base with a small line segment.

4   Use the Place Line tool (deselect all options), and Keypoint snap/data point on the rightmost end of each horizontal line (the base flange). Press the Reset button to terminate the dynamic line. The result is shown in the illustration at right.

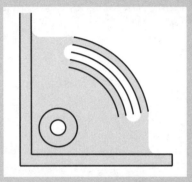

Base of bracket closed.

The next part of the bracket's construction is fun! In exercise 7-5, which follows, you are going to complete the adjustable slot's shoulder around the ends of the arcs you placed earlier.

EXERCISE 7-5:   CAPPING THE SLOT AND CREATING THE SHOULDER OF THE BRACKET

To cap the slot and create the shoulder of the bracket, perform the following steps. These components are shown in the following illustration.

1   Select the Place Arc tool and set the Method option to Center. Turn off all other options. You will not need them to construct the slot's end-cap arcs.

2   Snap from the innermost arc of the slot to the outermost arc (see following illustration).

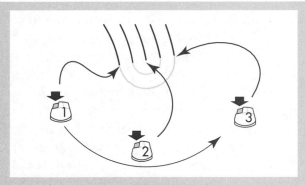

Slot to be capped and bracket shoulder to be created.

The order in which you tentative point/data point defines how the slot's end arc will appear. The first and second data points set the new arc's radius, as well as its centerpoint (the second data point).

3   Repeat the sequence for the inner edge of the slot.

You could go ahead and cap the other end of the slot using the operation just discussed. Do not! Later you will use the Mirror Copy tool to do this and more. For now, let's finish the bottom portion of the bracket by placing two lines that connect the slot and pivot to the base flange. These two lines are interesting in that they are both tangent to an arc or circle, yet are at the same time perpendicular to the base flange. Thanks to MicroStation's rich set of snap options, you can generate each of these lines in one operation.

4   Select the Place Line tool. Deselect all of its options. Next, select the Tangent snap (Settings menu > Snaps > Tangent).

You can set Tangent as your "permanent" snap by holding down the Control key while selecting it from the menu. Alternatively, you can select the Tangent snap from the Snap Mode button bar if you have it active (see the illustration at right), or from the snap field on the status bar (see the following illustration).

Tangent snap tool on the Snap Mode button bar.

5   With the Tangent snap selected, tentative point/data point on the small outer arc of the slot. A dynamic line will appear, and depending on where you clicked the tentative point on the arc, it may or may not be facing the right direction. In the following illustration, the line is definitely not facing the way you want. This is easily fixed: just perform a "loop-de-loop" counterclockwise around the right edge of the arc.

Adjusting the direction of the dynamic line.

A "loop-de-loop"? This is not exactly the most technical term used; however, it is the action performed to change the orientation of a tangent dynamic line. By crossing over the line's source element (the arc) and back out in the direction you want the tangent line to face, you direct MicroStation to change the orientation of the tangency, as shown in the following illustration.

Changing the orientation of the line's tangency.

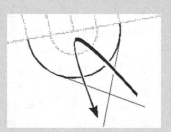

6   With the orientation of the tangent line set, all that remains is to make it perpendicular to the base. This is done with a different snap override. Select the Perpendicular snap from the Snaps menu or Snap Mode button bar. Tentative point once on the line indicating the base and data point to accept it. Repeat the process for the right side of the pivot. The results should be the two vertical lines added to your design.

With the shoulders on half your bracket, all that remains is the creation of the fillets and the mirror-copy operation.

7   Select the Circular Fillet tool (Main toolbox > Modify > Construct Circular Fillet). Set the Radius to 0.6. Finally, because the fillet to be applied (shown in the following illustration) modifies only one element, select the Truncate/First option.

8 Because the First option has been specified, the order in which you select the elements for the fillet operation is
important. Data point on the first shoulder line, placed in the previous step, and
then on the line indicating the base. The
location on the element you select the
baseline will establish the direction of the
resulting fillet. Data point on the side you
wish the fillet to face (in this case, to the
right of the shoulder line). Repeat for the
Pivot shoulder. The result of this step is
shown in the illustration at right.

Fillet to be applied.

In exercise 7-6, which follows, you will complete the design by performing the Mirror Copy operation.

Exercise 7-6: Mirroring Drawing Details

It is time to perform the mirror-copy operation and thus complete the design.
Prior to actually mirroring the elements, shown in the following illustration,
you must select them.

*Elements to be
mirrored.*

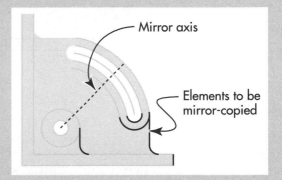

1 Select the PowerSelector tool (the Arrow icon). Identify the elements you
created for the lower part of the bracket (the slot arcs, the shoulder lines,
and the fillets). See the previous illustration for specific elements.

2   Select the Mirror Element tool (Main toolbox > Manipulate > Mirror Element). Select the Mirror About/Line option and the Make Copy option. The status bar should report the tool in use as Mirror Element About Line (Copy).

3   Set your snap temporarily to Midpoint, and tentative point/data point to one of the large arcs that constitute the adjustable slot.

4   Set your snap to Centerpoint, and tentative point/data point to one of the circles located at the pivot point of the bracket.

The mirror-copy operation results in the shoulder and the slot elements being mirrored about the imaginary axis you created between the midpoint of the adjustable arc and the origin of the pivot point. At this point, you have completed the project, as shown in the following illustration. Take a moment to review your results.

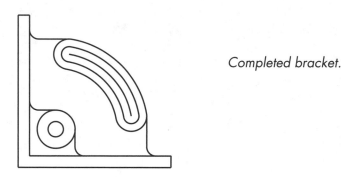

Completed bracket.

Summary

This series of exercises has shown you how to apply various MicroStation tools to create your finished design. With a little bit of practice, you should begin to see that there are many different combinations of tools you can use to get the same end results. Over time, you will develop your own style of construction. Some tools will become your favorites, and you will avoid others.

COMPLEX ELEMENTS

Working with Cells, Complex Shapes, Chains, and Other Sophisticated Elements

WHILE WORKING ON PRACTICALLY ANY DESIGN PROJECT there comes a point when you will want to use a particular drawing symbol many times within the same drawing. Examples of symbols include street signs on a signing and marking plan, a NAND gate on an electrical schematic, a north arrow on a site plan, callout bubbles on a floor plan, and so on.

If you were limited to using the MicroStation tools introduced thus far, you would have to use the Copy tool with a fence or selection set defined or (heaven forbid!) copy each element one at a time. This goes against the whole idea of making your job easier and more efficient, which means, of course, that MicroStation supports an efficient method for duplicating symbols. This is the subject of this section, MicroStation's Complex Element.

What Is a Complex Element?

Up to this point you have been dealing with what are commonly referred to in MicroStation programming parlance as primitive element types. MicroStation supports another object type called the *complex element*. Consisting of a varying quantity of primitive elements, a complex element is a method by which MicroStation can describe complex geometric shapes while manipulating them as single elements. Chief among this class of elements are the cell, the text node, the chain, and the shape. In this chapter, each of these element types is discussed in detail. Let's start with the cell.

What Is a Cell?

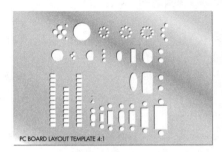

PC BOARD LAYOUT TEMPLATE 4:1

A cell is a named collection of one or more elements that are formally grouped together into a single component that as far as MicroStation is concerned *is* one element. Cells are a very efficient method for organizing and using your most commonly used symbols because they are initially stored in a separate file called a cell library (more on this later). It is from the cell library you extract a cell's definition and place it into your design file.

Once it is placed in your design file, you manipulate a cell in the same way you would any primitive element. Although you can snap to any part of a normal cell, you cannot modify individual elements contained within it. Think of a cell as a preprinted, adhesive-backed decal; for example, of a north arrow. Working with this "decal" (cell), you move, rotate, and copy it. You can even delete it (peel it off and throw it away).

The cell is also analogous to the drafting template. In fact, symbols found on such templates are definite candidates for conversion to cells. If you use a symbol more than a few times in a design, you should consider making it a cell. Examples of cells are shown in the following illustration.

Anatomy of a Cell

A cell also contains data beyond just its graphic components. For one thing, there is the cell origin, the location within the cell definition around which the cell is drawn. This origin can be snapped

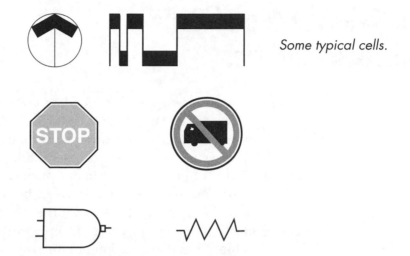

Some typical cells.

to using the (appropriately named) origin snap. A cell's origin is normally an important consideration when constructing the cell. It usually points to a common point of reference such as the center point of a callout bubble, the pivot point of a door, or some other obvious location. Cell origins do not have to share a location with any element contained within the cell, but it is a common practice to use a line or other element's endpoint as the origin point of the cell.

A cell also has a name. Limited to six characters in MicroStation/J and older versions (newer versions of MicroStation will eliminate this restriction), this name is a descriptive label of what the cell represents. For instance, a name such as VALVE is commonly used with a valve symbol. Because MicroStation limits the length of the cell name to six characters, it is difficult to get much more descriptive than DOOR or VALVE! Fortunately, when the cell is created and placed in its cell library, there is a description field of 27 characters, where you can provide a proper description of the symbol.

TIP: *Many companies have specific policies on cell names and organization. It is not unusual to find cells with names such as C1X5S, in which each character has specific meaning with respect to the design discipline or project type. It is a good idea to review any company cell definition policies before creating a cell for general consumption.*

The Relationship of Cells and Their Libraries

Before you can use a cell, you need to create its definition and store it somewhere. MicroStation uses a special file called a *cell library* to organize and store these definitions. You create a cell library file once and place your cell definitions in it one at a time. A cell library is nothing more than a purpose-built file that contains one or more cell definitions stored by cell name and organized for fast retrieval within the MicroStation design environment. There is no practical limit to the number of cells within a library, but there is a limit to the size (in bytes) that can be stored within each cell definition (65 Kbytes). It is not unusual to find cell libraries containing literally thousands of cell definitions.

As you would expect, you place cells in your design files more often than you generate the cell definitions themselves. In many companies the cell library is considered a corporate resource and is protected from unauthorized modification. You are generally given read access to cells in a library but are not allowed to create new cells. This makes sense as a way of maintaining drawing uniformity within a company or project. Using a cell from a cell library is a straightforward three-step process, as follows.

1   Attach the appropriate cell library to your active design file.

2   Identify the cell you want to use (the active cell).

3   Place the cell.

The Cell Library as a Post Office

A good analogy for how the library works is your local post office or, more specifically, the post office boxes located in the lobby. When you visit a post office, you can see hundreds of post office boxes lining numerous walls. What makes one unique from another is the unique label each box is given (usually a number). All post offices use the same basic numbering system for their P.O. boxes, some just have more boxes than others. When you observe the row upon row of P.O. boxes, you know each box potentially contains mail but, beyond that, you do not really know its content. Unless you look in a specific box, assuming you have the combination code or key, you may never know what a P.O. box has in it.

To correlate this postal analogy to cells, each cell library is like a post office. A cell is equivalent to an individual P.O. box. Each cell has a unique cell name; however, this uniqueness applies only within this cell library. Other cell libraries may have the same cell names, but may have totally different content.

The Cell Tools

There are several tools associated with cell creation and placement. These tools are located in the Cells toolbox, which is directly available from MicroStation's Main toolbox. The Cells toolbox is shown in the following illustration.

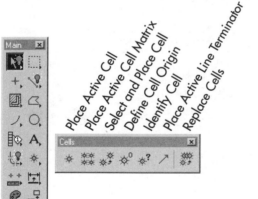

The Cells toolbox is used primarily to place cells in your design but also includes several tools used in the cell creation process.

In addition, there is a separate Cell Library dialog box (Element menu > Cells), in which you perform many of the cell library maintenance and placement operations. The Cell Library dialog box is shown in the following illustration.

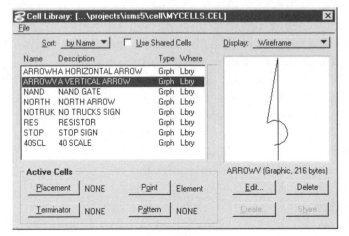

The Cell Library dialog box or window is often used to select the cells in the current cell library, as well as to create new cells.

The best way to understand how cells work is to use them. To see how simple it is, work through exercise 8-1, which follows. You will be attaching a cell library to your active design file, and placing a north arrow symbol from this library.

EXERCISE 8-1: PLACING A CELL

1   Open the *CELEX01.DGN* file, shown at right. Prior to using any cells, you need to attach an existing cell library. One has been provided for this exercise.

2   Open the Cells window (Element menu > Cells). This is the main user interface for working with cells. From this window, you attach cell libraries, activate specific cells within the current library, create cells, and perform other cell-related functions.

CELEX01.GGN file.

3   Attach the cell library MYCELLS.CEL (Cells File menu > Attach). Using the Open File dialog box, select MYCELLS.CEL. The cell library's name will appear in the title bar of the Cells window. A list of available cells is displayed in the Cells window. Clicking on a cell presents a preview of the cell's content in the preview pane (right side), as shown in the following illustration.

Preview pane.

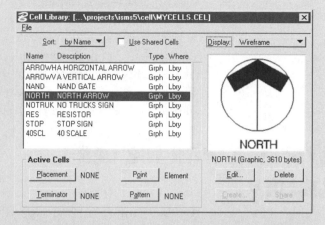

4   Select the cell NORTH. At this point all you can do is preview the image of the North symbol. To use it, you need to make it your active cell.

5   Click on the Placement button or double click on the North cell name. This makes the North cell your active cell. If you double click on the cell instead of the Placement button, the Place Cell tool is automatically activated and the North symbol appears on the cursor ready to be placed.

6   If Place Cell tool is not active, select it (Main toolbox > Cells toolbox > Place Active Cell). Now reorient the North symbol more toward the North-North-west.

7   In Place Cell tool's settings window, select the Active Angle data field and enter *15°*. The North arrow changes direction in the counterclockwise direction, as shown in the following illustration.

North arrow change of direction.

8   Data point once to place the North arrow.

9   Close the Cells window.

That is all there is to placing a cell. As you can see, there are other parameters that affect the appearance of the cell; in this case, the active angle. The active scale (*AS=*) will also affect the placement of the active cell. Active scale is easily set using the X, Y (and Z) Scale fields associated with the Place Cell tool settings. If you know the name of the cell you want to insert into your design, you can directly enter it via the Active Cell data field.

Working with the Cell Library

Unlike your regular design file, which you call up and work on directly, a cell library file is only accessible by indirect means. The graphic content of cells are drawn in an active design file. Then, using the Create Cell command (key-in shortcut: *CC=<cell-name>*), the identified elements are merged into the cell library.

Creating the Cell Library

Because the cell library must be ready to receive the new cell definition, you need to first create the cell library. This is accomplished within MicroStation's design environment using the New Cell Library command (Cell Library window > File > New). This brings up the familiar file navigation dialog box under the title Create Cell Library. Use of this dialog box is shown in the following illustration.

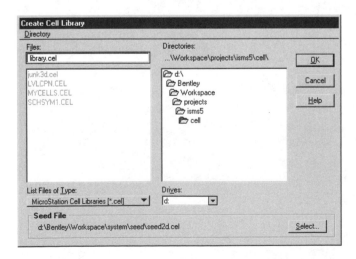

Type in the name of your cell file. In this example, LIBRARY.CEL is used.

MicroStation uses.CEL as the default cell library file name extension. It is highly recommended that you use this extension even if you do not use.DGN as your normal design file extension.

So, what does this new cell library file contain? At the moment, it contains the most basic cell library structure but no cell definitions (yet). As you create your cells, the cell library will hold the individual cell definitions, each of which may consist of hundreds of elements, up to 128 K of data per cell. It is not unusual to find cell libraries containing several hundred cell definitions. For this reason, it is a good idea to establish a concise cell-naming scheme.

A cell name can be up to six characters in length. The major reason for this limitation has to do with MicroStation's historical connection to the 1980s generation of CAD software and hardware and a data compression technology called RAD50 (Future versions of MicroStation will not have this limitation). True, you are not restricted to numbers only (letters work, but no spaces). To help alleviate this limitation, MicroStation provides a description field with each cell.

Here you can write a longer description of the cell. In the previous example, the description field described the cell NORTH as NORTH ARROW. With an obvious symbol like this, the description does not seem so important. But what about with a cell named P1THL-? Be wary of this mistake commonly made by first-time MicroStation users: Avoid cell libraries riddled with bizarre cell names and no descriptions, which makes their use difficult at best.

NOTE: *Another common mistake made by first-time users is the tendency to place all cells in one master library. At first this is not a problem. However, as you add more and more cells to it, you will find that you lose track of what the different cells represent. Now is a good time to organize your cell libraries by their function or content. It is much better to have multiple cell libraries to access than to have one huge central cell library. MicroStation even provides support for multiple cell libraries via a cell library search path [Workspace menu > Configuration > Cells category > Cell Library Directories (configuration variable: MS_CELL)].*

Creating a Cell

Creating a cell is easy. While working in design file, all you do is draw your proposed cell definition using the normal MicroStation drawing tools. The name of the design file in which you draw your cell is not critical; however, it should use the same working units as those design files with which you will be using the cell (working unit settings are not saved as part of a cell's definition). In this way, you will not have to deal with scaling the cell later to fit in your design.

Many users build their cells in the margins of a typical project design file and insert them directly into the attached cell library. You can do this as well; however, it is a better design practice to develop a separate "scratch" file for creating your cell library definitions. The following are the steps for creating a cell.

1   Draw the content of the new cell in a design file.

2   Attach the appropriate cell library. (This step can occur any time up to the last step.)

3   Identify the content of the cell using the Place Fence, Element Selection, or PowerSelector (preferred) tool.

4   Define the origin point of the new cell using the Define Origin tool.

5   Create the cell, providing a name, description, and cell type.

The cell creation process is not all that difficult to follow. The actual cell creation process is little more than a copy element operation, but instead of making the copy within your design file, the copy goes into the cell library, as indicated in the following illustration.

Cells consist of any elements supported by MicroStation. This includes all lines, arcs, circles, text, shapes, and so forth. All of the weights, line styles, colors, and so on are also allowed. Because the level information about each element is also stored, it is important to make sure your elements are on their correct level before creating the cell. There is nothing more frustrating than having an errant element of a common cell appearing on the wrong

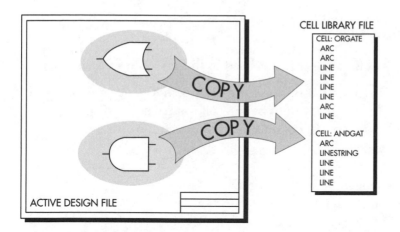

ACTIVE DESIGN FILE

CELL LIBRARY FILE

CELL: ORGATE
ARC
ARC
LINE
LINE
LINE
LINE
ARC
LINE

CELL: ANDGAT
ARC
LINESTRING
LINE
LINE
LINE

Cell creation in its most basic form is a Copy Element operation between your active design file and the cell library.

level. Of course, this usually happens just as you are ready to go to final plots.

Adding Your Cell to the Cell Library

With the cell library successfully attached, you need to identify the elements of your fledgling cell. This is done with the Place Fence tool or the Element Selection tool. In either case, you want to identify only those elements to be included in the new cell. One final item remains before you can create your new cell: establishing the cell origin.

Define Cell Origin

Every cell has an origin associated with it. The tool for setting your cell's origin is Define Cell Origin, found on the Cells toolbox. This origin point is crucial. Because it is the "handle" by which you place the cell, you need to make sure that the point you select for the origin is appropriate to the cell's usage. The Define Cell Origin tool is shown at left.

In exercise 8-2, which follows, you will build your first cell, a target symbol. You will do this by first placing a circle and a couple of line strings to give the new cell some content. By now you know how to place such elements, so the step-by-step approach to element placement is not provided. Instead, you will concentrate on the cell creation steps.

EXERCISE 8-2: BUILDING A CELL

1   Open the *CELEX01.DGN* file. This file was used in the previous exercise.

2   Attach the MYCELL.CEL library if it is not already attached (see previous exercise). You can verify the attachment by the cell library name displayed in the Cells window. Next, you will create the graphic elements that will make up the new target cell.

3   Using the dimensions shown, build the target with the Place Circle and Place SmartLine tools. The length of the line string's legs is not critical. Just make sure to place the middle data point of each line string at the centerpoint of the circle, as shown at right.

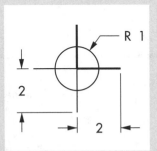

Placing the middle data point.

4   Place a fence around the elements just placed using the Place Fence Block tool. With the fence in place, you need to define the cell origin point at the center of the circle.

5   Select the Define Cell Origin tool from the Cells toolbox, shown in the following illustration.

Define Cell Origin tool.

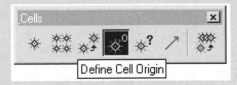

6   Tentative Point to the center of the circle (use Center snap to ensure this). Data point to set the origin point. An "O" character will appear at the origin's location, notifying you that MicroStation is ready for you to store the cell.

7   Open the Cells window (Element menu > Cells) and select the Create button. The Create New Cell dialog box appears, shown in the following illustration.

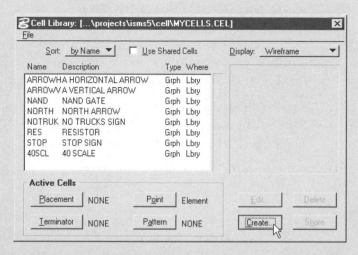

It is now okay to store the cell.

NOTE: *If the Create button is inactive (dimmed), you have not identified the target element with a fence (or selection set), or you have not defined the cell's origin. Once you have performed these steps, the Create button will activate.*

8   Enter the cell name *TGT* in the name field and *A simple target* in the Description field (as shown in the following illustration) and press the Create button. Cell names and descriptions are not case sensitive. A shortcut is available via the Key-in window, where you would type in *CC=TGT,A SIMPLE TARGET<Enter>*.

Name and Description field entries.

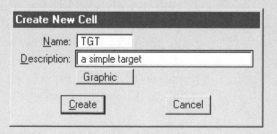

The TGT cell is now created, which MicroStation confirms by displaying a message on the status bar. The TGT cell appears in the Cells window in the cells list. You have now created your first cell. In exercise 8-3, which follows, you will place the TGT cell.

EXERCISE 8-3: PLACING THE TGT CELL

1   Select the TGT cell in the Cell Library window and click on the Placement button. This sets your new cell as the active cell. You can also type in *AC=TGT* in the Key-in window.

2   To place your cell, select the Place Cell tool from the Cells toolbox. Your TGT cell should now appear on your cursor. If it does not, simply type in *TGT* in the Place Cell tools settings' Cell field.

3   Data point once in View 1 to place the TGT cell in your active design file.

When you keyed in *CC=TGT,A SIMPLE TARGET* a moment ago, you actually invoked the Create Cell command. The actual name of the cell was TGT. A SIMPLE TARGET is a descriptive text label stored with the cell for listing purposes.

First-time users of MicroStation quite often make two mistakes with cell creation. First, their choice of cell names almost always includes MINE or XXX or A, all of which say nothing about what the cell represents.

The second mistake is, of course, not using the description field. When you can have literally hundreds, maybe even thousands, of cells in a library, this description field is important. That is why the description field can handle up to 27 characters. So, a good habit to get into, starting now, is using intelligent cell names. Use the descriptions to further define the content of the cell. It is always easier to scan a list of cell names and descriptions rather than study the graphic content of each cell within a library.

When Element Keypoints Get in the Way

If you were particularly observant you may have noticed that you created a normal cell in the Create Cell dialog box. This is the most common type of cell used. A normal cell contains the graphic elements, including their individual element attributes (color, level, line styles, and so on) and their snap points. This means you can snap to any part of the cell as if it were an individual element.

Sometimes the ability to snap to any element of a cell is a problem. For instance, when creating site plans, there is a real danger with using a normal cell that has more than one point to snap to. Depending on your cell, you could have serious problems with accuracy by inadvertently snapping to the wrong part of a cell, as indicated in the following illustration.

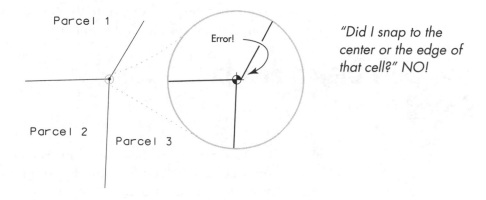

The Point Cell to the Rescue

MicroStation provides a special type of cell called the point cell. It differs in two important ways from a normal cell. First, a point cell takes on whatever element attributes you have active at the time you place the cell. For instance, if you have the color red, active level 11, and weight 5 as your active attributes, a point cell will take on these values when you insert it into your design. It makes no difference what the source elements were set to when you created the cell definition.

The other point cell distinction is its singular snap point: the origin point of the cell (thus its name). In the previous example, if the target or monument had been a point cell, that error condition could not have occurred.

You create a point cell in almost the same way as a normal cell. You set the fence and the origin in the same way. The difference comes when you select the type of cell. By setting the Type field on the Create New Cell dialog box to Point, you tell MicroStation to create a point cell. Using the *CC=* key-in, you append a comma and the letter P to designate the new cell as a point cell. For instance, to create the target cell as a point cell, you would enter the following.

```
CC=TGT,A POINT CELL TARGET,P
```

Note the additional comma and letter P. The *,P* informs MicroStation of your intention to make the cell a point cell. Outside of this difference, the point cell acts the same as all cells being affected by the active angle, active scale, and so forth.

The Cells Dialog Box in Detail

In the last two exercises you have been instructed to open and close the Cells Settings dialog box a number of times. By now you are probably curious about this dialog box and what all those buttons and menus do. The Cells Settings dialog box is shown in the following illustration.

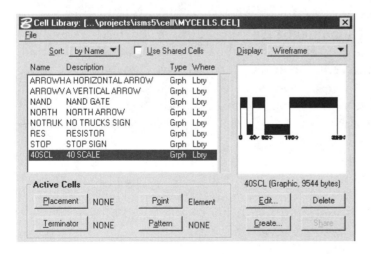

Most cell functions can be performed from the Cells Settings dialog box.

The first thing you will notice about this dialog box is the label in the window's title bar. Instead of a title such as Cells, the name of the currently attached cell library is displayed, as shown in the previous illustration.

You can sort the list of cells in your cell library via the Sort field, shown in the following illustration. Along with setting your current Placement cell, you can set the Point cell, the Terminator cell, and the Pattern cell via buttons in the lower left section of the settings box.

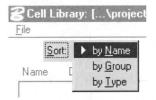

Display of currently attached cell library.

You can also perform maintenance on individual cells. Selecting the Edit button with a cell selected brings up the Edit Cell Information dialog box, shown in the following illustration. Here you can change the name and the description of your cell. Of course, there is also a Delete button for deleting a cell from your library. An alert box pops up when you select this command, confirming that you really do want to delete the selected cell.

Edit Cell Information dialog box.

Normally, when you place a cell, you are copying all the elements that constitute the cell into your design file every time you place a cell. In many cases, this means there is a lot of redundant data in your design file. Recognizing this, MicroStation gives you the option to share one occurrence of a cell's elements among many occurrences. By enabling the Use Shared Cells option on the Cells settings box, you can reduce the overall storage footprint of your design file. In addition, when you replace a cell (use the Replace Cell tool) that was placed with the Shared Cell option on, all occurrences of that cell are updated.

NOTE: *When using cells containing features such as enter data fields, text nodes, or dimension-driven definitions, you must turn the Shared Cell option off.*

Working with Cells

The sections that follow explore placing an active cell using the Relative and Interactive options, the cell matrix, and parameters that affect cell placement. The Select And Place Cell, Identify Cell, and Replace Cell tools are also discussed.

Place Active Cell

This is the primary tool used to place cells in the active design. The Place Cell tool, shown at left, provides two additional options that affect the behavior of this tool. These options are discussed in the sections that follow.

Place Cell: Relative Option

There are times when you need to place a cell on a level different than the one on which it was originally created. This is especially the case when using a level assignment scheme where levels are used to define different phases of construction or otherwise identify a subpart of the overall design. For instance, you may need to identify features within a rehab project where some doors and windows are identified as demolition and others are existing or new construction. In this scenario, you can use the same cell to represent any one of the three "states" of the construction object (the door or window) by using the Place Cell Relative option to place the cell on the appropriate level.

This command uses your active level as the starting point for the placement of the selected cell's elements. For all intents and purposes, the active level becomes level 1 to the cell. This means that any elements defined in the cell as level 1 will now appear on the active level. Elements on level 2 of the cell will appear on the next level higher than the active level (active level + 1). Elements on level 2 of the cell appear on the next level (active level + 2), and so on. The following illustration will help clarify how the relative placement operation works.

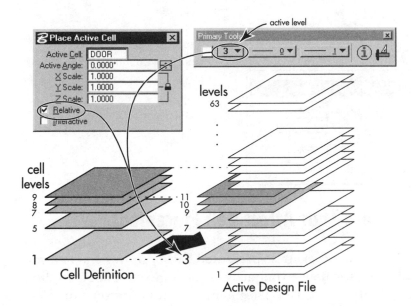

Cells placed as Relative start with elements on level 1 of the cell being placed on your active level in the design file.

The Interactive Option

Instead of using the current active angle and active scale, the Interactive option allows you to specify the scale and angle interactively. In keeping with the abilities of the previous Place Cell commands, the interactive command can place cells both absolutely and relative to the active level.

Cell Matrix

MicroStation includes a facility for placing a rectangular matrix of cells in your design file. This can be used to establish a structure for mapping projects, or a regular spacing of components for printed circuit board design. The Cell Matrix tool is at left.

Similar in operation to the Construct Array tool described earlier in the book, this tool requires you to set the parameters for the number of rows, the number of columns, and the spacing between each, as indicated in the following illustration. Once set, you data point at the lower left corner of where you want the matrix to appear.

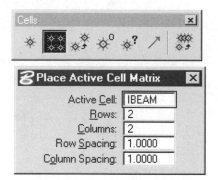

There are a number of options you must set in order to use this tool.

Parameters That Affect Cell Placement

In the North Arrow exercise presented earlier in this chapter, you were directed to change your active scale and active angle to show how these affect your cell placement. When laying out your cell, you should keep in mind that in all likelihood the cell will be inserted at various rotations and scales. Remembering the counterclockwise rule of angles, it is a good idea to build your cells in a predominantly horizontal direction, as indicated in the following illustration. If you do this, the active angle will affect the cell in a predictable manner.

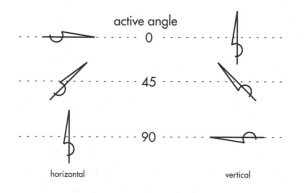

The cell on the left corresponds to the normal active angle rotation. The cell on the right could lead to some confusion.

Select Cell

More often than not, when you are editing an existing drawing, you come across a cell already placed that you would like to use elsewhere. You could use the Copy command, but if scale or angle has been used on the cell, you would have to undo these effects on the copied cell. Instead, use the Select And Place Cell tool, shown at left.

This tool allows you to click a data point on an existing cell, which then becomes your active cell. Select And Place Cell ignores the rotation and scaling of the identified cell. Thus, only the active scale and angle affect the placement of the new cell. As with the Place Cell tool, there is a Relative option.

Identify Cell

So how do you know what the name of a cell is once it has been placed in the design file a number of times? By using the Identify Cell tool, of course. Clicking on any cell returns its name in the status bar. The Identify Cell tool is shown at left.

Replace Cell

One of the disadvantages of the cell library approach adopted by MicroStation is the common situation you encounter when you want to perform revisions to previously placed cells with performing revisions to cells. It is easy enough to update the cell in the cell library, but what about all those cells you have already inserted into your design file? And, what if your original cell selection was incorrect and you want to update the placed cells with a different cell? MicroStation's Replace Cell tool to the rescue! Introduced with MicroStation/J, the Replace Cell tool, shown at left, performs the following functions.

- Updates one or more cells of the same name with the current definition in the cell library (names must be the same in the library and the design file)

- Replaces a single cell of one type with a new cell definition

- Globally replaces all cells of the same cell name with a new cell

By default, this tool performs a cell-library-to-cell-instance update (the first of the previous bulleted items). When the tool is selected, you simply data point on the cells you want to update to the current cell library definition. The cell must exist in the library for this to take place.

When you select the Replace method, the Replace Cell tool kicks into high gear. You have several new options available, most notably the Mode option. In Single mode operation, you select the source cell (the one you want to use as the new cell definition), followed by the specific cells you want to replace with the source cell. The "target" cells do not have to have the same name as the source cell, so it is a true replacement operation.

The Global mode also prompts you to select the source cell, but when you select your target cell, all cells with the same name in the active design file are replaced with the source cell definition. This can be an extremely powerful operation but should be used carefully (all global tools should be treated this way).

The Replace Attributes tool refers to this tool's ability to retain some of the information specific to each cell instance in the active design file. For instance, if your cell contains tag information (non-graphic database information), these attributes will be retained even though you have replaced the graphic representation of the cell.

The Use Active Cell option provides a method for selecting your cell definition from a cell library rather than a previously placed cell. This tool also works with a fence, which can be very useful when you do not want to quite do a global replacement but you also do not want to replace several instances within a small area of your design.

Reversing the Cell Creation Process

There are times when you may want to return a cell to its constituent components. For instance, you may want to create a new cell roughly based on an existing cell. Instead of recreating it from scratch, it would be easier if you could take a cell, undo its "cellness" (known better as its complex element status), and create your new cell from its parts. Well, you can. MicroStation supports just such a command: the Drop Element tool, shown in the following illustration. This tool is found in the Groups toolbox (Tools > Main > Groups).

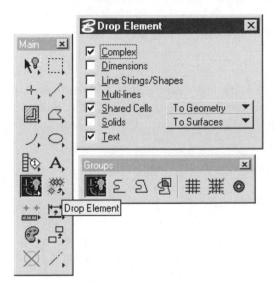

Drop Element tool.

Dropping a Cell Using Drop Complex Element

If you should accidentally delete a cell you absolutely need, there may be a way to recreate it. Earlier in this chapter it was mentioned how MicroStation copies the content of the cell into the design file from the cell library. This means that each cell occurrence within your design file either contains a full cell definition or points to one stored with the design file (the shared cells option). If you ever need to recreate a particular cell and do not have the original cell library handy, all you need to do is reverse, or "drop," a copy of one of the placed cells within your design, fence it, define the cell origin, and perform the Create Cell operation. This will regenerate that cell within your current cell library.

What does "drop cell" mean? Essentially, the Drop Element tool, shown at left, "explodes" a cell that was previously placed in a design file. This results in its constituent elements being returned to their non-cell status.

In earlier versions of MicroStation, this tool was called Drop Status because it simply deleted the cell header record in the design file "shopping list" (see following illustration), leaving behind the individual elements that constituted the cell. With the introduction of shared cells, this operation changed slightly. If a cell is placed with the shared cells option, a copy of the elements within the shared cell definition is placed at the end of the design file and the shared cell instance is deleted.

NOTE: *You can use the Drop Element tool (Complex option enabled) on any complex element, including text nodes, chains, and shapes, as well as cells.*

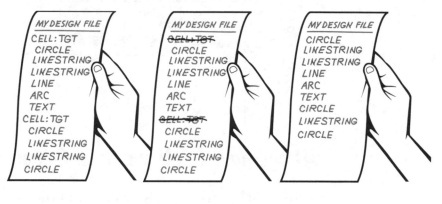

Start of design session...

...after Drop staus...

... results after compress design.

When you drop a cell's "status," the result is similar to using the Delete tool, except that the elements associated with the cell are left intact. Compress Design eliminates the deleted cell headers.

Convert Shared Cell to Unshared Cell

There are times when you want to release one instance of a cell you placed with the Use Shared option on. Maybe you are going to be replacing all but one occurrence of this cell with a new cell. To allow you to "unshare" a cell you have already placed, select the Shared Cells option in the Drop Element tool settings and set its option menu to Normal Cell. With this option configured, you data point a shared cell in your design file. This results in a single stand-alone instance (i.e., a normal cell) of this cell. There is no way, however, to turn an unshared cell back into a shared cell.

More Text Features

As mentioned earlier, cells are not the only type of complex elements supported by MicroStation. In fact, you may have already encountered one when editing or placing text: text node.

The Text Node

A common situation you run into when working with drawings is limited space to place text annotation within and around your design. One solution to this is obvious: use more than one line of text. Doing this is simple. During text input in the Text Editor window, you press the Enter key and continue to type in your next line of text. Behind the scenes, however, MicroStation actually groups each line of text you entered into a complex element called a text node.

Placing Multiple Lines of Text

Although there is a separate Place Text Node tool, most users generate the majority of text nodes using the aforementioned Enter key in the Place Text tool. Exercise 8-4, which follows, shows how you can verify the existence of text nodes in a design file.

EXERCISE 8-4: DISPLAYING TEXT NODES

Open *Text.dgn*. In View 1, several strings of text appear. Let's see which ones are associated with a text node.

1   From the View control menu (upper left corner of the window), select View Attributes. The View Attributes dialog box appears.

2   Turn on the Text Nodes option and click on the Apply button. Several text nodes appear in View 1. Note how several of the "paragraphs" of text do not have a text node associated with them, whereas others do.

3   From the Text toolbox, select Edit Text.

4   Click on the text starting with "Edit this single string." The string of text appears in the Text Editor window.

5   In the Text Editor window, place the cursor at the end of the line and press the Enter key followed by the string "This is a second line of text." Click on the Apply button. The text string updates and a text node appears. The Enter key signaled the placement of multiple lines of text. This feature works even with the Edit Text tool. You can even remove text nodes by removing Enter key strokes.

6   Continuing with the Edit Text tool, select the text string starting with "A long string." Two lines of text appear in the Text Editor.

7   Using the mouse, select the second line of text and delete it. Make sure you delete the Enter key denoted by the small dot at the end of the first line. Click on the Apply button. The text node previously associated with the target text string disappears.

As you just saw, MicroStation takes care of the entire process. The only indication that it has happened at all was the appearance of the text node in any views for which you have turned on text node display, as indicated in the following illustration.

 NOTE: *Although you may have inferred this from the fact that a view needs to be identified to apply the attribute to, it is worth noting that you can set the view attributes independently for each view window.*

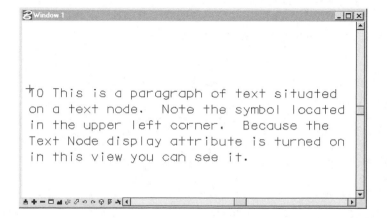

With Display of Text Nodes turned on, view windows show a text node symbol for multi-line text elements.

If you tried placing multiple lines of text where each line of text appeared too close to the one above or below it, you need to adjust the line spacing for that text node. This is the same text setting used in the text placements Above, Below, and Along, described in the previous chapter. The line spacing parameter is used to define the space between subsequent lines of text on a text node.

The Line Spacing value (*LS*=) found on the Text Settings box (Element > Text) specifies the distance in working units between each line of text on a text node. A good rule of thumb when starting out is to set your line spacing to the same value as your current text height, as indicated in the following illustration.

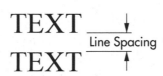

The Line Space parameter controls the vertical space between the top of one line of text to the bottom of the next line of text.

TIP: You can quickly adjust existing text node line spacing by invoking the Change Text Attributes tool (located in the Text toolbox), selecting the Line Spacing option, and identifying the offending text.

Text Node Justification

The other critical aspect of text node usage is its justification. This does not refer to whether you should use text nodes, but rather to the position the text takes up with respect to your data point. Text justification was covered in the earlier discussion of regular text. Text nodes have their own set of justification values. They include the nine associated with regular text, and six more, as indicated in the following illustration.

This text is left   This text is center   This text is right
top justified   top justified   top justified

This text is left   This text is center   This text is right
center justified   center justified   center justified

This text is left   This text is center   This text is right
bottom justified   bottom justified   bottom justified

The standard text node justification positions. Note the location of the text node "plus" sign in relation to its associated text.

Justification for text nodes is a lot more critical than it is for regular text. The reason is simple. You have a larger set of text to deal with, and selecting the wrong justification can push this text in all the wrong directions.

Place Text Node

You can also place text nodes that contain no text. The main use of an unpopulated text node is in annotation symbols. For instance, if you are creating a callout balloon for use in a detail drawing, it would be nice to construct it in a way that allows you to set the number or letter of the callout once the drawing is nearing completion.

A text node placed in your callout balloon allows you to do this. By selecting all of the various text parameters prior to placing the empty text node, all that remains is to place this node within your balloon graphics and create your balloon cell.

The Place Text Node tool, shown at left, does have one wrinkle. When you have identified your text node's location, you are prompted to define its angle using a second data point. This allows you to align the text node along an imaginary line from the text node data point and a second data point. If you wish not to do this, you can click on Reset, which instructs MicroStation to use the active angle parameter instead.

The Text Node Sequence Number

As you place each text node, a unique text node sequence number is assigned to each text node as it is placed. This number is the text node's address and is used by some advanced applications to place text onto, and read text from, text nodes. MicroStation is delivered with one example application that performs this operation, called Bulk Text. Database management applications also make extensive use of this feature.

There is a key-in parameter associated with the text node sequence number. Because the node number of the text node is automatically incremented each time you place a text node, there is an occasion when you may want to reset the value of the next node number.

For instance, you just deleted a series of text nodes via the Delete Fence Contents tool, and you want to start fresh with node number 1. You can do this by typing in *NN=1* in the Key-in window. Conversely, if you want to know what the next node number will be, you can key in *NN*. MicroStation responds with "Node Num = x."

Because a text node is a complex element, you can drop it using the Drop Complex Status tool, just as you would a cell. This results in each line of text becoming a separate (and unrelated) element.

Placing Text on a Text Node

Once you have placed a few text nodes, how do you get your text onto them? If you look at the tool settings for Place Text, you will find an option called Text Node Lock. Selecting this option forces all entered text to be placed on preexisting blank text nodes, as indicated in the following illustration. Clicking anywhere but on a previously placed text node results in no text being placed.

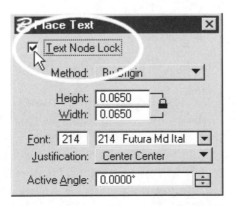

The Text Node Lock option must be selected to place text on a previously placed text node.

Fill-in-the-Blanks Text Feature

In addition to text nodes, MicroStation provides another form of reserving text for later placement. Called enter data fields, this useful feature of the text tool allows you to set aside individual characters for future updating.

When would you use *enter data fields*? How about part numbers on drawings. How about section callouts. How about specification numbers. In fact, any time you have an important string of text of a fixed format (for example, Part Number 360XXXX-001), an "enter data field" may be just the ticket.

"Enter data fields" can also be incorporated into cells and "filled in" once the cell has been placed in a drawing. As with the text node, the Shared Cells option must be turned off prior to placing these cells.

Creating an "enter data field" is easy. Using the standard Place Text tool, all you have to do is enter an underscore (_) character at each location where you want to reserve a character. When you type in a string of underbars, you create a single, fixed-length "enter data field" within your text string (you can put more than one "enter data field" in the same text string). For instance, all of the following are legitimate "enter data fields."

- Pursuant to Section __ of the Environmental Impact Law, no toxic gases greater than ___ PPB will be allowed.

- Refer to design guideline 123__-__ for more information.

- Date: __/__/__

- Part Number: 471____-__-00__

In each case, the text string contains more than one "enter data field." Each of these "enter data fields" is treated separately by specifically designed "enter data field" "fill-in" tools. Keep this in mind, as it can sometimes be more tedious to fill out multiple "enter data fields" than to key in repetitive data. However, the use of prefixes and suffixes with "enter data fields" ensures conformance to a part numbering standard, a common source of drawing errors. Once you have placed an "enter data field" or two, you have two tools you can use to directly fill them in. These are described in table 8-1, which follows.

Table 8-1: Enter Data Field "Fill-in" Tools

Tool	Description
Fill in Single Enter_Data Field	Identify a single "enter data field" and fill it in using the pop-up text editor.
Auto Fill in Enter_Data Fields	Identify a view. MicroStation then prompts you to fill in each empty "enter data field" displayed in the view (handy for bulk data entry).

Fill in Single Enter_Data Field

The Fill in Single Enter_Data Field tool, shown at left, allows you to manually identify each "enter data field" and enter its data. This tool works a little differently than other text placement tools in that you first select the "enter data field," then enter the text.

If you enter text that is longer than will fit in the "enter data field," the text will be truncated to fit the "enter data field." This truncation is an important characteristic of the "enter data field." Unlike the text node, which allows you to place a text string of any length on it, the "enter data field" is fixed in length.

Enter Data Field Justifications

By default, "enter data fields" are left-justified. This means if you place text shorter than the "enter data field," spaces will be added to the end of the text string. There are times, however, when you want to set the justification of the "enter data field" to something other than left-justified. To do this requires the use of the JUSTIFY key-in (there is no toolbox tool available for this "command"), the options for which follow.

- JUSTIFY LEFT
- JUSTIFY RIGHT
- JUSTIFY CENTER

Because the "enter data field" is fixed in length and size, center justification may not always give you the results you want. If you try to put a text string with an even number of characters into an odd-numbered "enter data field" with center justification, MicroStation will offset the text by one character. The same is true with odd-numbered text strings in even-numbered "enter data fields."

Auto Fill in Enter_Data Fields

A neat labor-saving tool is the Auto Fill in Enter_Data Fields, shown at left. By selecting a specific view, MicroStation will highlight each empty "enter data field" in turn, at which time you key in the text.

The order of highlight is the order in which the "enter data fields" were placed in the design file. This tool skips any "enter data field" that already contains characters.

Copy Enter_Data Field

If you have an "enter data field" already filled out and want to transfer this data to another "enter data field," MicroStation provides the Copy Enter_Data Field tool, shown at left. By selecting the first "enter data field" that contains the text you want to copy, you can select any number of other fields. MicroStation will copy this text into the selected "enter data fields" one by one.

NOTE: *If you use the Fill in Single Enter_Data Field or Copy Enter_Data Field tool and select a previously filled-in field, the text you key in will replace the text already there. Only one string can reside in an "enter data field" at a time.*

Copy and Increment Enter_Data Field

The Copy and Increment Enter_Data Field tool, shown at left, is related to the Copy and Increment Text tool previously introduced. In this case, you select the source data field with a data point and identify subsequent fields into which you want the source data field's text to be incremented and inserted. As with the other data field tools, any data in the target data field will be overwritten by the incremented data.

Viewing Enter Data Fields

You can make the underscore character associated with the "enter data field" invisible. This is done by turning off the Data Fields option by removing the check mark next to this option in the View Attributes settings window, as shown in the following illustration.

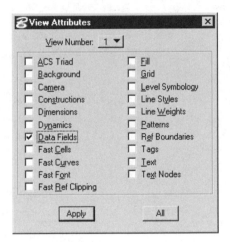

Making the underscore character invisible requires you to turn off the Data Fields attributes (shown activated here).

An Exercise in Cells and Text

To show how cells and text work hand-in-hand, a drawing exercise is in order. An electrical schematic diagram is an ideal candidate for both cell usage and fill-in-the-blank text placement. In exercise 8-5, which follows, you will work with cells, text, and "enter data fields."

EXERCISE 8-5: USING CELLS, TEXT, AND "ENTER DATA FIELDS"

1   Open the *SCHEM.DGN* design file. Once you have this file active, you will need to create a cell library for your electronic symbols.

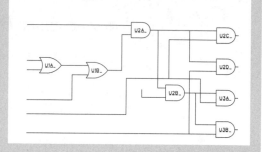

2   Create a new cell library named ELE-SYM.CEL. Use the Cell dialog box (Element menu > Cells) to create this cell library.

3   Set your grid to.05 (Settings menu > Design File > Grid category, Grid Master: 0.05); set the grid reference to 2 (Grid Reference: 2); and turn on your grid lock (select Grid Lock on).

4   Using the Place Arc, Place Line String tools, create the ORGATE symbol as shown in the following illustration. Make sure to keep the input lines (left side of the symbol) on the reference grid (the large tic marks on the screen).

Creating the ORGATE symbol.

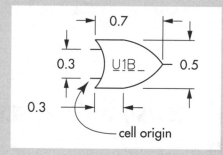

Most gate symbols include a text designator unique to each symbol occurrence. To provide for this, you will place a blank "enter data field" in your new symbol.

5   Select the Place Text tool and enter four underscore (_) characters. Place the text in the center of the symbol. Adjust the text justification using the Text dialog box or Place Text tool settings.

6   Create a cell with the name ORGATE, and with the description A TWO INPUT OR GATE. To review the cell creation process, use a fence to identify the elements for your new cell, use the Define Cell Origin tool to locate the new cell's origin, and then click on the Create button in the Cells dialog window.

7   Repeat the construction process for the AND gate symbol.

8   Create the cell ANDGAT, with the description A TWO INPUT AND GATE. Once you have created your two cells, delete the elements you used to create them, as shown in the following illustration.

Deleting elements used to create cells.

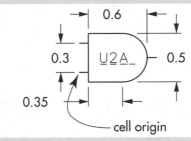

9   In the Cells dialog window, select the ORGATE cell and click on the Placement button. ORGATE is now your active cell.

10   Using the figure at the beginning of this exercise as a guide, place two instances of the ORGATE using the Place Cell tool.

11   Select ANDGAT as your active cell and repeat the placement process for the six instances shown.

12   Interconnect the cells using Place SmartLine. Make sure your grid lock is on during this phase of the work.

13   Use the Fill in Single Enter_Data Field tool from the Text toolbox to enter the text shown in the following illustration. You can also use the Auto Fill in Enter_Data Fields tool to automatically select each cell's "enter data field" in the order you placed them.

Text to be entered.

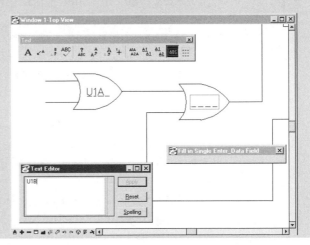

If you do not see the enter data fields displayed, the Use Shared Cells option in the Cells dialog box was probably turned on. Unfortunately, MicroStation cannot give you "enter data fields" (or text nodes) within cells if the cells were placed with this option turned on. Just set the Shared Cells option to off and repeat the cell placement, or use the Drop Complex Element tool with Shared Cells: Normal option selected and try the Fill in Enter Data Fields tool again.

More Complex Elements

So far, you have been introduced to two complex element types. In the case of the cell, you had to complete a certain number of separate operations in order to create and use your cell. When you place a cell in your design file, there is little you can do to its individual elements. You can, of course, change the overall size and attributes of the cell, but you cannot change a single vertex of any of its elements.

There are times, however, when you may want to modify vertices of a complex element. Take a road centerline, for instance. On a traditional hand-drawn plan, this centerline is treated as one object. In a design file, the centerline consists of lines, line strings, tangent arcs, and even spiral curves. These are, in turn, grouped as a complex element called, appropriately enough, a Complex Chain.

Along with its close cousin, the Complex Shape, these elements allow the user to take a series of connected elements and link them as a single contiguous object. Unlike a cell, the other major type of complex element, the vertices of the individual elements within a complex chain or shape are modifiable. You can even insert extra vertices within a complex chain/shape. In fact, all of the element modification tools work on these complex element types as if they were simple lines and arcs.

Imagine drawing a centerline of a road and with one operation creating the two edges of pavement, complete with varying arc radii as the centerline "bends" along its path. If you have created that centerline as a complex chain, you can use the Copy Parallel tool in just that way. Many experienced MicroStation users (including the author of this book) even develop their own drawing strategies to exploit this feature and enhance their drawing productivity.

The Complex Chain Versus the Complex Shape

The main difference between the complex chain and complex shape is closure. With a complex chain you create an open element with two endpoints. The complex shape is a closed element that has no endpoints (like a shape or block). As a result of its closed nature, a shape encompasses a fixed area of the design plane, which in turn can be measured using the Measure Area tool. It can also be filled or patterned using the appropriate attribute tool.

Groups Toolbox

The tools for creating complex chains and shapes are located in the Groups toolbox (Main toolbox > Groups). This is the same toolbox (shown at right) in which the Drop Complex Element tool resides.

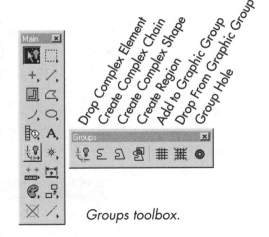

Groups toolbox.

Create Complex Chain

Used to link previously created elements into a single complex chain, the Create Complex Chain tool, shown at left, has two distinctly different operating modes.

With the manual method, an example of which is shown in the following illustration, you identify each element you want in your new chain with data points. A final data point in space (meaning not on any element) accepts the creation of the chain, at which point MicroStation creates the chain.

TIP: *It is always a good idea to click on Reset after the acceptance data point.*

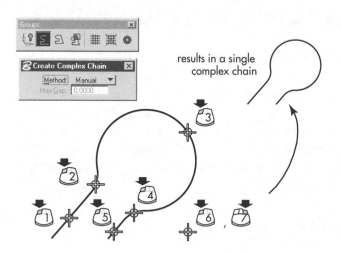

results in a single complex chain

Creating a chain is simply a case of selecting the elements to make up the target shape with a series of data points. In this example, the manual method is being demonstrated.

The automatic method, on the other hand, uses your first element selection as the starting point for a search for additional elements that touch. In other words, you identify an element in your proposed chain, and MicroStation will, it is hoped, find all of the associated elements for the chain, as indicated in the following illustration.

Automatic method of complex chain creation.

Fortunately, this process does require your approval for each element added. MicroStation highlights each element it finds and, if it is the right one, you data point to accept it. In case you meet a "fork in the road," you can click on Reset to deselect an element, and MicroStation will look for the next element sharing the same location. The following illustration depicts the sequence of events involved in the automatic method of complex chain creation.

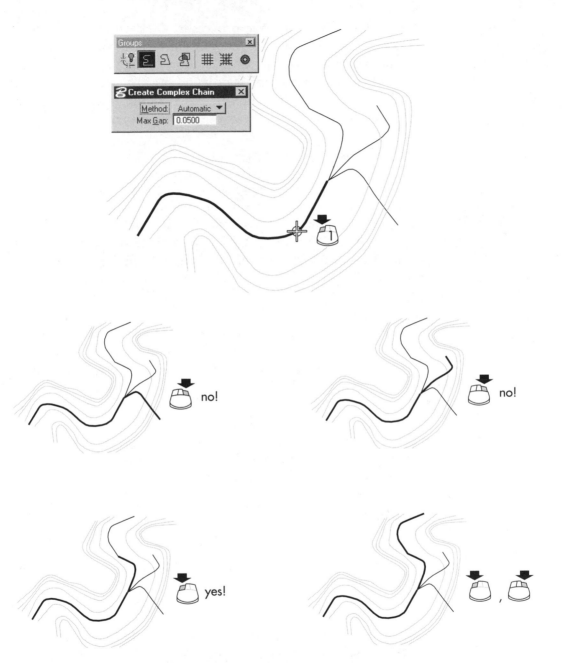

In this example, the automatic option for creating a complex chain has been selected. Once the first segment of a series of contours is selected with a data point, the Create Complex Chain tool highlights the next segment and waits for a data point to move to the next one, or for a reset to search for a more appropriate candidate.

Of course, for the automatic process to work, the elements must share common endpoints, right? Not necessarily. As an option to the automatic method, you can select the air gap between the end of one element and how far MicroStation will search for the next element. By default, this value is small enough to be treated as zero. However, you can change this value and span short distances.

Create Complex Shape

The procedure for creating a shape is almost the same as that for creating a chain. When you want to close the shape, you can place your final data point on the element with which you started your shape, or just data point once in space. MicroStation will then provide closure for this shape. The Create Complex Shape tool is shown at left.

As with the Chain, you can set the method to Automatic and have MicroStation make the selection of elements. The Create Complex Shape tool is smart enough to know when it has closed upon itself, so no further action needs to be taken by you to close the shape.

You may have noticed how the Complex Shape tool provides options for setting the Area attributes of the new shape. This is consistent with the complex shape's newly acquired area characteristic. As with polygons and circles, you can set the fill color and type by selecting the appropriate settings.

Create Region

One of the more powerful tools for creating a complex shape, Create Region performs Boolean operations with existing elements to generate the new shape. The Create Region tool is shown at left.

Instead of limiting the creation of a complex shape to using existing linear elements, Create Region computes a new shape from the relationship of two or more elements. For instance, Create Region can take two overlapping polygons and create a shape consisting of only that area where the two shapes overlap. Or you can create a shape by subtracting one polygon from another. There are four possible methods for creating a region (i.e., a complex shape) from two or more closed shapes, as indicated in the following illustration.

The four methods that control how the Create Region tool will interpret your data points.

From Element Intersection

Selecting the Intersection method allows you to select two or more closed shapes from which the Create Region tool will compute the exact area these elements co-occupy (thus the term *intersect*) and create an appropriate complex shape of this region, as indicated in the following illustration. Selecting the first two elements will display the computed intersection. A third data point accepts this shape. If this third data point falls on yet another element, the computed shape will incorporate this element as well.

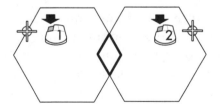

Use of "From Element Intersection."

From Element Union

The complement of the intersection, the Union method, computes a complex shape by computing the perimeter around both elements. If the two elements you select do not touch, no action is taken. As with the Intersection method, you can continue to add elements to the new complex shape. Use of the "From Element Union" method is shown in the following illustration.

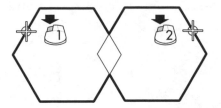

Use of "From Element Union."

From Element Difference

If intersection computes the overlapping area and union computes the combined area, the Difference method should subtract one area from another. That is precisely what Difference does. The first data point selects the source element, the second data point the element whose area is to be subtracted. Use of the "From Element Difference" method is shown in the following illustration.

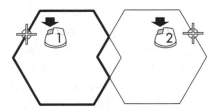

Use of "From Element Difference."

From Area Enclosing Point (Flood)

The only method that does not require closed shapes, the Flood option is also one of the more interesting operations performed by MicroStation. By striking lines, arcs, or any element supported by MicroStation that encloses an area of the design file, Create Region From Area Enclosing Point, shown in the following illustration, will calculate the complex shape that matches this enclosed area.

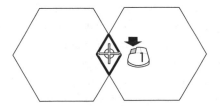

Create Region From Area Enclosing Point tool.

This has real design potential. Now you do not have to be concerned with cleaning up the intersections of your linework, as indicated in the following illustration. Instead, the judicious use of this tool can speed things up.

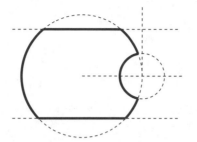

Using the Flood option, a single data point within a closed region can clean up the construction linework as shown.

Keep Original Option

One important option associated with the Create Region tool is Keep Original. When this option is not set (the default), MicroStation deletes the elements used to create the new complex element. This may or may not be your intention. Selecting this option ensures the continued existence of your original design elements.

You may want to turn the Keep Original option off when using the Flood method. In this way, you create your objects and clean up at the same time. If you inadvertently "flood" an area, just use the Undo command (Edit > Undo) to return your elements.

Exercise: Using the Create Region Tool

In exercise 8-6, which follows, you will use the Create Region tool to create a series of complex shapes based on the relationship of various elements. The various elements you will use are provided in the *REGION.DGN* design file. From these you will create the shapes shown in the following illustration.

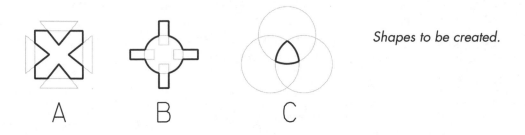

Shapes to be created.

EXERCISE 8-6: USING THE CREATE REGION TOOL

In this exercise you will be placing a series of multi-line elements with a variety of multi-line definitions.

1   Open *REGION.DGN*. Three separate sets of elements labeled A, B, and C will be used with different Create Region methods to generate a variety of complex shapes.

2   Using Window Area, zoom in on element set A. The Difference method will be used to create the "X" shape shown in this exercise's opening illustration.

3   Select the Create Region tool (Main toolbox > Groups). Set the Method to Difference, as shown in the following illustration. Deselect (turn off) the Keep Original option.

Method set to Difference in the Create Region dialog.

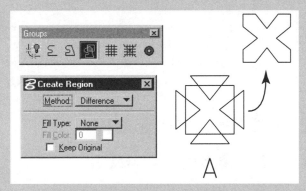

4   Select the square block with your first data point, and select each of the four triangles in turn. Create Region previews the complex shape candidate constructed from the elements you have selected. Note how the "square" now has four triangular indentations.

5   Place a final data point "in space" (anywhere in the design except on an element) to accept the shape, followed by a Reset. Create Region generates the complex shape as previewed.

6 Scroll your view to element set B. You can use the scroll bars or, as a short-cut, you can use saved view B (Utilities menu > Saved Views > B > Attach) in View 1. You will use the circle and rectangles to generate a complex shape combining all of the shapes into a single shape, as shown in the following illustration.

Complex shape to be created.

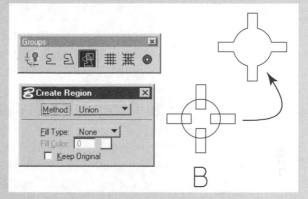

7 With the Create Region tool still active, set Method to Union.

8 Select each shape in element set B.

9 Data point in space to accept the selected shapes. The new complex shape is the combination of all elements in element set B.

10 Scroll your view to element set C. As a shortcut you can use saved view C (Utilities menu > Saved Views > C > Attach) in View 1. In this example you will be creating a shape of the intersection of these three circles, as shown in the following illustration.

Shape as the result of the intersection of three circles.

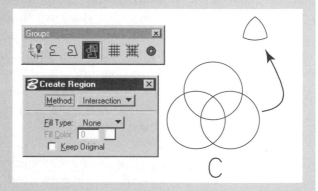

11 Set Create Region Method to Intersection.

12   Data point on each circle. The preview shows only that portion of each circle that overlaps its neighbor.

13   Data point in space to accept the previewed shape.

In the last part of the exercise you could also have chosen to use the Flood method to generate the intersection of the three circles. However, this is not always the case.

The Composite Curve

Before leaving the world of complex chains and shapes, you need to look at one more tool: Place Composite Curve. Not found on the Groups toolbox, but every bit a complex chain/shape tool, the Place Composite Curve option does not require existing elements in order to generate a chain or shape. Instead, it provides options for creating such an element on the fly via a number of options. Found on the Curves toolbox, this is a powerful tool you will find many uses for.

Place Composite Curve

By definition, a complex chain can consist of a variety of element types. A tool designed to create such a chain should, in turn, support the creation of such elements, and Place Composite Curve, shown at left, does just that.

Providing an array of options, this tool allows you to adjust the type and parameters of each element created as part of the Composite Curve (Tools menu > B-spline Curves > Create Curves toolbox) tool operation. The Composite Curve tool is shown in the following illustration.

The most important option found here is the Mode field. By selecting the appropriate element type from this field, the Place Composite Curve tool presents you with this element as you click your way across the design plane.

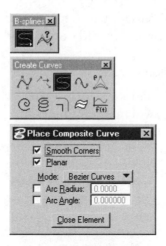

The Composite Curve tool can be found on the Create Curves toolbox. The Mode field controls the type of element currently under construction.

Mode: Line Segments

The simplest of the modes, Line Segments, results in a line string being generated as you data point.

Mode: Arcs By Edge

If, during your placement of line segments, you want to transition into an arc, this is one of two options that allow you to do just that. With Smooth Corners selected (the default), this transition will also remain tangent to the last line segment. This is extremely useful, especially in design situations where such transitions are critical. (Pipes and road alignments immediately come to mind.) You can also control the resulting arc's radius and even its sweep (arc angle) by selecting the appropriate options. This, again, is an important consideration when working with known pipe sizes and minimum bends.

Mode: Arcs By Center

Similar to the previous option, the Arcs By Center option gives you the ability to set the centerpoint of the arc while maintaining tangency with your previously created linework. Being able to set the radius and angle is especially helpful with known construction details such as cul-de-sacs and the like.

Mode: Bezier Curves

An element type not yet discussed, the Bezier curve (i.e., B-Spline) is a sophisticated, mathematically generated, free-form curve used extensively in design situations where absolute curve definition is needed (e.g., aerodynamics, mold design, and boat design).

Smooth Corners Option

As previously mentioned, Smooth Corners controls whether arcs created with this tool will maintain tangency with the previous element segment. Turning it off results in a sharp corner at the point of transition from one element to another. There are times when you may want such a transition, thus this option.

Close Element

The Composite Curve tool can create either a chain or a shape. This button selects which of the two element types you want. By clicking on the Close Element button, you create a shape. The element under construction is immediately created, with a final segment placed between your last data point and the beginning point. Regardless of which mode you have selected, the final segment will be a line.

NOTE: *The Place SmartLine tool discussed in Chapter 4 also creates chains and shapes. However, it is more geared to creating line strings and lines combined with circular arcs, with the ability to automatically generate fillets and chamfers, than to composite curves that maintain tangency.*

Grouping Elements

A complex chain or shape, consisting of variety of primitive element types, requires an end-point to end-point spatial arrangement. There are times, however, when you may want to informally group a set of design elements that are not connected to one another in this way; something like a cell, only less formal in organization, and which allows you to still modify the individual members of the group.

Element "Groups"

Enter MicroStation's graphic group feature. Whereas cells or complex chains and shapes always behave as a group (unless you drop them), graphic groups can act as a group or as individual elements, depending on whether or not you have the graphic group lock enabled. For instance, a steel plate layout complete with drill holes and annotation text may need to be copied a number of times and then modified. By first creating a graphic group of the elements that constitute the plate, you can quickly and easily copy this group using the Copy Element tool, provided you have turned on the Graphic Group Lock.

The graphic group allows you to modify any member of the group using the appropriate modify command. In fact, you can drop individual elements from the graphic group if they are no longer needed. You can also add elements to the individual graphic group as necessary. The key to the graphic group is *flexibility*.

Creating a Graphic Group

The tools for graphic group maintenance are found on the Groups toolbox. These are shown in the following illustration.

Found on the Groups toolbox, two tools control the "care and feeding" of your graphic groups.

Add to Graphic Group

The Add To Graphic Group tool, shown at left, is used to create new graphic groups or add to existing graphic groups. You are prompted to identify an element to add to a graphic group. Once an element has been identified, MicroStation will first check to see if the element already belongs to a graphic group. If it does, you are prompted to select the next element to add to the existing group. If the element is not a member of a group, a new graphic group will automatically be created. There can be up to 65,536 graphic groups in a design file.

Once your graphic group is started, you select each element in turn to add to the group. Selecting each new element also accepts the last element in the group. This is important. Many first-time users of the Add To Graphic Group tool will find that the last element they chose did not end up in the group. This is because the last element was highlighted but not accepted. Remember to data point "in space" one last time to accept the last element selected.

The Graphic Group Lock

Once you have a graphic group, you control how MicroStation's other tools act upon it using the Graphic Group Lock, shown in the following illustration. The following are three different ways of accessing this lock.

- Access the Full or Toggle Locks dialog box (Settings > Locks > Full or Toggles)

- Key-in *LOCK GGROUP*

- Access the Lock options menu (located under the Lock icon on MicroStation's status bar)

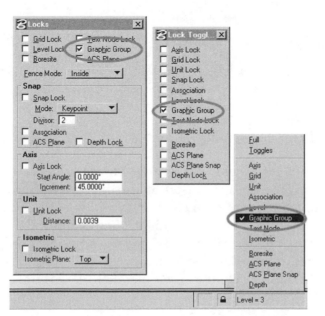

Graphic Group lock.

With the Graphic Group Lock on, all standard element manipulation tools act upon all members of the graphic group. When you select an element for, say, copying, and it is a member of a graphic group, instead of telling you it is a line or arc, MicroStation will inform you on the status bar that you have selected a graphic group.

It is important to read your status bar when working with graphic groups. You can unintentionally alter your design if Graphic Group Lock is on and you meant to manipulate only a single element.

WARNING: *ALWAYS READ YOUR PROMPTS when dealing with graphic groups. You can destroy your design without knowing it when elements in a graphic group are out of view. The only indication when selecting elements that something might be amiss is the (GG) following a description of the highlighted element type on the status bar when you select an element for manipulation.*

Graphic Group is a powerful and extremely useful feature of MicroStation. In fact, many of MicroStation's tools automatically generate graphic groups as part of their operation. For instance, the Text Along text placement method generates a graphic group of every character it places along an element.

Drop from Graphic Group

Part of the flexibility of the graphic group is the ease with which you can drop a single element from it. You can also drop the entire group, thus releasing all elements from it. This is done with the Drop from Graphic group tool, shown at left.

The effect of this tool depends on the current state of the Graphic Group Lock. If it is on, and you select an element in a graphic group, you will be prompted to drop the entire group. If the Graphic Group Lock is off, the single element is released.

The Select Element Tool Revisited

By now you have probably seen the Group and Ungroup commands found under the Edit pull-down menu. Not to be confused with the graphic group tools just discussed, these two commands work in conjunction with the Select Element tool to create yet another type of element association, the orphan cell.

The Group and Ungroup Commands

There is one additional complex element creation tool in Micro-Station: the Group command (Edit menu > Group). Used in conjunction with a selection set, this command generates an object called an "orphan cell." This is defined as a cell with no name, or one that originates from a cell library. You can think of this as the ability to "group" a set of elements into a cell. However, unlike a normal cell, the source elements you select become the unnamed cell. As the Group command's complement, the Ungroup command (Edit menu > Ungroup) does just as you would expect; it breaks apart a previously grouped set of elements. You can also use the Drop Complex Element tool on "groups."

Locking Elements

While on the subject of working with the Element Selection tool, there are two interesting commands: Lock (Edit menu > Lock) and Unlock (Edit menu > Unlock). There are times when working with complex drawings that you wish you could lock a set of elements in place where they are "out of harm's way." When you select an element and lock it, no element modification or manipulation command will work on it.

For instance, say you have a complex drawing and want to globally delete a large area. However, in the middle of the "deletion" area there is a set of elements you need to leave intact. By first selecting the elements with the Element Selection tool, followed by the Lock command (Edit menu > Lock), you can proceed with the global deletion operation with no worries. The Unlock command (Edit menu > Unlock) releases the elements from their locked state.

Summary

By now you have realized that MicroStation really does provide you with more than one way to perform your tasks. Many of these features may seem redundant in operation. However, when you start using them on a daily basis, you will find there are times when cells are appropriate, other times when creating a complex shape is the right thing to do, and yet other times when a combination does what you want. Experimenting with these various options, combined with experience, will help you work out their place in your design.

PART TWO

2

CREATING A FINISHED DESIGN

REFERENCE FILES

Bringing the Work Group Together

IN THIS DAY AND AGE OF PROJECT COLLABORATION it is rare to see one person create a design from start to finish, totally unassisted by other individuals. Instead, most designs, whether a new hospital or a microelectronic gadget, involve teams of skilled people, each responsible for a specific portion of the overall design. You may have architectural, structural, mechanical, and process designers working simultaneously on a new building project. In such an environment, the need to share information, especially as it might affect others, is crucial to keeping the project on track.

Using Reference Files

To support project collaboration, MicroStation is equipped with the Reference File system, which provides you with the ability to attach and display the design content of external design files within your active design file. This ability will have a very positive and fundamental impact on how you use MicroStation in your design project.

Sharing Files in a Workgroup

Suppose you are a structural engineer charged with creating a rebar plan. Using the reference file capability of MicroStation, you call up your design and attach the architectural plan to it. By using this architectural reference file as a background to your design, you can quickly pull off the information needed to get started with your design process. The following illustration shows an example of cross-discipline information sharing.

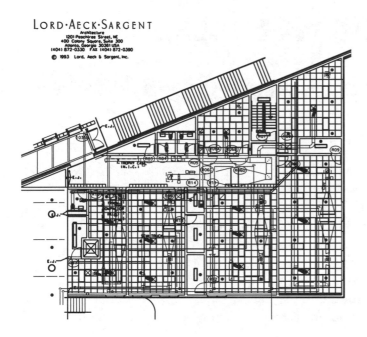

An example of an architect and engineer sharing information. (Courtesy of Lord Aeck Sargent.)

"Aha!" you say. "I can do the same thing by starting with a copy of the original design file and simply draw my rebar layout on one level." Not so fast! Suppose the architect in the next cubicle/office/building/state/country (pick one, all are possible), the one who is responsible for the architectural plan, has a last-minute change that involves moving several walls within the design.

If you had copied the architectural plan into your rebar plan, at best you would be required to copy the changed elements into your drawing or, worst, you might have to start all over again. On the other hand, if you had been using reference files in MicroStation, a simple reference file update would immediately show the architect's modification in your rebar plan, where you would only have to deal with the design impact and not the drawing formalities. An example of this approach is shown in the following illustration.

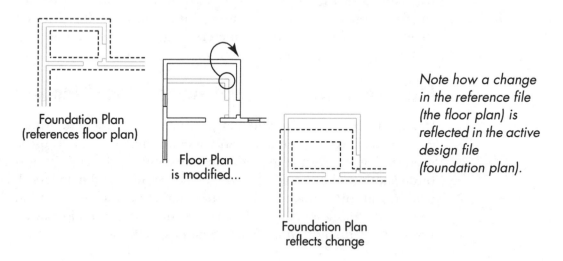

Foundation Plan
(references floor plan)

Floor Plan
is modified...

Foundation Plan
reflects change

Note how a change in the reference file (the floor plan) is reflected in the active design file (foundation plan).

This immediate feedback has many benefits. First, it cuts down on the number of reworks of duplicate information required because of design changes, and the inherent ripple effect through all parts of the design process. Second, it cuts down on potential errors because a change in one critical drawing was not incorporated into a different but related drawing (the relocated wall never gets updated in the separate rebar plan, for instance).

Create Drawings from a Model

Another use of MicroStation's reference file system is in the area of document preparation (plotting). With it you can set your various plotted drawing scales independent from the full-size model of your design.

From the beginning, this book has stressed the importance of creating your designs at full scale. However, when it comes time to annotate the final design, the plot scale you will use to create paper plots determines the size of text and symbols you place in the design file. Thus, if you were to decide to change the plot scale after all text and annotation symbols were in place, you would have to change their size throughout the design file.

If, on the other hand, you were to create your model in one design file and your individual drawing sheets containing all the annotation in another, you could then attach the design model to your sheets at the appropriate scale. You could also place your callouts, notes, and other detailing information in the drawing sheet file with consistent parameters (e.g., text size, patterning, spacing, scale, and so on).

Multiple Drawings from a Single "Base" Model

When you design a road, you may produce hundreds of drawings, each with its own orientation and specific plot scale. Because a road design can be a large and time-consuming project, it is most often supported by a staff of engineers, designers, and detailers all working at the same time. As the engineers create the road design, the designers and detailers are given the task of drawing up the individual drawing sheets to meet specific standards.

All of the drawings have one thing in common—the stretch of road being designed. Using the reference file approach, drawing sheets reference the main road model in sections appropriate to the final drawing scale (1:20, 1:50, 1:200, and so on).

By aligning the sheets along the road's main axis, you also generate uniform, detailed drawings and still maintain the modeling approach no matter which way the road curves. Sometimes design changes make subtle shifts in the overall alignment of the road. With traditional hand drawings, such a subtle change is difficult to incorporate across every sheet without serious impact to the project schedule. However, when using MicroStation's reference file system, such changes are immediately incorporated into all detail drawings, thus eliminating the need to "touch" every individual plan and profile sheet. An example of this is shown in the following illustration.

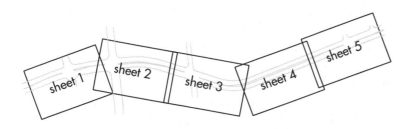

An example of how a reference file can "travel" along a road.

Extending the Design Plane for Large Areas

No matter how capable a design product is, there comes a time when you must break up a drawing into more manageable parts. When you do this, you must deal with those features that cross from one drawing into another. In most cases, this is not a critical junction. In an architectural design, for instance, you can plan logical breaking points that minimize such crossovers.

Mapping, on the other hand, deals with cartographic features crossing from drawing to drawing almost at random. Physical land features cannot be controlled in the sense of designed layout. In such cases, the mapping professional must deal with matchlines and closely scrutinize those elements that cross between adjacent drawings. In addition, some sort of spatial check must be possible to ensure that the maps match the real world.

This is where MicroStation comes into its own. By allowing you to connect several design files to your active design file, adjacent linework is available for immediate reference and identification. In addition, a master gridwork of monuments, carefully laid out, can provide a useful framework upon which to calibrate the individual map sheets (see following illustration).

Edge matching between adjacent map drawings is easy with reference files.

No matter how large your design plane is, there will come a time when you exceed it. Because mapping involves large areas and tight dimensional control, you are always running the risk of falling off the edge, so to speak.

However, with proper project planning, you can use the reference file capability of MicroStation to extend this design plane. Because MicroStation allows you to move a reference file's point of origin with respect to the active design file, you can maintain a "traveling canvas" of design files.

The Capabilities of Reference Files

The previous discussion about the various scenarios may have stimulated some thoughts about how reference files can be used in your operation. With this in mind, in this section you will take a detailed look at how MicroStation's reference file system works.

In its most basic form, a reference file facility is nothing more than a means of displaying the content of another design file on top of (or under) your active design file. You can have numerous reference files attached (up to 255). Each reference file can be rotated, scaled, clipped, and moved with respect to your active design file. You can do all of this without affecting the actual content of the reference file itself.

V8: MicroStation version 8 will allow you an unlimited number of reference files!

Once you have attached a reference file to your active design, you can use any of the Copy tools to copy selected elements of the reference file into your design file. Of course, this defeats one of the advantages of the reference file concept. However, there are many times when you will need to do this. You can also restrict the ability to copy on a reference file-by-reference file basis.

Another important aspect of the reference file is its "snapability." When enabled, the snap feature of the reference file facility allows you to use the tentative point snap functions with elements in the reference file. In this way, you can use the reference file as a starting point for your active design.

In addition, you can set the reference file to be viewable only—no interaction allowed. This keeps you from accidentally copying elements from the reference file into your active design file. Even the fence copy tool will not work on a reference file whose elements are not "located." This restriction is not available with normal elements in a design file, unless, of course, you have locked them using the Lock command (Edit > Lock) on elements in a selection set.

Also of special note: There is nothing to stop you from attaching a design file to itself, an example of which is shown in the following illustration. Why would you do this? By scaling and clipping the self-referenced file, you can highlight an area of the main design and add text, arrows, dimensions, and so on, to better illustrate the object of interest. As you make changes to the main design, this "live" detail will be updated as well.

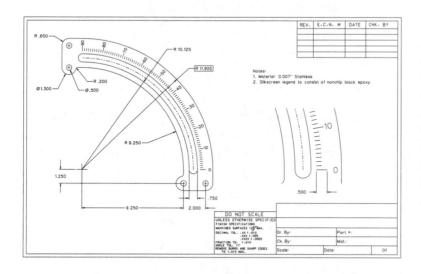

A design file referenced to itself is an excellent method of detailing a small section of a design without duplicating the design subject's linework (example from MicroStation for AutoCAD Users).

Design Files Versus Raster Reference Files

MicroStation supports two distinctly different types of reference files. The first type, which you are already familiar with, is MicroStation's design file; this is the file format used to store your drawing data. The other type is the raster reference file. Raster refers to the pixels used to describe the graphics stored within this type of file.

There are many forms of raster data in use today, some of which you have already encountered. For instance, JPEG images are a common image type used for web-page graphics. The subject of raster data formats warrants an entire book, which is beyond the scope of this one, so to keep it simple: think of MicroStation's raster reference file support as a means of overlaying (or underlaying) a raster image in your design file.

MicroStation supports a wide variety of raster data formats. Raster data falls into three main categories of types, and its most likely usage is within the design environment. Table 9-1, which follows, provides a summary of each data format.

Table 9-1: Three Main Raster Data Formats

Raster Type	Typical Examples
Monochrome (1-bit) imagery	Scanned images of existing engineering drawings
Mapped (4- or 8-bit) color imagery	Continuous-tone aerial surveys (scanned or direct digital photography)
Full (24-bit) color imagery	Scanned or digitally captured photographs

The decision as to which of these raster types to use depends a lot on the source materials (why use color to represent a monochrome engineering drawing?), as well as optimizing the storage and overall system performance (8-bit images are almost always easier to manipulate than 24-bit images). MicroStation provides several raster manipulation solutions, depending on the raster type and its usage. In previous versions of MicroStation you had to choose between several raster-editing products.

Descartes

Developed and initially marketed by HMR of Canada (now, a wholly owned part of Bentley), this full-featured raster-editing product was designed to provide powerful tools for continuous-tone color images for the professional photogrammetrist. Descartes provides several specialty tools for geo-correcting imagery from aerial photos to satellite data, so that they map properly onto geo-centric ground features. In addition, Descartes provides a very fast raster display capability, so that the extremely large (in byte count) raster files normally associated with this type of work can be quickly and efficiently manipulated and displayed.

Reprographics

Co-developed by Bentley and HMR, Reprographics is primarily aimed at the scanned engineering drawing user. Using 1-bit black-and-white images, this product provides extended support for hybrid drawings, where a scanned-in image of an existing drawing is used as a background with additional MicroStation design elements overlaid on this image.

Providing several options for optimizing this process, this tool provides a reasonable solution to the problem of decades worth of engineering drawings (non-CAD) that still need to be updated, but only as needed and where needed. Utility companies with incredible archives of paper drawings are prime users of Reprographics.

Third-party Raster-editing Products

Over the years, several additional raster-related products have been offered for use with MicroStation. Examples include Intergraph's BRAS and BRAS/C (binary raster, binary raster continuous), as well as others. However, with the integration of raster reference support into MicroStation itself, as well as the recent acquisition of HMR by Bentley, the market for these third-party products would appear to be on the decline.

MicroStation's Raster Reference Support

In addition to these add-in raster-editing products, MicroStation itself has incorporated a limited editing capability since the release of MicroStation 95. With MicroStation SE, an additional image display manager was introduced, called Image Manager. Designed primarily as a display tool for Descartes-generated raster data, Image Manager provides several image manipulation capabilities not found in the foundation MicroStation raster support.

NOTE: *In late 2000, MicroStation will incorporate a consolidated raster editing facility that integrates several discrete raster functions into a feature set called Raster Manager. Designed to be extensible by other products, Raster Manager will provide the user interface "front end" to all raster tools. This promises to bring a consistency of operation to an area of MicroStation that, from a user's perspective, has been "problematic."*

The Reference File Tools

Before you explore how to work with a reference design file, you should review the reference file tools. The sections that follow discuss the various components of the Reference File tool set.

The Reference File Settings Window

All reference file operations can be performed from the Reference Files settings window (File menu > Reference), shown in the following illustration. This is also where you review the current setting of your attached reference files.

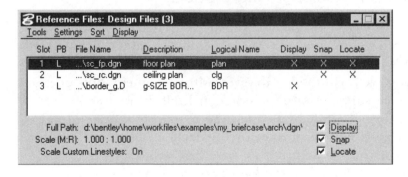

The Reference Files window is the primary point of contact with your reference files. Here you set the various parameters associated with each reference file attachment.

When using the reference file commands from this window, you must select the reference files first, and then the command (most reference file commands can work on several files at a time). When you select a reference file, details about it are displayed in the lower section of the Reference Files window. These include the absolute location where the reference file is located, and its current display, snap, and locate settings. To change a reference file's settings, double click on its entry in the reference file list.

Table 9-2, which follows, presents the commands available in the Reference Files window under the Tools, Settings, and Display pull-down menus. The Tools pull-down includes these commands.

Table 9-2: Reference File Tools Menu

Command Name	Description
Attach	Attaches a new reference file
Attach URL	Attaches a new reference file over the Web (*http* and *ftp* protocols supported)
Detach	Deactivates a previously attached reference file
Detach All	Deactivates all previously attached reference files
Ref Agent	Opens the Reference File Agent window used to manage URL-attached reference files
Exchange	Opens the highlighted reference file as the active file
Reload	Updates the reference file image stored in the computer's memory
Move	Repositions the reference file with respect to your active design file
Scale	Changes the working-unit to working-unit ratio between the reference file and active design file
Rotate	Changes the angle at which the reference file is displayed
Copy Attachment	Copies the selected reference file attachments as new reference file attachments
Mirror Horizontal	Mirrors the display of the reference file about the horizontal axis
Mirror Vertical	Mirrors the display of the reference file about the vertical axis
Clip Boundary	Clips the area within a fence to display from a reference file
Clip Mask	Clips the area within a fence to hide from a reference file
Clip Mask Delete	Deletes the clipping mask for a reference file
Clip Front (3D only)	Clips the reference file's display on the front of the Z axis
Clip Back (3D only)	Clips the reference file's display on the back of the Z axis

As you can see, there are quite a few commands associated with reference files. There are also a number of settings associated with each reference file attachment. These settings, described in table 9-3, are managed by the commands found under the Settings menu.

Table 9-3: Reference File Settings Pull-down Menu

Command Name	Description
Attachment	Modifies settings of previously attached reference files. Includes Display.
Levels	Controls the display of reference file levels in each view.
Level Symbology	Controls the level symbology settings associated with each reference file.
Update Sequence	Invokes the Update Sequence dialog box that allows you to change the sequence in which Micro-Station displays attached raster and design files.
View Reference	Provides a preview pane displaying the content of the selected reference file.

Because the tools associated with referencing raster files are very different from those associated with design files, the Display menu provides the switch between the design file and raster reference file categories. Table 9-4, which follows, describes the options under the Reference File Display menu.

Table 9-4: Reference File Display Pull-down Menu

Command Name	Description
Design	Enables controls in the window for manipulating reference design files
Raster	Enables controls in the window for manipulating raster reference files

The Reference Files Toolbox

Many the commands found under the Reference Files window's Tools menu are also available from the Reference Files toolbox (Tools > Reference Files), shown in the following illustration. However, at first you will find the Reference File menu version of these tools a little easier to follow. As you become proficient in manipulating reference files, you will find the toolbox version of the commands to be very useful for faster changes to individual reference files.

Reference Files toolbox.

The Reference File Tools

Let's look at the tools used to manipulate and control the operation of your reference design files. The discussion that follows specifically deals with design files attached as a reference.

Attach Reference File

To use the reference file system, you need to attach a reference file using the Attach Reference File tool, shown at left. As mentioned earlier, this can be accomplished a number of ways.

- From the Reference Files window invoked from the MicroStation main menu (File > Reference)

- From the Reference Files toolbox (Tools > Reference Files)

- Typing in the command *REFERENCE ATTACH (RF=)* via the Key-in window

All three of these methods invoke the same command. When you select the Attach Reference File command/tool, you are presented with the Attach Reference File dialog box (closely resembles the Open Design File dialog box), shown in the following illustration. Here you identify the file to attach as your reference file. Once selected, a second dialog box is presented, in which you specify key information about the soon-to-be attached reference file.

Although you need not enter any information in this dialog box to create an attachment, unless the selected file is already attached as a reference, it is a good idea to fill in the fields provided. Table 9-5, which follows, describes the fields within the Reference File dialog box.

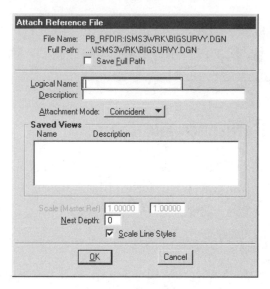

The Attach Reference File dialog box associated with the Attach Reference File tool. Here you define key information about the new reference file.

Table 9-5: Reference File Dialog Box Fields

Field Label	Description
Save Full Path	This checkbox, when enabled, saves the directory path prefix with the reference file name in the design file.
Logical Name	The shorthand name by which you can identify the reference file.
Description	A short description of the attached reference file.
Attachment Mode	Coincident (matching the design plane of active design file) or Saved View (attach using parameters defined in view).
Saved Views	Used to select a specific saved view (SV=).
Scale (Master:Ref)	The relationship of Master Units to Master Units between the active design file and the reference file.
Nest Depth	The depth of reference file nesting (how far to "dig" into the reference file for attaching its referenced files to your active design).
Scale Line Styles	This checkbox, when enabled, scales for display the custom line styles in the reference file using the Scale (Master:Ref) value previously described.

Logical names are usually short (one or two characters) mnemonic strings that you can use in many reference file commands to quickly identify the reference file. Logical names are especially useful when you have multiple attachments of the same reference file.

In many companies, the use of logical names is standardized as part of the CAD usage guidelines. One such standard is to use names that describe the reference file's usage within the project: R1 for general reference file first attachment, D1 for DTM file 1, and so on.

Other engineering projects might have their own unique reference file logical names. For instance, when working with edge-matched map sheets, adjacent referenced sheets can be identified by their relationship to the active design file (e.g., logical reference file names N, NE, E, SE, and so on). Using the Exchange tool, you can quickly navigate through a fairly large set of edge-match design files using these reference file logical names.

TIP: *You can change a reference file's logical name at any time using the Settings menu > Attachment command.*

Reference File Dialog Box Fields: Use Those Descriptions!

The description part of the Attach Reference command is an informational field that is displayed when performing a directory of the design files attached. This field is often overlooked by even long-time users of MicroStation. This is unfortunate because it is a built-in means of communicating your intention with the reference file, and other details that are not obvious. At the very least, it should be used to define the relationship of the logical name to the design file (e.g., B = BASELINE).

The Coincident attachment mode means the design plane of the reference file matches your design file exactly. The absolute X and Y coordinates of the two files are precisely the same unless the two files are using different working units. MicroStation does not pay attention to the working units. Instead, it uses the positional units (the 2^{32} number or 4.2 billion points) for alignment.

If, on the other hand, you want to attach the reference file at a scale or with part of it clipped, you can key in a named view you previously created in the reference file using MicroStation's Saved

Views facility (Utilities > Saved Views). This view defines how much of the reference file you wish to see in your active design file. This essentially presets what is called the "clip boundary" of your reference file as part of the attachment process.

NOTE: *You can always set the clip boundary of a reference file using the Clip Boundaries command (Tools menu > Clip Boundary).*

If you have such a view, and MicroStation finds it, the next question is, at what scale do you wish it attached? Unlike the scales you have used so far, the scale factor this time consists of setting a ratio between the master units of the active design file and the master units of the reference file. This relationship is always master unit to master unit, regardless of the subunits or the positional unit values. This is one of the most important aspects of the reference file system, and can be exploited to great benefit by the experienced user.

As mentioned earlier in the book, text and symbols must be defined based on the scale of the final plotted drawing. Many common plot scales are very misleading to the first-time CAD user. For instance, the U.S. architectural scale 8:1, commonly referred to as "eighth scale," is actually a true scale of 1/96 (8 feet X 12 inches per foot equals 96).

As a result, having to remember obscure text and symbol sizes as they relate to the final plotted scale can be prone to drawing mistakes. If, on the other hand, you use this Master Unit to Master Unit feature of the reference file system, you can avoid most annotation scale errors by organizing your annotation in a separate "sheet file," which references the model at the correct plot ratio.

Attach URL

One of the more recent enhancements to MicroStation's reference file system is its ability to access files directly from the Web using the Attach URL command. Instead of specifying a file name, you enter a URL (Uniform Resource Locator) string, as indicated in the following illustration.

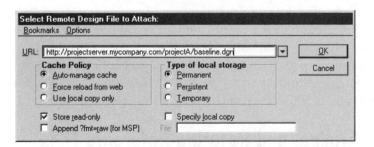

You specify the full URL path to the remote file. Note the various options related to the temporary storage of the remote file.

MicroStation will then locate the file specified over your web connection and download a locally cached copy of the file to your system (the location of the local directory is controlled by the MicroStation configuration variable *MS_WEBFILES_DIR*, which can be reviewed via Workspace menu > Configuration > Engineering Links). Once downloaded, this command operates the same as the Attach Design File command just described.

Detach Reference File

This command is self-explanatory. Selecting Detach Reference File, shown at left, from the Tools pull-down menu brings up an alert box for confirmation of the operation. The Tool Box version requires you to identify the unwanted reference file by clicking on an element within that file or by keying in its logical name.

WARNING: *The Detachment operation cannot be undone! Once you have identified the reference file, you lose all of the clipping, movement, and scalar information you may have painstakingly generated. Turn off the display of your reference file prior to detachment, which provides a preview of what the design will look like without your reference file.*

Reload Reference File

One of the strengths of the reference file is its ability to let you look over the shoulder of another person's design session. For performance reasons, reference files are cached in local memory only at the time you open your design file or first attach the reference file. In the meantime, if the reference file source is modified, your on-screen version of the reference file will not reflect these changes. The Reload Reference File tool is shown at left.

The Reload Reference File command forces MicroStation to refresh the reference file content it holds in memory from the original file on the disk or from the network. You will not want to do this very often with large reference files or over a slow network connection; however, if you need to update the reference file, simply invoke this command to see the most current version of the reference file.

Moving, Scaling, and Rotating Reference Files

The three tools discussed in the following sections are used to reposition the reference file with respect to your active file's design plane.

Move Reference File

The Move Reference File tool, shown at left, allows you to reposition the entire reference file with respect to your design plane. When selected, this command prompts you for a starting point and a finishing point to shift the file. This command has a similar "feel" to a fence move or selection set move operation.

Scale Reference File

This tool allows you to change the ratio of active design file master units to reference file master units. The Scale Reference File tool, shown at left, performs the same function as the Scale option in the Attach Reference File dialog box.

NOTE: *This command does not operate in the same manner as MicroStation's Move Element tool. The ratio between the reference file and the active design file is not cumulative. For instance, if you perform two Scale Reference File commands, entering a scale of 1:5 followed by a 1:2, the final result will be 1:2.*

Rotate Reference File

When you select the Rotate Reference File tool, shown at left, the tool settings window prompts you for the angular rotation about each of the axes (one for 2D, three for 3D) you want the reference file rotated (shown in the following illustration). You are also prompted in the Status Bar for the point of rotation.

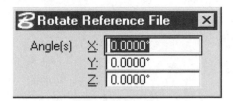

Angular rotation prompt box.

Mirror Horizontal and Mirror Vertical

The Mirror Horizontal and Mirror Vertical tools, shown at left, are provided for mirroring the entire content of the reference file. Once you have identified the reference file in question, you are prompted for the mirror axis. As with other mirror tools, this data point defines the axis, not the direction of the mirror operation.

Setting the Bounds of Your Reference File

In many instances, you may want to display only a portion of a reference file. This ability is controlled by the various Clip tools, discussed in the sections that follow.

Clip Boundary

The first of these tools, and probably the most used, is Clip Boundary, shown at left. This tool defines the outer boundary of your reference file. Any elements that fall outside this boundary, or any portion of elements that cross the boundary and lie outside it, will not be displayed.

To define the boundary, you must first place a fence. You can use either Place Fence Block or Place Fence Shape to define the clip boundary. In either case, when the tool boundary has been defined, and the Clip Boundary tool is invoked, the result is the disappearance of the elements outside the fence, as shown in the following companion illustrations.

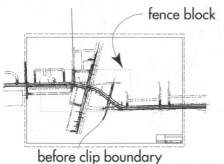

fence block

before clip boundary

after clip boundary

A "before" and "after" picture of the reference file clip boundary operation. Note how the operation only affected the roadway baseline and not the

Any type of fence shape is allowed. You can place complex fence shapes and display specific portions of the reference file. However, it should be noted that complex shapes affect the various window commands. To maximize the performance of MicroStation, you have the option of setting the reference file display as fast or slow.

View Reference Boundaries

A display attribute tool related to the reference file clip boundary is Ref Boundaries (Settings > View Attributes > Ref Boundaries), shown in the following illustration. You have the option to display the boundary on the screen and plot it as well. This is useful when determining where a reference file stops and your active design file begins.

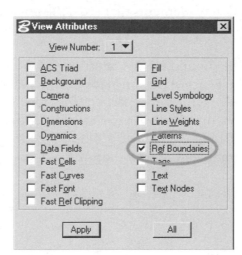

Ref Boundaries tool.

Set Reference Clipping Mask

Working like an inside-out boundary, the reference file Set Reference Clipping Mask tool, shown at left, allows you to specify areas within a reference file you do not want to display. Following the same rules as the normal clip bounds, this command is useful for exploring options with an existing design. For instance, a rehab job on a building would use an as-built plan as a reference file, using the reference file clip masking to "white out" those parts of the design that are to be replaced by the new design shown in the active file.

NOTE: *The total number of masks allowed depends on the total number of vertices those masks contain. The total number of vertices cannot exceed 101. In addition, each mask takes up one "point." This means you can have a maximum of 25 three-vertice masks (25 x 3 vertices + 25 masks = 100).*

The Reference File Settings Pull-down Menu Functions

The following are functions you may want to perform when working with reference files during the design process.

- Display or not display entire reference files
- Enable or disable the ability to snap to reference elements
- Enable or disable the ability to locate elements in a reference file for use with various copy tools
- Turn specific levels on and off for display
- Set a reference file's level symbology

MicroStation supplies you with commands for performing all of these functions. They can be found on the Settings pull-down menu of the Reference Files window. The sections that follow describe these commands.

Attachment Settings

Once you have attached a file as a reference file, there are a number of settings associated with its operation. The three most important of these (the Display, Snap, and Locate options) are defined via the Reference Files (File > Reference) window, shown in the following illustration.

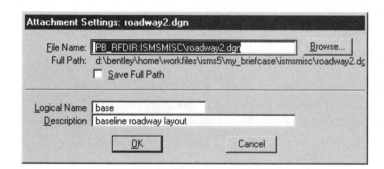

Selecting Reference from the File menu brings up the Attachment Settings dialog box.

The three check boxes (Display, Snap, and Locate) control the fundamental relationship between the selected reference file and your active design file. Table 9-6, which follows, describes the attachment settings options.

Table 9-6: Attachment Settings Options

Option	Description
Display	Controls the display of the entire reference file on all views
Snap	Enables/disables the tentative point snap capability to the selected reference file
Locate	Enables MicroStation to copy one or more elements from a reference file into the active design file

Selectively Displaying (or Not Displaying) a Reference File

Sometimes you may want to temporarily suspend the display of a reference file. Maybe you have a complicated active file and an equally complicated reference file, and you cannot tell what is in which file. The Display option lets you toggle the display of a reference file without actually detaching it permanently.

Locating Reference File Elements

With the Locate option, MicroStation's fence manipulation and selection set tools can be used to copy elements from a reference file. When you set a reference file's Locate setting to off, its elements can still be displayed but cannot be copied into your active design file.

Snapping to Reference Elements

Just as you can locate an element, you can snap to it using the tentative point snap. Enabling the Snap option of a reference file allows you to snap to any visible element within the reference file.

Reference Levels

The Reference Levels window (Reference Files > Settings > Levels) allows you to select which levels you want displayed (or not) for each of your views. However, unlike View Levels, you can only indicate your levels by number, not by name. You can also access a reference file's display levels via the standard Level Manager dialog box (Settings menu > Level > Manager).

Exercises: A Site Plan

Now that you have been shown some scenarios in which reference files are especially helpful, and have been shown the basic set of commands for manipulating them, you can try using them. The online companion website contains a set of design files specifically for this exercise. Table 9-7, which follows, describes the design files for use with the exercises.

Table 9-7: Reference File Exercise Design Files

File Name	Description
myhouse.dgn	A single-story residential floor plan
mysite.dgn	Site plan to which the other files will be attached
asize.dgn	An A-size drawing sheet

These files will be used in the following interactive sessions, which are designed to help you understand the reference file capabilities of MicroStation. In this series of exercises, you will locate the house plan on the site plan and place an A-size border around the entire.

EXERCISE 9-1: REVIEWING A FLOOR PLAN

Before you start using the reference file tools to lay out the site plan, you should first become familiar with the files contained therein, by performing the following steps.

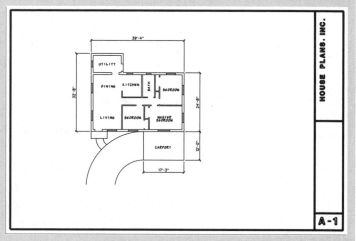

1   Open the design file *MYHOUSE.DGN.* This is the floor plan, shown at right, you will be attaching as a reference file to the site plan.

The MYHOUSE.DGN design file on screen.

2   Review the working units of this file (Settings menu > Design File > Working Units). Note the dimension elements and the other annotation items in this file.

Now that you have reviewed the floor plan, you can attach that floor plan to a site plan using MicroStation's reference file tools. Exercise 9-2, which follows, takes you through this process.

EXERCISE 9-2: ATTACHING THE HOUSE PLAN

In this exercise you will use MicroStation's reference file tools to attach the floor plan just reviewed to a "typical" site plan.

1 Open the design file *MYSITE.DGN*. This file consists of a surveyed lot outline and dimensions. To this file you will attach and manipulate the floor plan file you reviewed in the previous exercise.

2 Open the Reference Files window (File > Reference). Select the Attach tool (Reference Files > Tools menu > Attach). Select the file *MYHOUSE.DGN* and click on OK. This brings up the reference file attachment dialog box labeled Attach Reference File, shown in the following illustration. Here you will enter specific information about how you want this reference file to be attached and described.

Attach Reference File dialog box.

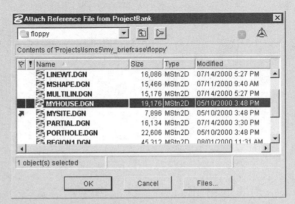

3 Enter the Logical Name and Description, the dialog box for which is shown in the following illustration. Click on the OK button.

Logical Name and Description fields.

An indication that something happened is a confirmation message in the status bar and the appearance of the reference file name in the Reference

Files window. Although the house plan may not be visible within the extents of your view window, the house plan is indeed attached to your site plan. To verify this, you need to use the Fit View command, which will give you the complete picture.

4   Select the Fit View icon in View 1's view control tool bar. Verify that the All option is selected in the Tool Settings window. This command performs the Fit command on all elements, including those active and those in all attached reference files.

The result was not what you expected, was it? Instead of seeing the site plan, you now see the floor plan. If you look in the lower right corner of the view, you will see the site plan smaller than life. The reason for this size discrepancy is in the different working units with which these two designs were made (i.e., feet/inches/8000 positional units in the floor plan versus feet/hundredths of a foot/10 positional units in the site plan). You will need to "set" the scale of master units between the site plan and the referenced floor plan.

5   In the Reference Files window, highlight the HOUSE reference file entry and select the Scale tool (Reference Files > Tools > Scale). The Scale Reference tool settings appears with the "coincident" plane value.

6   Enter a 1:1 ratio (Master:Ref) in the Tool Settings window. MicroStation prompts you in the status bar for the point to scale the reference file about.

7   Click a data point once at the lower right corner of the house drawing border (under the A-1 label).

The house will now reappear at the same scale as the site plan. Although the working units between the two files are dissimilar (FT:IN versus FT:tenths), MicroStation successfully reconciles the working units between the two design files.

At this point you may want to use MicroStation's view controls to look over the house plan. You may even want to try to delete some of its elements; however, you will not be successful. Remember, you cannot delete elements from a reference file. You can, however, copy elements from a reference file.

The next step in this process is to clip the floor plan down to size. This means eliminating the surrounding material, such as the drawing sheet border incorporated into the floor plan design file, as shown in the following illustration. This will be performed using the Reference Clip command. Exercise 9-3, which follows, takes you through this process.

Eliminating the unwanted portion of the floor plan for the site plan preparation.

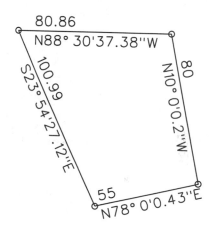

EXERCISE 9-3: CLIPPING THE FLOOR PLAN TO SIZE

1   Use the Place Fence tool with the Shape option to place a fence around the house, as close as possible without touching the actual walls.

2   With the house reference file selected in the Reference Files window, select the Clip Boundary tool (Reference Files > Tools menu > Clip Boundary).

This results in the floor plan drawing being clipped to your fence shape. However, if you look closely at the house plan, you will see part of the witness lines of the dimensions around the house, as shown at right. This will not do for a site plan. To get rid of those unsightly lines, you need to turn off the dimensions level.

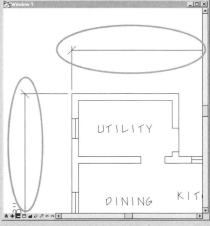

Witness lines (top and left in image) to be removed.

3   From the Reference Files window, select the Levels command (Reference Files > Settings menu > Levels). Deselect level 50, the dimensions level for the floor plan reference file (turns off display of level 50), and click on the All button to apply the change to all view windows.

The witness lines disappear, leaving only the walls of the house itself. If you want, turn off other levels of the reference file. (Level 11 is room names, and level 10 is interior walls.) The house is now ready to be placed on the property in its proper location and orientation.

Positioning the House on the Lot

The shape of the site plan chosen for this exercise is irregular and requires you to turn the house at least 20 degrees to get it situated appropriately on the property, as shown in the following illustration. The Move Reference File and Rotate Reference File commands will be used to accomplish this. Exercise 9-4, which follows, takes you through this process.

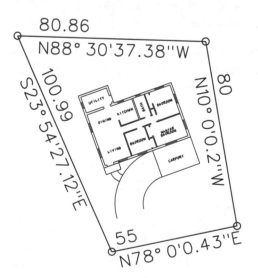

A picture of the house sitting on the site plan.

EXERCISE 9-4: MOVING AND ROTATING THE REFERENCE FILE

The first step is to move the house within the site plan property boundary.

1   Select the Move Reference File tool (Reference Files > Tools > Move).

2   Data point once within the house for the "from" location, and place another data point in the middle of the site layout for the "to" location, as shown in the following illustration.

Data points for "from" and "to" locations.

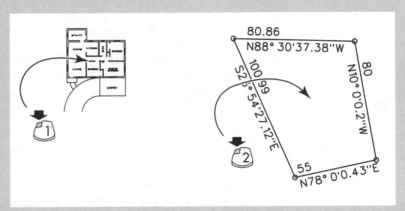

Next, you need to rotate the reference file to better position the house on the property.

3   Select the Rotate Reference File tool (Reference Files > Tools > Rotate). Enter the rotation angle (*20*) to reorient the house to the front of the property.

You can try other rotation values. The reference file's boundary polygon is dynamically active at this point.

4   Using the cursor, position the rotated house within the property boundary and data point once.

At this point, you can use the Move Reference File tool to move the rotated house around on the property to get the best fit. If you want, you can add line work in the site plane to complete the driveway to the property line. Turn the house's reference file snap option on so that you can use the tentative point snap with the Place Arc by Edge tool.

Placing the Drawing Sheet Format

Before continuing with the exercise, you need to understand use of the border as a reference file. Quite often, first-time users of MicroStation start a new drawing within a project by creating the drawing sheet border. However, this may not be the best way to start a design.

First, a drawing sheet border tends to go through many revisions during a company's life (for instance, a merger changes the name of your company). By their very nature, such drawing sheet borders should be the same across an entire project, if not the entire company. Finally, when you start your design with a sheet border embedded in the drawing, its elements will be the first highlighted should you stray too close to them while using an element manipulation tool such as Delete Element. This can lead to accidental deletion of key drawing sheet information.

One solution is to use a cell for the drawing border, but you would still have to deal with accidental manipulations, and there is always the problem of globally updating your company standard sheet borders within your current project. The other solution is to attach your drawing sheet border file as a reference file. The major advantage of using a reference file is the ability to update the sheet border used within a project by simply updating a single design file. The positioned border for the house in this exercise is shown in the following illustration. Exercise 9-5, which follows, takes you through the process of attaching the drawing format.

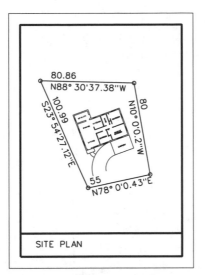

The final drawing of the house with the border positioned.

EXERCISE 9-5: ATTACHING THE DRAWING FORMAT

To complete the site plan exercise, you need to incorporate a drawing sheet border. The target plot scale of our site plan is 1 inch (on the plot) = 20 feet (1 inch = 20 feet x 12 inches/1 foot, or 1/240 scale). However, as you will see shortly that you do not have to worry about the 12-inch to 1-foot math. All you need to do is attach the A-size drawing sheet border *(ASIZE.DGN)* to the site plan at the 20:1 ratio.

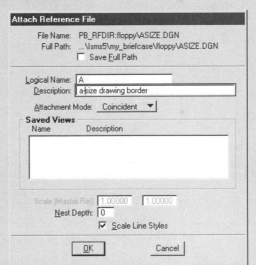

Logical Name and Description entries.

1   Attach the design file *ASIZE.DGN* to the site plan using the Attach tool (Reference Files > Tools > Attach). Give it the logical name *border* and a short description, as shown in the illustration at right. The drawing border is now attached to the site plan.

2   Use the Fit All view control to view both the drawing border and your site plan. You may want to use the Zoom Out tool to further shrink the image in the view window. This will give you some elbow room when it comes time to move the drawing border reference file.

3   With the *ASIZE* reference file highlighted on the Reference Files window, select the Scale tool (Reference Files > Tools > Scale).

4   Enter a 20:1 ratio between your master file and this reference file. This defines 20 feet of the site plane as equal to 1 inch within the drawing sheet border. Click near the lower right corner of the *ASIZE* reference file. The result should be an overlap of the drawing on your site plan, as shown at right.

Overlap of drawing and site plan.

5   Using the Move Reference File tool (Reference Files > Tools menu > Move), reposition the drawing sheet over the site plan. Make sure the *ASIZE* reference file is highlighted in the Reference Files window before invoking this command, or you may be moving the house instead!

Summary

In this series of exercises you used the Reference File facility of MicroStation to compose a site plan. Many companies use this technique to develop drawings. With a real project you would now add the various notes and other annotation information to the site plan and then plot the final results to your printer or plotter.

The subject of the next two chapters is getting to the final plotted product. In the next chapter you will learn how to take your design model and turn it into a finished set of drawings by providing all of the traditional drawing constructs of patterns and dimensions. Combined with the various text annotations and reference file features discussed so far, you are now well on your way to mastering MicroStation in a production environment!

DETAILING TOOLS

Finishing a Design with Meaningful Details

CREATING THE COMPUTER MODEL IS ONLY HALF of any project. True, there are methods available to go directly from a computer model to final product (for instance, numerical controlled machining), but in most cases you must generate detailed "paper" drawings to ensure the final product matches your original design.

The process of converting your model into usable drawings is better known by its drafting function: detailing. Starting with your computer model, you add drawing-specific details such as hatching, sectional details, dimensions, callouts, notes, and the like. The end result of the detailing process will be drawings with enough information for the manufacture of your design to your specifications.

In most design disciplines, such drawing details must conform to drafting standards originally developed for manual drafting. This can sometimes lead to compromises between what is efficient for the computer design process and what is acceptable as a final drawing. In many cases, to get the details "just right" you have to generate special detail-only features in your design files. For this reason, many of the final drawing details are created toward the end of the design process. This way, if there are any major changes in the design, rework of detail-oriented drawing features is minimized.

Dimensioning Your Drawings

Paramount in any design project is the need to convey the accurate information about the measurements of the model. One of the first subjects you learn in any first-year drafting class is the importance of dimensioning a drawing. Of course, MicroStation provides a host of dimensioning capabilities flexible enough to handle most dimensioning requirements. Before continuing with the initial setup and use of MicroStation's dimensioning tools, a review of what constitutes a dimension is in order.

The Anatomy of a Dimension

A dimension is essentially text that conveys spatial or dimensional information about a specific component or feature of your design. In other words, a dimension tells you how long something is or where it is located and whether it is in units of measure or degrees of arc. Additional information may also be included in a dimension, such as the acceptable tolerance (how much the physical dimension can be off from the ideal) and what the measurement is in an alternate measurement system (e.g., metric/English).

MicroStation enables you to adjust the various features of a dimension to best match your design standards. To accommodate this, each part of a dimension is referred to by name within MicroStation, as indicated in the following illustration. This allows you to customize practically everything about these components, even eliminate them altogether. For this reason it is a good idea to take a moment to familiarize yourself with the various parts of a dimension.

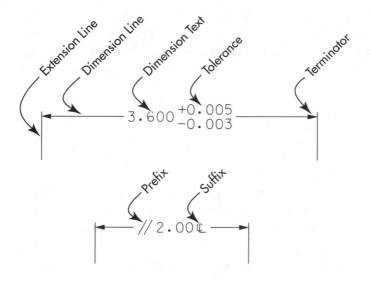

The parts of the typical dimension as defined by MicroStation's dimension settings.

Dimension Line

The most obvious part of the dimension is the dimension line. Defining the measured distance, this line may be a single line with the dimension text resting above it, or a broken line with the dimension text sandwiched between its two segments.

Dimension Text

This is probably the most important part of the dimension. Used to document a measurement of the design object, this text has all of the standard text attributes plus additional dimension-specific parameters (accuracy, format, and measurement system, to name a few).

Extension Lines

Extension lines (i.e., witness lines) are used to indicate the point in the design where the dimension originates and ends. This way, there is no question as to what is being measured. Again, there are options associated with this dimension component; most important, selectively turning it on and off.

Terminator Symbol

Always located at the ends of the dimension line, terminator symbols (also known as arrowheads or terminators) serve the purpose of identifying the text and linework of a dimension as a dimension. Without terminators, dimensions would get lost in the maze of elements that constitute the drawing. Arrowheads come in a variety of styles and types to support the various dimensioning standards.

Prefixes, Suffixes, and Tolerances

Additional information critical to interpreting the dimension, these components can be activated and adjusted as needed. These, too, warrant their own settings boxes, and provide MicroStation the ability to adhere to some of the more rigorous dimensioning standards (e.g., ANSI Y14.2).

All told, there are 11 dimension setting categories dedicated to your dimensions, far more than any other element class within MicroStation. This fact serves to point out the importance of dimensioning within the design community.

Setting Up Dimensions

Because MicroStation tries to be all things to all people, when it comes to dimensions, it can require a fair amount of setup prior to actually placing your first dimension. The dimensioning options are set via the Dimension Settings dialog box, shown in the following illustration.

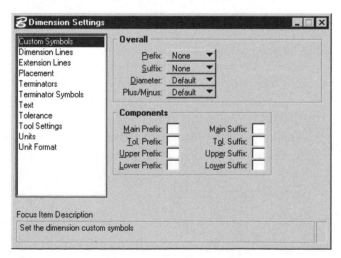

The Dimension Settings dialog box is used to control every aspect of the dimensioning process.

Accessed via the Element pull-down menu (Element > Dimensions), the Dimension Settings dialog box offers a list of categories on the left, each of which controls different dimension attributes. Table 10-1, which follows, describes the commands within the Dimension Settings dialog box.

Table 10-1: Dimension Settings Categories

Command Name	Description
Custom Symbols	Controls the creation of suffix/prefix annotation
Dimension Lines	Controls the level and symbology settings for dimension lines
Extension Lines	Controls the level and symbology settings for extension lines
Placement	Controls the alignment of linear dimensions relative to the element, location of text, and other placement parameters
Terminators	Controls the geometry, symbology, and orientation of arrowheads
Terminator Symbols	Allows you to select a symbol or a cell for any of the four terminator types: Arrow, Stroke, Origin, or Dot
Text	Used to select text placement and symbology parameters
Tolerance	Controls the geometry and attributes for dimension tolerance text
Tool Settings	Allows adjustment of settings for individual dimension tools
Units	Controls the dimension format (mechanical or AEC), and whether or not to place the dimension in two units, primary and secondary
Unit Format	Sets the format for the units selected above

The sections that follow examine each of these dimension settings categories in detail.

Custom Symbols

MicroStation provides the Custom Symbols option for adding a prefix and a suffix character symbol to each dimension you place. This category gives you control over what MicroStation places. By default, no suffix or prefix is placed; however, you can opt to display either a symbol from a text font or a cell from your active design file (shared cells only). The Custom Symbols option is shown in the following illustration.

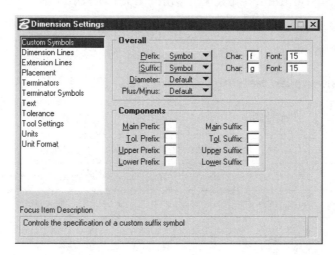

Custom Symbols gives you options for setting dimension prefixes and suffixes. More importantly, this is where you change the default diameter and plus/minus symbols.

In addition, you can change the default diameter symbol and/or the plus/minus symbol used with all dimension tools. In this case, however, you can only specify an alternative font and character (no cells).

In addition to suffixes and prefixes for the entire dimension text string, you can set single-character text for the individual text parts associated with toleranced dimensions. As you can see, this can get very complicated, so the recommendation here is to tread lightly through this settings box, and only use those settings you absolutely need.

Dimension Lines

At first glance, a dimension can be thought of as a series of lines and text. In fact, prior to MicroStation version 4, this was precisely how MicroStation treated the entire dimensioning routine. Now, however, dimensions are seen as elements in and of themselves, subject to the same element attributes as all other elements. However, the individual parts of the dimension can have their own attributes, subject to the settings you select in the Dimension Lines category, shown in the following illustration.

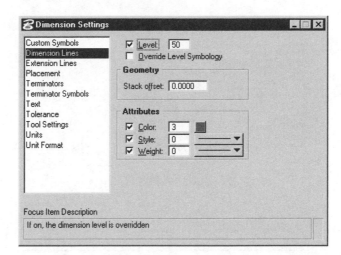

The Dimension Lines category is used to override the active element attributes for dimension lines placed.

By clicking on the checkboxes of the attributes you want to override, you ensure uniform dimensional appearances. This is an important aspect of the dimensioning system. Standardizing the color, weight, and especially the level of dimensions, will lead to a better-looking final product.

TIP: *The Dimension Level override associated with the Dimension Attributes settings box also has a shortcut key-in. Use* LD=50 *in the Command window to set the level to which all of your dimensions will go.*

Extension Lines

This category is similar to Dimension Lines for independently setting the color, style, and weight for the extension lines. Whereas dimension lines are always generated when using a dimensioning tool, the Extension Lines checkbox, shown in the following illustration, allows you to control whether or not extension lines are placed when dimension lines are placed.

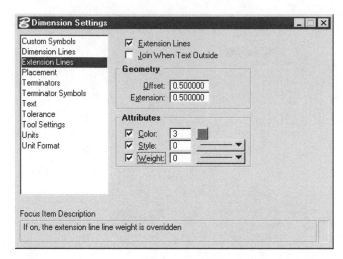

The Extension Lines category is used to specify extension line parameters.

Normally, when dimension text is longer than the gap between extension lines, the text is placed outside, with a gap between the extension lines. When you enable the checkbox labeled Join When Text Outside, the gap between the extension lines is closed with a line.

The Offset parameter under Geometry specifies the distance, in text height units (TH=), between the start of the extension line and the element being dimensioned. This measure is a percentage of the active text height associated with the dimension. The default is 0.5000 (or 50%), which means that the space between the witness line and the object is half that of the current dimension text height. The Extension parameter specifies the distance, again in text height units, the extension line should extend beyond the dimension line.

Placement

The Placement category on the Dimension Settings dialog box, shown in the following illustration, is used to set dimension placement parameters. These include the orientation of the dimension text to the dimension line, its justification, whether MicroStation automatically generates the dimension text position or prompts the user for input, and other parameters.

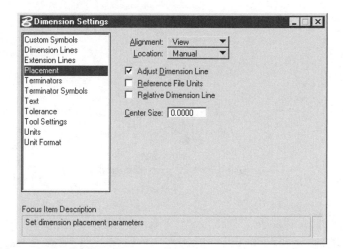

Dimension Settings dialog box.

Alignment Options

By default, dimensions are placed normal to your current view. This means they align along the X and Y axes of the view, which is why this type of placement is referred to as View alignment. There are three additional dimension alignment options supported by MicroStation. Drawing alignment uses the drawing's absolute XYZ axes for calculating a distance to be dimensioned, regardless of the view's orientation. True alignment computes the dimension alignment directly from the dimension tool's data points and is most often used to dimension an oblique face of a 3D object.

Arbitrary alignment allows you to generate a dimension without keeping the extension lines perpendicular to the object being measured. This is especially helpful when dimensioning 2D, isometric drawings. Use of the Alignment control is shown in the following illustration.

Location

The Location option sets how the dimension text is placed with respect to the dimension line. With *Automatic* location, the text is automatically placed by MicroStation using the justification option set in the Text category (discussed in material to follow). *Manual* location prompts you to manually identify the point along the dimension line where you want the dimension text to be placed.

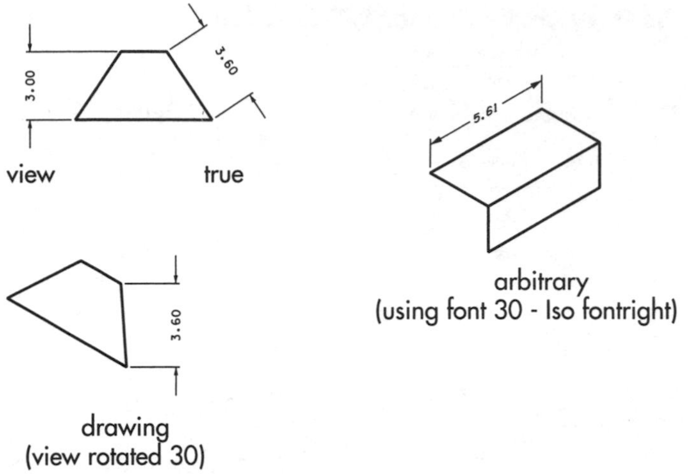

view   true

drawing
(view rotated 30)

arbitrary
(using font 30 - Iso fontright)

The Alignment control gives you control over how MicroStation calculates the distances for the active dimension tool.

Semi-auto location is a combination of the manual and automatic options. When a dimension fits between the witness lines, Micro-Station automatically places the dimension. If the dimension text is too large, you are prompted to manually place the text. This last option is probably one of the better compromises between human- and machine-generated dimensions.

Manual location is most often used when you are working with small distances that require you to place dimensions very close to one another. MicroStation's automatic dimensioning feature is good, but it does have a tendency to place the dimension string in an arrows-out configuration before the distance being dimen-sioned really requires it.

The following illustration points this out. The automatic place-ment results in the dimension at the top of the object. By selecting the manual placement option, the additional step of placing the dimension text allows you to place the text between the arrow-heads.

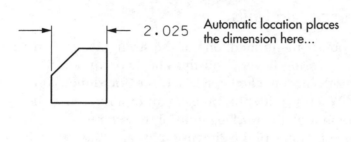

2.025   Automatic location places
         the dimension here...

*Dimensioning differences
between manual and
automatic.*

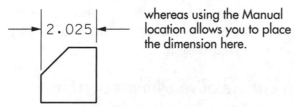

2.025   whereas using the Manual
         location allows you to place
         the dimension here.

Adjust Dimension Line

This option, shown in the following illustration, gives the Place
Dimension tools some latitude when placing adjacent dimensions.
With the Adjust Dimension Line option selected, a second dimension's text, terminators, and arrowheads will be shifted up to clear
the first dimension's text.

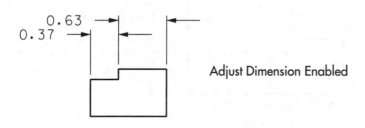

0.63
0.37   Adjust Dimension Enabled

*Using this option results
in a cleaner finished
dimension.*

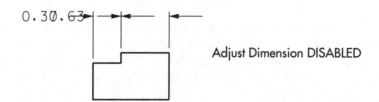

0.30.63   Adjust Dimension DISABLED

Reference File Units

As previously mentioned, one of the strengths of MicroStation is its reference file facility, and this checkbox option allows you to look inside an attached reference file for dimension measurements. When you identify an element of a reference file with the dimension tool, the working units of the reference file are used to compute the value of the dimension. By attaching your design to a border sheet, you can set up your text and other drawing-related functions based on the final drawing size, while still being able to accurately dimension your design.

Modify Element: Relative Dimension Line

This option, shown in the following illustration, controls how the Modify Element tool affects the dimension text string whenever you modify the witness line location of a dimension.

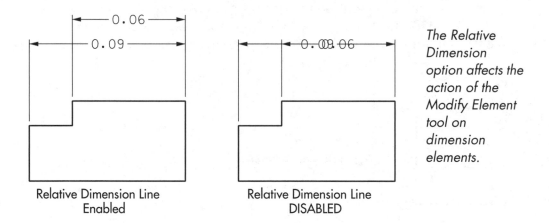

Relative Dimension Line
Enabled

Relative Dimension Line
DISABLED

The Relative Dimension option affects the action of the Modify Element tool on dimension elements.

As you can see, when a dimension is placed with the Relative Dimension Line option selected, any changes to the dimension's extension line keeps the dimension line at the same relative distance from the extension line's endpoint. With the Relative Dimension Line option turned off, the dimension line keeps its original location and only changes its length in response to changes to the extension line location.

NOTE: *The Relative Dimension Line option is effective only at the time you place the original dimension. Changing the setting while using the Modify Element tool has no effect on existing dimension elements. You must use the Change Dimension to Active Settings tool to update an existing dimension if you change this setting. Keep in mind, though, that this tool changes all parameters of the chosen dimension to the current settings.*

Center Size

This text field sets the size of the center mark that is used when placing radial dimensions. The center mark size is specified in the design file's working units.

Terminators

This category, shown in the following illustration, controls the geometry, orientation, and display attributes of a dimension's terminators. The sections that follow describe the options within this category.

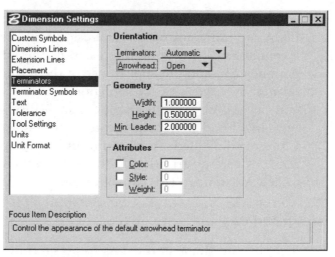

Terminators category on the Dimension Settings box.

Orientation

The terminator orientation refers to how MicroStation places the terminator with respect to the dimension line and text. *Automatic* terminators mean MicroStation will place the terminators inside the witness lines whenever possible. However, in the event there is not enough room for them, they are placed on the outside of the witness lines facing in.

Setting Terminators to *Inside* forces MicroStation to always orient the terminators on the inside of the dimension, along the dimension line. Conversely, setting Terminators to *Outside* forces the terminators to the outside of the extension lines. *Reversed* simply flips the terminators from where they would normally be placed in Automatic mode, which is handy when MicroStation appears to be placing the terminators in exactly the opposite place you expect. These settings are shown in the following illustration.

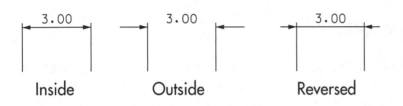

Three of the four terminator orientation options. The Automatic option uses the Inside option unless the terminators and text do not fit the distance being dimensioned.

If you are using any of the dimension tools that place arrowhead terminators, you may be interested in the effect the Arrowhead parameter has on the appearance of the terminators. The default terminator used with MicroStation is a simple open line string. You can choose two other styles of arrowheads using this parameter. The Closed arrowhead creates a three-sided shape with no fill. Choosing the Filled arrowhead results in a filled arrowhead terminator. To see the arrowhead's fill, turn on the Fill display (Settings > View Attributes > Fill) for your views.

Geometry

The size and shape of the dimension's terminators is governed by the settings in the Geometry section of the Terminators category and the current text height and width setting used for the dimension. By default, the width of the terminator is 1:1 with the text (1.0), and the height is 50% of the text height (0.5). Both of these parameters are adjustable in the Geometry section.

TIP: *By selecting the Filled Arrowhead option and adjusting the Width-to-Height ratio in the Geometry section, MicroStation gives you some of the best-looking arrows available in any CAD program.*

Attributes

As with the other components of the dimension element, you can override the current active element attributes by setting the Color, Style, and Weight fields in the Terminators' Attributes section.

Custom Terminators

If MicroStation's selection of terminator styles does not meet your requirements, you can replace them with one of your own design or one found in any text font installed within MicroStation. Using the Custom Terminators category, shown in the following illustration, you can selectively replace the Arrow, Stroke, Origin, or Dot dimension graphic with either a symbol from a font or a shared cell.

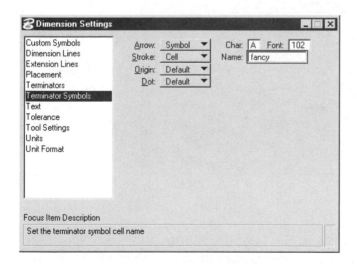

The Custom Terminators category provides options for replacing the default terminators with those of your own choosing. Note the use of a text font symbol for the arrowhead, and the use of a cell for the stroke terminator. Defaults refer to MicroStation's default terminator setup.

The symbol font numbered 102 is a font resource delivered with MicroStation that contains nothing but arrowheads perfect for use as replacement (custom) terminators. Symbol font 102 is shown in the following illustration.

Font 102

A	B	C	D	E
➢	❯	▶	▷	➢

F	G	H	I	J
▶	▷	➢	≀	∅

K	L	M	N
⌐	⌐	⋊	•

The terminators found in font 102 are designed for use with the Custom Terminators settings box. Note that all of these symbols are uppercase.

Text

This dimension settings category, shown in the following illustration, controls both the attributes and placement options for dimension text.

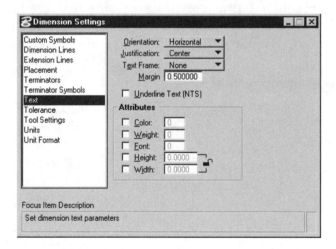

The Text category controls the attributes and placement parameters of dimension text.

Orientation

This option, shown in the following illustration, gives you three choices: Above, Inline, and Horizontal. Above places the dimension text above the dimension line, Inline breaks the dimension line and places the text midstream, and Horizontal breaks the line, but also forces the text in all cases to appear horizontal to the screen.

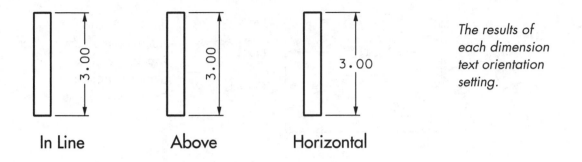

In Line　　　Above　　　Horizontal

The results of each dimension text orientation setting.

Justification

Justification, shown in the following illustration, controls where the text appears in relation to the dimension line. The default, Center, places the text in the center of the line. The Right and Left options place it to either side of the line midpoint. You usually use the latter two settings when stacking dimensions and staggering the text for better appearance.

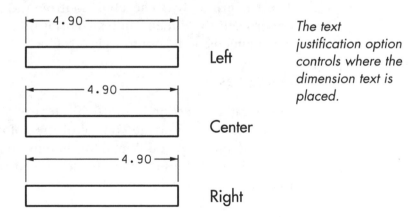

The text justification option controls where the dimension text is placed.

Text Frame

The Text Frame option, shown in the following illustration, controls the addition of graphic elements around the dimension text. The text frame default, None, represents normal text, the most common form of the dimension element. The other two selections, Box and Capsule, provide a graphic box or capsule-like construct around the dimension. These are usually associated with reference or quality control dimensions.

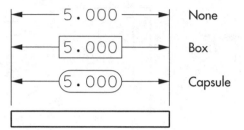

None

Box

Capsule

The Text Frame option controls the appearance of graphics around the dimension text.

Margin

The Margin field sets the space between the leader line and the dimension text. The unit of measure is relative to the dimension text height, which as a default is the active text height. Thus, the default value of 0.5 means that the space between the dimension text and the leader line is half of the dimension text height.

Underline Text (NTS)

The Underline Text checkbox controls the placement of a line underneath the dimension text. This is normally used to indicate that a dimension is "not to scale" (NTS).

Attributes

The Attributes section in the Text category of the Dimension Settings box offers override control of the color and weight of the dimension text. In addition, if for dimension text you need to use a font or size different from the active text setting, you have the font, height, and width checkboxes.

Tolerance

In mechanical design it is common to find tolerances associated with one or more dimensions on a drawing. MicroStation supports the display of dimension tolerances using the Tolerance dimension setting category, shown in the following illustration. When you select the Tolerance Generation checkbox, all dimensions created from the time you set the option will include the tolerance as configured in the Dimension Tolerance settings category.

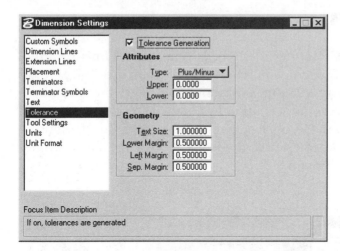

The Tolerance dimension setting category controls all aspects of the tolerance function. Selecting the Tolerance Generation checkbox activates the use of tolerance values with your current Place Dimension tools.

The Type field provides you with two options: Plus/Minus and Limit. The default shows the tolerance as a plus and minus text string appended to the dimension text. Limit is a minimum/maximum configuration where the dimension is two lines of text, the top being the maximum dimension allowed and the bottom the minimum dimension allowed.

Both of these tolerance types use the Upper and Lower field for calculating the tolerances. The Text Size option in the Geometry section of the setting box is used with the Limits option to set the height of the two strings of text. The default of 1 means the text is the full height of the current dimension text value.

NOTE: *Tolerance can only be used with the mechanical format dimensions.*

The three fields for the Lower, Left, and Sep margins set, respectively, the space between the dimension line and the bottom of the dimension text, the horizontal space between the dimension text and the tolerance text, and the vertical space between the tolerance values. Again, the unit of measure of the margins is a ratio of the dimension text.

Tool Settings

In addition to the settings described so far, each dimension tool has its own set of parameters. These are set using the Tool Settings category's Dimension Settings window. You select the tool whose parameters you want to set, either by name or by selecting the tool icon adjacent to the Tool label, as shown in the following illustration. Unique in MicroStation, the icons pop up just like any other text-based option menu.

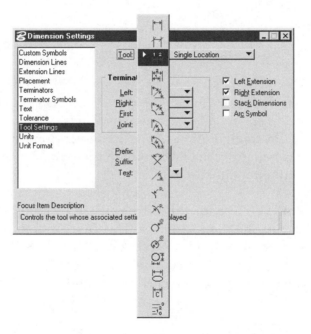

Selecting the tool icon presents you with the entire list of tools for use with the Tool Settings box.

Earlier, the Extension Lines category discussion showed how you can control the appearance of a dimension's entire set of extension lines. There are times when you may want explicit control over the individual extension lines (and terminators) associated with a dimension. For instance, you are placing a dimension from the face of your design object, a situation where one of the normal dimension extension lines would not be appropriate. At this point you would turn to the dimension tool's Tool Settings category in the Dimension Settings window.

Terminators

In the Terminators section of the Tool Settings category you select the type of terminator used by the specified dimension tool. You have independent control over each terminator associated with the current tool. In each case, you select from the option field the type of terminator you want to use.

There are four fields listed under the Terminators section. The Left and Right fields are self-explanatory. However, the First and Joint fields require a short explanation. When you put in a string of dimensions with a dimension tool such as Single Location, the first terminator placed is different from the rest. (In this case, a dot signifies the starting datum.) In MicroStation, this is referred to as the *First* terminator. The *Joint* terminator is used at the shared junction between two dimensions. This occurs when there are two or more dimensioned distances in the same dimension string.

Prefix and Suffix

Although similar to the global custom symbols discussion, the prefix and suffix used here are more specific and are used with each tool. For instance, the prefix for the Diameter Extended tool consists of the diameter symbol, as shown in the following illustration.

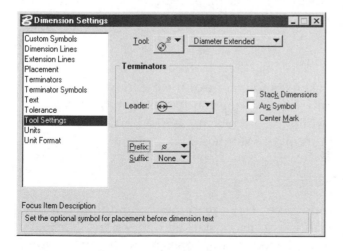

The settings associated with the Diameter Extended tool include the diameter symbol as its prefix. Note the use of special terminators with this tool.

Text

The Text field, shown in the following illustration, controls the orientation of the text in relation to the dimension line. In most cases you will leave it at the Standard setting. However, there may be occasions when you may want to change this.

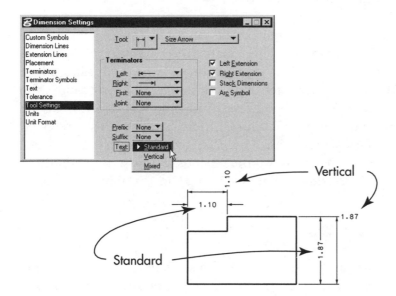

The Text option controls the orientation of the text with respect to the dimension line, not the drawing.

The Mixed option will normally place the text in standard mode. However, in cases where the text does not fit the space (e.g., in the case of small dimensions), this option allows MicroStation to place the text in the vertical format.

Extensions

There are two checkboxes that control the creation of extension lines. The Left and Right Extension boxes can be selectively turned on and off while you are placing dimensions. This is one of the most-used options on this settings box.

Stack Dimensions

This checkbox option controls the placement of multiple options. If selected, the result is a series of stacked dimensions. Normally this option is already selected for the appropriate dimension tool, such as Dimension Location (Stacked).

Arc Symbol

This checkbox option controls the placement of an arc symbol above arc dimensions.

Center Mark

This checkbox option controls the placement of a center mark graphic at the center of a radial dimension. The size of this center mark is controlled by the Center Size field in the Placement category on the settings box.

Units

When it comes time to dimension your drawing, you must choose whether you will be using a mechanical or architectural dimension style. This is usually a straightforward decision based on your design discipline (you normally know whether you are a mechanical designer or an architectural designer). The Units option is shown in the following illustration.

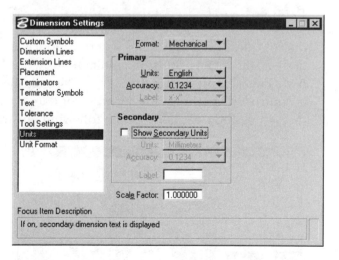

Units option of the Dimension Settings dialog box.

However, just for the record, one difference between the AEC dimensioning format and the mechanical format is AEC's use of master units and subunits versus Mechanical's use of master units only. Formatting the feet and inches with the appropriate "gingerbread" (i.e., ft-in, or ft' in", and so on) used with AEC dimensions is the other main distinction. These differences are shown in the following illustration.

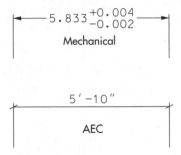

$$5.833\,{}^{+0.004}_{-0.002}$$

Mechanical

$5'-10''$

AEC

The differences between the Mechanical (top) and AEC (bottom) dimensioning styles are very apparent.

Format Field

The Format field in the Dimension Units settings box gives you the two dimensions. The Dimension Units Settings box gives you the two dimensioning options just described, AEC or Mechanical. When you select AEC format, the Labels field for both the primary and secondary dimension sections are activated.

Primary and Secondary Dimension Control

MicroStation's dimensioning system supports dual dimension standards. This means you can simultaneously display a dimension in both English and metric units, either singularly or combined. This is controlled via the primary and secondary dimension sections of this settings box, as shown in the following illustration.

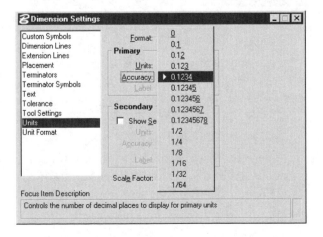

Primary and secondary dimension settings.

In addition to the units of measure, you control the amount of accuracy for each dimension via the Accuracy field. Note how you can set the accuracy for the secondary dimension independent of that of the primary dimension. Clicking on either will present an array of options, ranging from eight decimal places of accuracy to half a unit or a whole unit.

Scale Factor

The Scale Factor field allows you to enter a multiplier that is applied to all linear dimension values prior to their placement in the drawing. This is handy if you want to enlarge a detail on a drawing, but want the dimension to still measure out at the model's true units. For instance, if a detail is drawn at twice the scale of the design file, setting this value to 0.5 (or 1/2) results in the correct dimension text.

Unit Format

The Unit Format category on the Dimension Settings box controls the format for angular and metric units, and whether or not to display leading/trailing zeros for primary or secondary units.

Angle Format

The other major unit of measure you use in dimensioning your drawing is the Angle format. These settings are used with the Angular Dimension tools. MicroStation gives you control over how the angular information is displayed. Table 10-2, which follows, describes the various angle parameters.

Table 10-2: Angle Parameters

Angle Parameter	Value	Description
Units	Length	Measured distance along the arc curve.
Degrees	0 to 360 degrees	Sweep of the angle measured in degrees.
Accuracy	0 through 0.0000	Degree of accuracy to display for each angular value.
Display (option 1)	Decimal degrees	Degrees are shown in decimal format.
Display (option 2)	DD^MM'SS"	Degrees shown in degrees/minutes/seconds format.

NOTE: *The Dimension Settings window can stay open or be mini-mized in your MicroStation design session (it is a non-modal dialog box). This allows you to adjust the various options during the place-ment of the your dimensions. As a result, you can control the accuracy and the selective display of witness lines and arrowheads during the dimensioning operation.*

Metric Format

Although the metric system is in wide use worldwide, the charac-ter used as a decimal separator is not the same everywhere. For instance, in Europe the comma character is used to delineate between the whole part of a number and the fractional portion, whereas most of Asia and the Americas use the period character to perform this same function. In addition, it is the convention in many parts of the world to use a space after the thousand and mil-lion places. For this reason, MicroStation provides settings for adjusting the readout format in your dimensions.

Use Comma for Decimal

This option swaps the period character used by default as the dec-imal separator with the comma character.

Unit Separation

This option, when enabled, leaves a space after the million and thousand places in a number.

Other Dimension-related Options

There are a number of common checkbox options found on Unit Format. Each controls the display of specific dimension text fea-tures related to units. These features are applicable to both pri-mary and secondary units.

Show Leading Zero

When a dimension is less than one unit, selecting this option results in a zero being placed in front of the decimal point of the dimension.

Show Trailing Zeroes

When selected, this option pads the dimension value with zeroes out to the number of places set by the accuracy value for the Primary and/or Secondary dimension field. For example, if a dimensioned distance is 1.5 and the accuracy is set to 0.12345678, the resulting dimension text will be 1.50000000.

Dimensioning Tools

You might very well say at some point, "With all of the setting up involved with dimensions, there had better be a rich selection of dimensioning tools." Yes, there is a wide variety of dimensioning tools. In fact, there are no fewer than 33 dimensioning tools at your disposal, as shown in the following illustration.

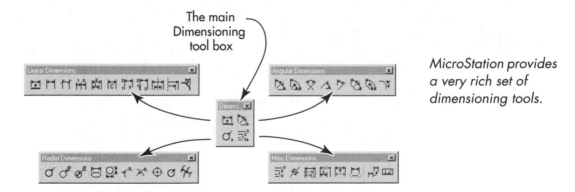

The main Dimensioning tool box

MicroStation provides a very rich set of dimensioning tools.

The dimensioning tools can be accessed from the main Dimensioning tool box (Tools menu > Dimensioning > Dimensioning). This tool box is used to access the four dimensioning tool boxes, which are organized around the type of dimension you are going to create (linear, radial, angular, and miscellaneous).

Linear Dimensioning Tools

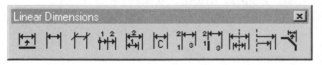

Linear dimensioning is usually the first type of dimensioning you try. Simply put, a linear dimension measures the straight-line distance between two given points. MicroStation supports normal, stacked, ordinate, and datum methods of linear dimensioning.

Dimension Size Tools

Measuring a distance between two points and creating a dimension is the most common dimensioning style used. MicroStation calls this the size method of dimensioning. When you use the Dimension Size tools, you get two witness lines, a dimension line, a text string showing the measured distance, and two arrowheads or slashes. The type of terminator generated depends on which of the two tools you select. The tools are shown at left.

These two tools operate in exactly the same fashion, but the terminators used are different. Depending on the dimension control settings, some or all of the components that constitute the dimension may appear. Use of these two tools is shown in the following illustration.

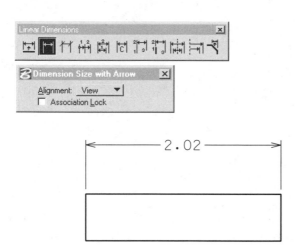

The simplest of the dimension tools, the Dimension Size with Arrows or Dimension Size with Strokes tool gives you straight-line dimensioning between two points.

The Dimension Size tools can also create a string of dimensions, with the end of one dimension serving as the beginning point of the next dimension. Two Resets will start a new chain of dimensions. One Reset tells MicroStation of a new location for the next dimension string. This is helpful when you want to turn a corner and dimension a second side of an object, as shown in the following illustration.

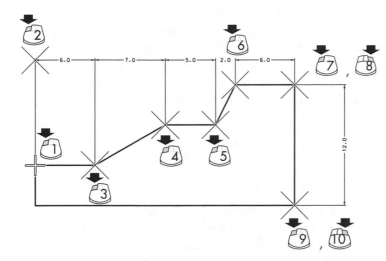

Dimensioning with a continuous chain of dimensions. Note that when you turn the corner for the second string of dimensions, this becomes a separate dimension.

Dimension Location Tools

The other common linear dimension used in mechanical design is the datum, or location, dimension. This dimensioning method uses a common starting point (the datum) from which all dimensions are referenced. MicroStation provides two tools, shown at left, for performing this type of dimension: Dimension Location and Dimension Location (Stacked).

The Dimension Location tool creates one linear dimension line containing dimensions for each data point entered. The Dimension Location (Stacked) tool, on the other hand, creates a new dimension line and text for each additional data point. The following illustration shows the operation of each tool.

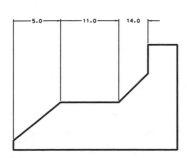

Dimension Location

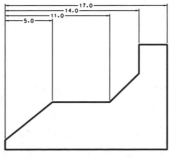

Dimension Location (stacked)

The results of the Dimension Location tools. Note the different appearances between the Stacked version and the normal version of this tool.

Before proceeding with more dimensioning tools, try your hand at creating some dimensions. In exercise 10-1, which follows, you will be using the widget drawing you created in a previous chapter. This time, however, you will document it with dimensions. At the conclusion of this exercise, and those that follow, you will find out if your previous work was accurate.

Editing a Dimension's Text

When you place a dimension, you have the option of modifying the text string associated with it. This is done just after you have established the endpoint of the dimension. If you look at the status bar after picking the endpoint of a dimension, you will notice the message "Press Return to edit dimension text."

After pressing the Enter key (just make sure the focus is on the Key-in window when you press Enter), you are presented with a dialog box, shown in the following illustration, in which you can change the values of the text or add to them as needed. Why would you want to do this? The answer is simple. In many instances you will want to add suffixes such as REF or TYP. This feature allows you to do this without compromising the measurement text of the dimension.

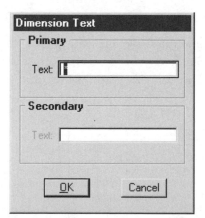

Pressing the Enter (Return) key at the right time in the dimensioning process will bring up this dialog box. Here you can append to or change the text of the dimension. The asterisk represents the dimension value.

NOTE: *The Key-in window must be the active window for this feature to work. If the Tool Settings window is active, the result will be the selection of one of its options. Use the Utilities menu to open the Key-in window. If the Tool Settings window has the focus, and the Key-in window is open, pressing the Esc key makes the Key-in window active.*

TIP: *You can also use MicroStation's Edit Text tool to change a dimension's text string. When you select a dimension's text with this tool, the dialog box previously shown is displayed. You can replace the auto-generated dimension (designated by the asterisk in the text field) with any text, or prepend/append text to the dimension string. If at any time you want the dimension text to display its true measurement value, simply put an asterisk back in the text string.*

EXERCISE 10-1: LINEAR DIMENSIONING A MECHANICAL PART

1   Open *BRACKET.DGN*. To get started, you need to select the default dimensioning settings.

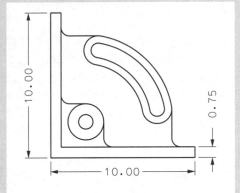

2   Open the Dimension Settings window (Element menu > Dimensions). Set/ verify the Dimension Lines settings to the following values.

- Dimension Lines category: Level enabled, set to 50. Color Attribute enabled, set to 3 (red)

- Terminators category: Arrowhead set to Filled

- Text category: Orientation set to In Line

- Units category: Format set to Mechanical, Units to English, Accuracy to 0.12

3   Select the Dimension Size with Arrows tool (Dimensions tool frame > Linear Dimensions toolbox > Dimension Size with Arrows). Because dimensioning tools have specific steps to follow, you should pay close attention to the prompts in the status bar. At this point, the status bar should be prompting you to "Select start of dimension." You will be placing the horizontal (*x*) 10-inch dimension along the bottom edge of the bracket, as shown at right.

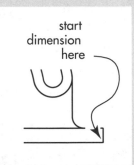

Placing the horizontal dimension.

4   Select the lower right corner of the bracket with a tentative point/data point. The message "Define length of extension line" appears in MicroStation's status bar. You must select the location for the dimension line and text. This would be where the 10-inch dimension's rightmost arrowhead is located. This location does not need to be precise.

 NOTE: *To avoid confusing the dimension command as to which direction your dimension string is going to face, always pick the point in line with the starting point of the dimension.*

5   Place a data point below the bracket's lower right corner. The dynamic display of the dimension x appears. The message "Select dimension endpoint" appears in MicroStation's status bar.

snap to corner

At this point you need to "turn the corner" to continue dimensioning the left edge of the bracket.

6   Place a tentative point/data point at the lower left corner of the bracket's base, as shown at right. The Dimension Size tool will prompt you to place more dimensions along the same line.

7   Press Enter once. MicroStation prompts you for the new dimension line and text string location.

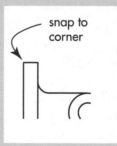

snap to corner

Vertical dimension placed.

8   Place a data point just left of the vertical face of the bracket at the location you want the vertical dimension line to run.

9   Place a tentative point/data point on the top left corner of the bracket. This places the vertical dimension, as shown at right.

The last dimension to be placed is the 0.75-inch one in the lower right corner of the drawing. The problem is that the dimension text is too large to fit within the space to be dimensioned. This is a case in which you need to intercede in the dimensioning process and manually place the dimension text. To do this, you need to turn Dimension Text Location to Manual.

10 Returning to the Dimension Settings window (Element > Dimensions > Text category), set the Location option to Manual, as shown in the following illustration.

Location option set to Manual.

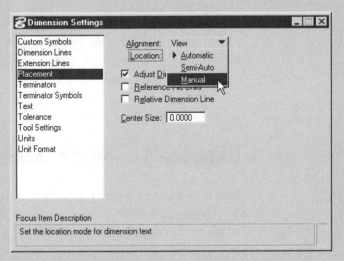

11 Select the Dimension Size tool again. If the tool is still active from the last dimension placement, you must press Enter twice to return the tool's operation to its starting point. The message "Select start of dimension" is displayed in MicroStation's status bar.

12 Place a tentative point/data point on the lower right corner of the bracket. Next, you will define the location of the dimension line and text string.

13 Place a data point just to the right of the previous data point. Remember that this step sets how far away from your design the dimension string will appear.

14 Define the endpoint for the dimension by selecting the top edge of the bracket base with a tentative point/data point, as shown at right. The message "Place dimensioning text" appears in MicroStation's status bar.

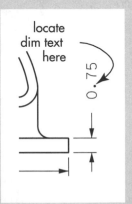

Defining the dimension's endpoint.

15 Place a data point either above or below the dynamic dimension. You will not have room between the extension lines, so pick a point outside the dimension.

16 Press Enter twice to complete the dimension placement.

This simple exercise showed you one method for placing dimensions around your drawing. As you can see, you need to be aware of how the various dimension options are set. If your results were not precisely the same as shown, it is probably due to an incorrect setting of these values. You might want to try some variations to the values in the Dimension Settings box and see how they affect your dimension placement commands.

Using the Association Lock

You may have noticed the presence of a checkbox option on the Tool Settings window called Association Lock. When enabled, a tentative point/data point sequence attaches the dimension to whatever element was highlighted as part of the tentative point. If, in the future, this element is moved, or the selected point is modified, the dimension associated with the point is automatically updated to reflect the change, as indicated in the following illustration.

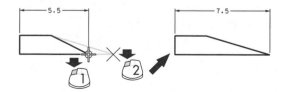

The Association Lock adds a degree of intelligence to a dimension by establishing a link between it and a specific design element. In this example, Modify Element was used to stretch the tip of this shape, resulting in the dimension change.

NOTE: *If you decide to use associative dimensioning, be sure to use it throughout your design. Later, you will assume all dimensions in a drawing are associative; thus, mixing non-associative dimensions may lead to faulty results.*

Angular Dimension Tools

Another type of dimension is the angular dimension. This dimensioning tool type documents the angle (or "sweep") of arcs. The sections that follow describe the angular dimensioning tools.

Angular Dimensioning Setup

As with linear dimensioning, you need to set up some parameters before you can begin using the angular dimensioning tools. In the Dimension Settings window, you set the angular readout settings in the Unit Format category. For Units, you have the following two choices.

- *Degrees*: Show angle measurements in degrees

- *Length*: Show angle's arc-length value

If you choose units of degrees, you can further decide to show the angle in decimal degrees (more properly known as sexagesimal), degrees-minutes-seconds, or centesimal (right angle divided into 100 equal parts or grads, only available in MicroStation SE or newer) by selecting the appropriate setting from the Display option menu. The angle format control of the Dimension Settings window is shown in the following illustration.

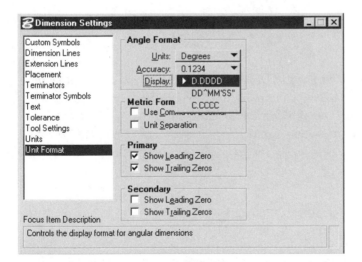

The angle format used with the Angular dimensioning tools is controlled via the Dimension Settings window. You select both the format and measurement technique (degrees or arc length).

MicroStation also lets you specify the number of places to display after the decimal point. You can set between zero and four decimal places. Both the Show Leading Zero and Show Trailing Zeros checkboxes also affect the appearance of angular dimensions.

Dimension Angle Size

The Dimension Angle Size tool, shown at left, does for angles what the Dimension Size with Arrows/Strokes did for linear distances. As indicated in the following illustration, the tool prompts you for a starting point, the extension line location, the center-point for the angle, and the endpoint of the angle.

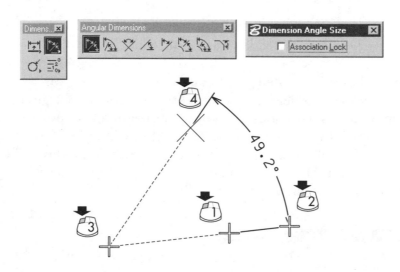

The sequence of data points required by the Dimension Angle Size tool.

NOTE: *First-time users of Dimension Angle tools are confused by the order in which the program prompts for the location of various points. Getting the order wrong sometimes results in unexpected results (usually a large, circular dimension). Keep an eye on the prompts in the status bar and follow them. If you get confused, just press Enter a few times and start over.*

Dimension Angle Location

The Dimension Angle Location tool, shown at left, works much like its linear cousin. You select a starting point, dimension line location, centerpoint, and then a series of endpoints. Each dimension will be stacked one above another.

Dimension Angle Between Lines

The Dimension Angle Between Lines tool, shown at left, takes some of the drudgery out of dimensioning angles. By selecting the two elements between which you want a dimension, MicroStation skips the requests for centerpoint and starting/ending points. All you have to do is supply the location of the dimension.

Dimension Angle from X-Axis and Dimension Angle from Y-Axis

The "dimension from axes" tools, shown at left, are a slight modification of the previous Angle Between Lines tool. Instead of selecting two elements, you identify only one. MicroStation then calculates the angle from that line to the axis you specified by the specific tool you selected.

Radial Dimension Tools

Angles and arc lengths are not the only dimensions of interest when detailing arcs and circles on a drawing. Many times you need to provide diameters and radius information. MicroStation provides another set of dimensioning tools (Dimensioning tool frame > Radial Dimensions) to address radial dimension requirements. The sections that follow describe the radial dimensioning tools.

Dimension Radial

The primary radial dimensioning tool available in earlier versions of MicroStation was Dimension Radial, shown at left. This single tool could be used to generate most radial-style dimensions by simply selecting the appropriate mode. Recognizing that in some instances it is more efficient to provide separate tools to access specific radial dimensioning modes, the most current release of MicroStation provides discrete tools for each radial dimension, as shown in the following illustration. This is in addition to the all-purpose Dimension Radial tool. In the following Dimension Radial tool description, the related discrete tool is included.

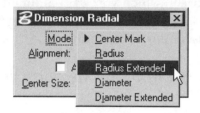

The Mode button lets you select a radial dimensioning option.

Dimension Diameter and Dimension Diameter (Extended Leader)

These two tools, shown at left, provide a means of dimensioning circles by their diameter. By selecting the target circle, you are prompted for the final location of the dimension text. Depending on the Location setting on the Text category in the Dimension Settings box, this may take one or more data points, as indicated in the following illustration.

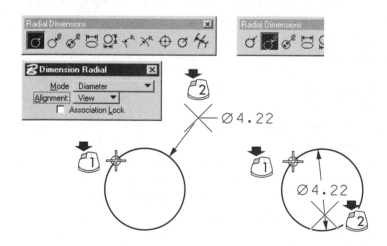

The Dimension Diameter tool in action. By setting the Text Orientation to Horizontal in the Dimension Placement settings box, the resulting dimension always appears level. Note how the radial dimension's final appearance is highly dependent on the location of your second data point.

A final Enter creates the dimension. By default, the dimension text includes the diameter symbol, as shown in the following illustration.

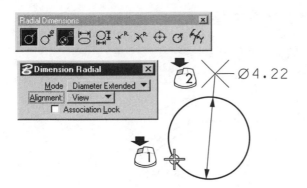

The Dimension Diameter Extension mode changes the dimension line style to incorporate two arrows across the diameter of the target circle.

Dimension Radius and Dimension Radius (Extended)

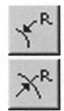

The Dimension Radius tool, shown at left (top), gives you a leader line, with text denoting the radius of the arc or circle selected. The Dimension Radius (Extended) tool, shown at left (bottom), gives you the same leader and text, except that it also extends the leader to the center of the arc or circle.

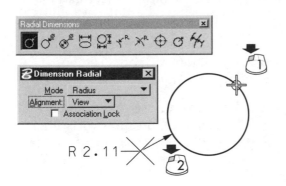

The Dimension Radius tool behaves very similar to Dimension Diameter.

Place Center Mark

The Place Center Mark tool, shown at left, prompts you to select an arc or circle and proceeds to place a center mark at the center of the selected element. The tool setting's Center Size field lets you control the size of the center mark placed by the tool. In exercise 10-2, which follows, you will practice adding an angular dimension, using the bracket drawing of exercise 10-1.

EXERCISE 10-2: DIMENSIONING THE BRACKET DRAWING

Add an angular dimension to the bracket design by performing the following steps.

1   Continuing with
    BRACKET.DGN, open the
    Dimension Settings window
    (Element menu > Dimen-
    sions). Under the Unit For-
    mat category, set the Angle
    Format values as follows.

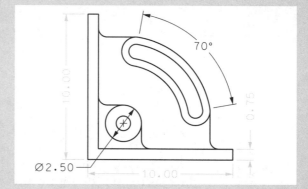

- Units: Degrees

- Accuracy: 0

- Display: D.DDDD

2   Select the Dimension Angle Size tool from the Angular Dimensions tool-
    box. MicroStation prompts you with "Dimension Angle Size > Select start of
    dimension" in the status bar.

3   Snap to the lower right endpoint of the major arc in the design with a tenta-
    tive point/data point sequence. MicroStation prompts you with "Dimen-
    sion Angle Size > Define length of extension line" in the status bar.

4   Select a point to the outside of the arcs (to the right), where you want to
    place the dimension line with a data point. MicroStation prompts you with
    "Dimension Angle Size > Enter point on Axis" in the status bar.

5   Snap to the centerpoint of the arc (the pivot point of the bracket) with a
    tentative point/data point. The angular dimension appears in dynamics.
    All that is needed is the final endpoint to tie the dimension down.

6   Snap to the opposite end of the arc angle with a tentative point/data point.
    The 70-degree dimension appears.

    If you were dimensioning several arc features, you could continue to select
    additional endpoints and MicroStation would add the arc dimensions to
    the 70-degree one just placed. However, you need to move on and dimen-
    sion the diameter of the bracket's pivot shoulder (the outer circle that
    shares the centerpoint of the arc you just dimensioned). This is accom-
    plished using the Dimension Radial tool. Proceed with the following steps.

7   Select the Dimension Radial tool from the Radial Dimensions toolbox. Select the Diameter Extended mode in the tool settings window. MicroStation prompts "Dimension Diameter (Extended Leader) > Identify Element" in the status bar.

8   Identify the larger of the two concentric circles with a data point. MicroStation prompts "Dimension Diameter (Extended Leader) > Select Dimension Endpoint" in the status bar, and displays the dimension in dynamics mode, as shown in the following illustration.

9   Place a data point just to the left and below the bracket. The dimension appears, complete with the diameter symbol prepended to the dimension value.

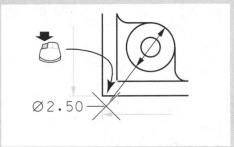

10   Change the Dimension Radial tool setting Mode to Center Mark. MicroStation prompts "Place Center Mark > Identify Element" in the status bar.

Dimension displayed in dynamics mode.

11   Identify one of the pivot point circles with a data point. A centerpoint marker appears at the pivot point (and the object's circles and arcs shared centerpoint).

If you want, you can continue to use the dimensioning tools just discussed to further annotate this drawing.

Additional Dimensioning Tools

MicroStation organizes most of its dimensioning tools in the three categories of Linear, Angular, and Radial dimensions, each with its own toolbox. In addition, a general-purpose toolbox is available directly from the Main tool frame. This toolbox, named Dimension (see following illustration), contains the most popular dimensions from the other dimension toolboxes.

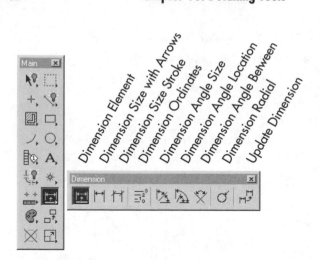

The Dimension toolbox is a collection of the several types of dimensioning tools, including several special-purpose tools.

Dimension Element

Now that you have experienced several of MicroStation's dimensioning tools, you can now utilize the Dimension Element tool, shown at left. The "Swiss army knife" of dimensioning, Dimension Element provides access to the major types of dimensions with a press of the Enter key. First, Dimension Element prompts you to select an element for dimensioning.

Once an element has been selected, you have the option to choose the type of dimension you want generated. This is done by selecting the Next button, or by pressing either the Enter key or the space bar on your keyboard. Dimension Element will step through the types of dimensions available for the element chosen. The illustrations that follow show the Dimension Element options for linear and circular elements.

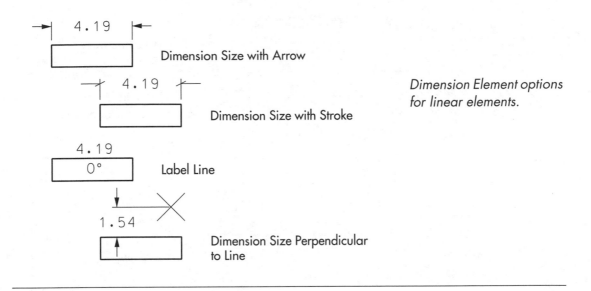

Dimension Size with Arrow

Dimension Size with Stroke

Label Line

Dimension Size Perpendicular to Line

Dimension Element options for linear elements.

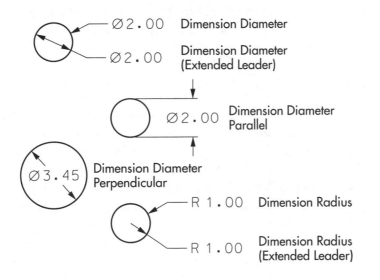

Dimension Diameter

Dimension Diameter (Extended Leader)

Dimension Diameter Parallel

Dimension Diameter Perpendicular

Dimension Radius

Dimension Radius (Extended Leader)

Dimension Element options for circular elements.

Dimension Ordinates

An alternative to the linear dimensions discussed, the Dimension Ordinates tool, shown at left, labels distances as values along a specified axis, as indicated in the following illustration. The result is a simple annotation of distances from a fixed datum. This is quite effective in some design disciplines in which the drawing would otherwise be overwhelmed with dimensions.

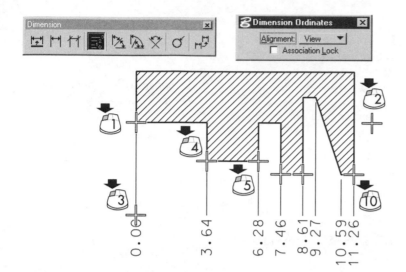

The second data point in Dimension Ordinates is used to set the major axis to be dimensioned; in this case, the horizontal axis.

This tool is available from both the Dimension toolbox (Main tool frame > Dimension) and in the Misc Dimensions toolbox (Dimensions tool frame > Misc Dimensions).

Label Line

A simple dimensioning tool, Label Line, shown at left, is located on the Misc Dimensions toolbox, or as one of the options of the Dimension Element tool. It is particularly useful to civil engineers, who need to work with bearings and lengths of lines. When you select Label Line, MicroStation prompts you to select a line. Once identified, two strings of text are displayed along it: the text above the line representing its length, and the text below the line representing its angle. You accept this dynamic display with a data point to place the label for the line. If Dimension Text Location is set to Manual, you must provide an extra data point to locate the position of the text.

Modifying Dimensions

Once you have placed your dimensions, you may need to modify them. As mentioned earlier, MicroStation treats dimensions as fundamental (primitive) elements. This means that the various tools provided to modify such elements work just as well on dimension elements. The secret to their use is where you select the dimension for modification. The sections that follow explore some of these tools and situations.

Modify Element

Used to change the vertices of lines and the diameters of circles, the Modify Element tool can be used to change key features of a previously placed dimension. Table 10-3, which follows, describes the effect of selecting various dimension components using Modify Element.

Table 10-3: Effect of Selecting Specific Parts of a Dimension with Modify Element

Dimension Component Selected	Modify Element's Effect
Extension Line	Modifies the endpoint of the dimension
Dimension Line	Repositions the dimension line and text without affecting the endpoints
Dimension Text	Repositions the text along the dimension line

Insert Vertex

Selecting an extension line of a dimension with Insert Vertex results in a dimension being appended to the dimension string. Selecting a dimension along its dimension line results in the dimension being divided into two separate segments. In either case, your second data point defines the location of the new extension line.

Delete Vertex

The inverse of the Insert Vertex tool, Delete Vertex removes an internal dimension of a dimension string and recreates a single dimension of the combined values. It should be noted that both vertex modification tools work even if the chosen dimension was placed with the association lock turned on.

Changing a Dimension's Settings

As with other elements in MicroStation, dimensions may need to have their parameters adjusted after they have been placed in the design. Fortunately, there are two tools to help accomplish this. These tools are described in the sections that follow.

Match Dimension Attributes

Due to the wealth of options associated with each dimension, when you are ready to change a dimension's parameters it is best to first set the current dimension settings to the target dimension. In this way, you can change only those parameters you need without affecting other settings. To do this, use the Match Dimension Attributes tool (Tools > Match > Match Dimension), shown at left. This sets the current dimension settings to match those of the chosen element. Through its SmartMatch tool on the Change Attributes toolbox, MicroStation 95 offers yet another way of matching these settings.

Change Dimension to Active Settings

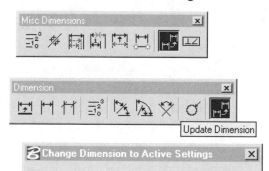

Once you have adjusted the parameters the way you want them, you use the Change Dimension to Active Settings tool (Dimension tool frame > Change Dimension to Active Settings), shown at left, to identify those dimension elements you want updated to the current dimension settings.

Saving Dimension Settings

There is no question that MicroStation's dimensioning system requires close watch of many parameters. It gets even more difficult when you try to maintain uniformity across multiple drawings. Some would say it is impossible. Fortunately, there is a way to save your dimension settings and reuse them in all of your drawings.

A special facility within MicroStation called the Settings Manager (Settings menu > Manage) is designed to capture sets of complicated tools settings such as those associated with dimensioning (multi-line settings are the other major category). These settings can then be saved to an external file, where you can then access them from any design file.

NOTE: *The Settings Manager and its functions are not for the faint of heart. It was designed to enforce a company's design standards. As such, it supports more than just dimension settings. As a result, the procedure for using it can be somewhat confusing. However, if you keep in mind that you are only using one small part of this facility, you should be able to follow its operation.*

Accessing Settings Groups

When you open the Settings Manager window (Settings menu > Manage), the name of the current settings file is displayed in the Select Settings title bar. In MicroStation's default workspace, the name of this file is *STYLES.STG*. Looking at the settings box, shown in the following illustration, you will see two major sections: Group and Component.

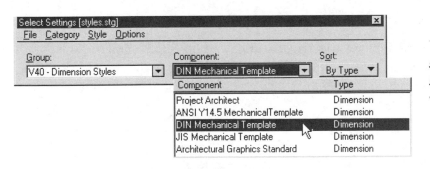

The Settings Manager showing the default settings associated with STYLES.STG.

You can see that there are two distinct groups, whose content is readily identified by their names: dimensions and multi-lines. As you may remember, multi-lines also suffer from too many parameters to be manipulated, and are thus good candidates for the settings file.

With the V40 - Dimension Styles group selected, you will see a number of items listed in the Component Section of the settings box. These components are a variety of dimension settings that have been saved under descriptive names for recall at a later time. Here you can see there are dimension settings for ANSI Y14.5, DIN, JIS, and Architectural. To set your active dimension settings to any one of these, all you have to do is click on the one of interest, but not just yet.

When you select a component from the Components list, the settings associated with that component will override your current settings. This has a major impact on configuring your dimension settings. You can have several components set up for different types of dimensions. You could have one component for detailed, four-decimal-places accuracy, and another for a "reference purposes only" dimension.

Managing the Settings Manager: An Exercise

Managing the Settings Manager is a subject worthy of a chapter in itself, which is beyond the scope of this book. However, if you want to try your hand at creating your own settings file and component, go ahead and try exercise 10-3, which follows.

EXERCISE 10-3: CREATING YOUR OWN DIMENSION COMPONENT

1   Open *BRACKET.DGN*.

2   Open the Settings Manager window (Settings > Manage). By default, the content of *STYLES.STG* will appear. To avoid modifying this system file and potentially upsetting your CAD administrator, you should create your own settings file.

3   From the Settings Manager menu bar, select File > Edit. The Edit Settings dialog box appears. Here you can modify existing settings components or create your own.

4   From the Edit Settings dialog box, select File > New. In the Create New Settings File dialog box, shown in the following illustration, enter the name *INSIDEMS.stg* and click on OK. You now have an empty settings file to work with. Next you need to create a "group" for organizing your various components within the settings file. A group can be all dimension settings components, or a set of related standards (Civil group or section detail group, and

so on), or whatever other organizational grouping you want to use. For this exercise, keep is simple; name the group Dimensions.

Create New Settings File dialog box.

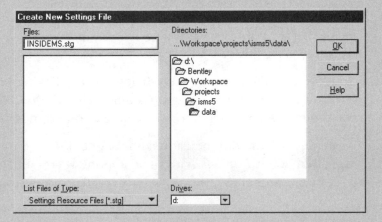

5   Select Edit > Create > Group. An "Unnamed" group appears in the middle display pane. Next you need to give it a name.

6   With the "Unnamed" group highlighted, select the name field in the middle of the dialog box (just beneath the Group list field). Type in Dimensions and press Enter. The Unnamed changes to Dimensions. Next, you will add a dimension "style" to this group. A style is the actual dimension settings captured from your current active design session. Styles are useful because you can have a few standard dimension styles that are accessed by many more dimension components.

7   Select Style menu > Dimension. The Edit Dimension Styles dialog box appears, shown in the following illustration. First you need to "capture" the current dimension settings.

Edit Dimension Styles dialog box.

8 Click on the Get Active button. The dimension "style" named *Dim0* appears in the list. As earlier, you will need to rename the style and give it a description.

9 Double click on the *Dim0* text in the Name field, change it to *Linear,* and press Enter. Enter a description such as *Default Linear Dimension values.* Note that there is no OK or Cancel button on this dialog box.

10 Close the Edit Dimension Styles dialog box by clicking on the Close Window icon [X] in the upper right corner of the window. With a dimension style established, you need to create at least one dimension component.

11 Create an Unnamed dimension component (Edit menu > Create > Dimension). Rename it *Linear1.* With the component created, all that remains is to set its parameters to use your newly created dimension style.

12 Double click on the *Linear1* component. The Modify [Linear1] dialog box appears, which is shown in the following illustration. Here, you set all of the parameters associated with this specific component. Note how you can even override the active element attributes or even invoke special commands (the key-in field). For this exercise, you will just set the dimension style.

Modify dialog box.

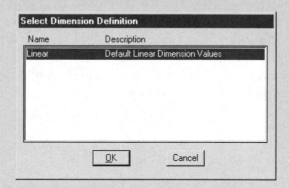

13 Click on the Select button in the Dimension Setting section of the dialog box and select the Linear dimension style you created earlier (click on the OK button!).

14 In the Modify [Linear1] dialog box, click on the Save button to save your changes. At this point, you have one component defined. You can now use this component to reset your dimension settings to the values captured in your Linear dimension style.

Obviously, you would need to create several dimension styles and components to flesh out your system. This exercise simply showed you the steps involved in creating your own settings file and components. This procedure can also be used to generate specific settings for multi-lines, cells, text, patterns, and other features specific to your drawing standards. In fact, Settings Manager can also be used to establish working units in new design files and even drawing "scales" as they apply to final plots.

Area Patterning

In addition to placing dimensions on your drawing, the other typical detailing operations in regard to drawings are applying hatching and crosshatching patterns or otherwise applying patterns to specific features of your design. For instance, in a wall section detail, you show insulation batts as curves, in a topo map you show marshy areas with a grassy detail, and in a cross section of a machine part you show hatch patterns to represent the different types of materials from which the part is constructed. Creating these drawing details is referred to as *area patterning* in MicroStation. An example of the use of patterning is shown in the following illustration.

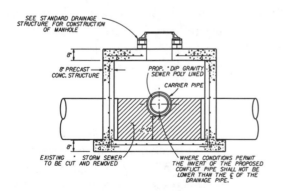

In this typical drawing detail, you can see the use of several drawing patterns representing everything from concrete aggregate to steel piping.

The Area Patterning Process

Area patterning takes a selected cell or linear line definition (hatch or crosshatch) and "fills" a defined area of the drawing with this "pattern." By trimming off those parts of the cell that lie outside the boundary (sort of like a Fence Clip command) and repeating the process row after row, the final effect is an area filled with drawing texture.

Obviously, there are several parameters associated with the patterning process. You must set the type of pattern to be used, the row and column spacing, and the angle at which you want the patterning to be applied. This is done with the Pattern Area tool settings, shown in the following illustration.

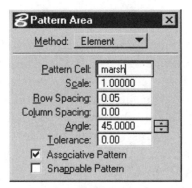

The Pattern Area tool settings illustrate the degree of control you have over the patterning process.

Once the patterning parameters have been established, the area to be patterned has to be defined, an example of which is shown in the following illustration. This is typically done by identifying a closed shape (circle, ellipse, complex shape, block, or shape element). In addition, there are powerful methods for specifying the area to be patterned without identifying a source element (more on this later). In most company operations, however, patterning is performed with a closed element.

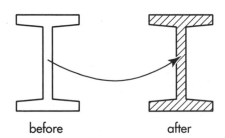

before   after

With all of the patterning parameters set, the appropriate patterning tool and "target" area will result in a finished pattern. In this case, the Hatch Area tool was used to pattern a shape representing a cross section of an I-beam.

Patterning Parameters

Each of the patterning tools provides a data field in the Tool Settings window for entering these critical patterning parameters. The patterning settings are described in table 10-4, which follows.

Table 10-4: Patterning Tool Settings

Setting	Description
Pattern Cell	Name of the cell to be used as the pattern
Scale	Scale to apply to the pattern cell during the pattern process
Row Spacing	Distance between each row of pattern cells (same as Spacing)
Column Spacing	Distance between each cell along a row
Angle	Angle of the pattern rows or hatch lines
Tolerance	How close to a curved element a pattern is calculated
Spacing	Distance between each hatch line

The Row and Column Spacing settings control how far apart each cell is placed with respect to each row and column. By controlling this inter-cell spacing, you can generate dense or sparse area fills with relatively simple patterns. The Hatch and Crosshatch Area tools use a related parameter simply called "Spacing." Because the lines used to hatch and crosshatch are continuous, there is no "column" spacing per se, thus the parameter name change. To paraphrase: Spacing is spacing.

The parameters that most affect the outcome of the patterning process are the Spacing and Angle. In the previous illustration, you can see how the patterning process repeats in a specific direction. When the patterning process encounters an obstacle (the edge of the pattern element), it clips the current pattern occurrence and resets the pattern spacing distance, and starts the process over.

The distance between the cells is expressed in the actual distance between one cell's extents (the area of the design file it occupies) and the next cell's extents. In this way, you do not need to know how "fat" a cell is to specify the spacing. An example of the manipulation of cells is shown in the following illustration.

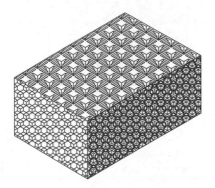

Using the Pattern Area command is a time-consuming process, especially if the pattern cell is complicated. The cells used in this fanciful example are from the GEOMPAT.CEL cell library delivered with MicroStation.

The Tolerance setting may be a bit confusing. It refers to a mathematical dilemma. Because a curved element's (circle, arc, curve) edge is only an approximation of the curve geometry (thanks to pi) you must tell MicroStation how fine to calculate the cutoff point for the pattern. The finer the tolerance, the longer it takes to complete the pattern.

Associative Pattern

Pattern association "connects" the resulting pattern with the source boundary element. This means that if you change the boundary element later, the area pattern will automatically update to reflect this change. This works the same way as associated dimensions. With pattern association turned on, you can easily modify patterned elements without a second thought. Because MicroStation stores the pattern with intelligence, any changes to the underlying master element will result in a recomputation of the pattern, as indicated in the following illustration. Associative patterns also take up much less space.

Snappable Pattern

Another feature of patterns is snappability. This refers to the ability to use the tentative point snap with a pattern. If selected at the time you pattern an element, MicroStation will allow future tentative points to find the pattern elements.

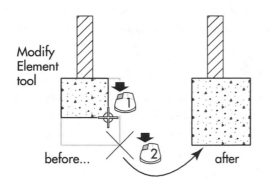

Modify
Element
tool

before...   after

When you modify an element with associative patterning, the pattern takes care of itself.

On the face of it, this sounds good. However, there are many times you do not want to snap to a pattern. In fact, if there are small elements within the pattern close to a desired vertex, the result may be an inadvertent snap to said pattern components. This option should be used with great care.

Pattern Methods

In MicroStation you are not limited to just patterning a closed element or shape; you can use Boolean operations on closed elements to define a virtual area that should be patterned. Table 10-5, which follows, describes various patterning options.

Table 10-5: Patterning Methods

Setting	Description
Element	Pattern a selected closed element
Fence	Pattern the current fence
Intersection	Pattern the intersection area of two or more closed elements
Union	Pattern the combined area to selected closed elements
Difference	Pattern area of element 1 less area of element 2
Flood	Find an area enclosed by elements
Points	Pattern area enclosed by a series of user-entered data points

If these methods sound familiar, it is probably because they are the same ones described when automatically creating a complex shape.

Area Patterning Tools

MicroStation's patterning tools are accessed from the Patterns toolbox, which can also be torn away from the Main toolbox (Tools menu > Main > Patterns). The sections that follow examine the tools you use to create patterns. The Patterns toolbox is shown in the following illustration.

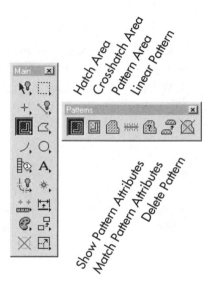

Patterns toolbox.

Hatch Area

Hatch Area, shown at left, is the simplest of the area pattern tools. Instead of a user-defined cell, a line is repeated at the specified angle and offset to fill an area, an example of which is shown in the following illustration. Hatch Area requires you to enter a Spacing and Angle in order to perform its function. In most instances, you will want to select the Associative Pattern.

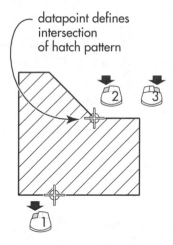

datapoint defines
intersection
of hatch pattern

*Setting the angle to
45 degrees results
in the hatched
pattern shown.*

Crosshatch Area

The major difference between the Crosshatch Area tool, shown at left, and the Hatch Area tool is the addition of the second spacing and angle fields. This is necessary to set the angle of the second set of hatch lines, an example of which is shown in the following illustration. If you leave the second fields blank, MicroStation defaults to a right-angle pattern to the first angle at the same spacing.

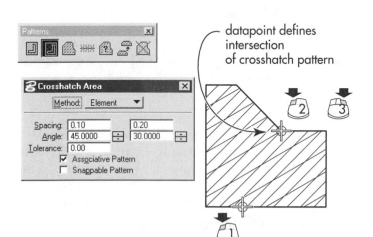

datapoint defines
intersection
of crosshatch pattern

*Varying the value of
the two hatches
changes the final
crosshatch
appearance.*

NOTE: *If you pattern an element that was previously patterned, the new pattern replaces the previous one. This is handy when experimenting with the settings.*

Pattern Area

The workhorse area-patterning tool, Pattern Area (shown at left), uses a shared cell as its pattern element, as indicated in the following illustration. Using the Scale, Angle, Row, and Column spacing, the Pattern Area tool computes a final pattern shape based on the method chosen.

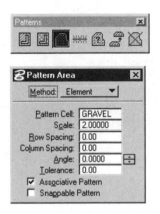

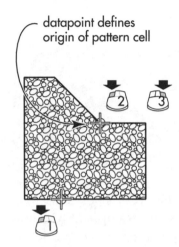

datapoint defines origin of pattern cell

The Pattern Area tool relies on the cell definition to generate the pattern's content. Note the use of a pattern scale to enlarge the results from the default cell's normal size.

Punching Holes in a Pattern

One of the typical situations that arises with patterning is the creation of voids within the pattern's field. Take a plate, for instance, drilled with holes. How do you pattern the plate but leave the holes clear?

There are two methods for performing this feat. In both cases, however, the result is a static pattern (one that is not easily edited). This harks back to the less convenient days of patterning (version 3.x and IGDS).

First Method: Patterning by Difference

The first method involves the difference method. With this method chosen, select the main element in which the holes are to be "drilled." Proceed with the following steps.

1   Select the Pattern Area tool and set the Method to Difference.

2   Identify the main object with a data point.

3   Identify each element representing a hole in the main object.

4   Data point once, clear of any element.

5   Press Enter once to create the pattern.

If you take a close look at the result, you may be surprised at what you find. Your original outline is still a separate entity, but all of the elements you chose as holes, a copy of the outline element, and the pattern itself have been turned into a single orphan cell. This happens regardless of the alignment setting.

Second Method: Create a Group Hole Cell

This act of creating an orphan cell is similar to the other method for patterning with voids. Found under the Groups toolbox, the Group Hole tool lets you manually create the orphan cell before patterning it. The procedure follows.

1   Select the Group Hole tool (Tools > Main > Groups > Group Hole).

2   Identify the main object with a data point.

3   Identify each element representing a hole in the main object.

4   Data point once, clear of any element.

5   Press Enter once to create the orphan cell.

6   Select the appropriate Patterning tool and use the Element Method to select the freshly created cell.

V8: MicroStation version 8 will introduce true associative patterning with "flood fill." This means that you will be able to move the interior elements and the pattern will automatically update!

Line Patterning Versus Custom Line Styles

Before you tackle linear patterning, some differences between linear patterning and custom line styles should be discussed. Prior to version 5 of MicroStation, linear patterning was the only tool available for creating a custom line style beyond the eight standard line styles (*LC=0-7*). With the advent of custom line styles, this is no longer true. You can create and use complex custom line styles. Although the process of creating such line styles is a little tricky, their use is not.

By selecting a line style from the Line Styles settings box (Primary Tool Box > Line Style > Custom) at the time you place the original element, you do not need to go back and pattern it. Because these custom line styles are stored external to your design file, it also provides greater consistency between drawings than with linear patterning.

However, due to historical reasons (custom line styles were implemented for the first time in Version 5), you may still encounter in your day-to-day operations files that use linear patterning. For this reason, you should be familiar with this tool.

More Pattern Features

Basically, that is all there is to patterning. However, recognizing the complex nature of the patterning function, MicroStation is equipped with a set of helpful tools and commands that make working with patterns a little easier. These are discussed in the sections that follow.

Turning the Pattern Display On and Off

The patterning elements created by MicroStation include a special marker that provides a method for turning on and off their display without affecting other view attributes. This is very helpful when you have large areas of a drawing patterned with very complex patterns. The Patterns display option is located in the View Attributes window and can be toggled on and off just like any other display attribute, as indicated in the following illustration.

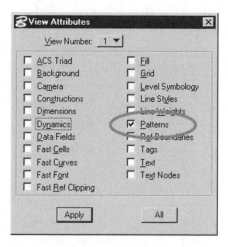

Toggling the Pattern attribute to off (removing the check mark, which indicates the active state) suppresses the display of all patterned elements in the selected view.

Show Pattern Attributes

The Show Pattern Attributes tool, shown at left, is used to identify a previously placed pattern element. Once a patterned element is selected, MicroStation displays the pattern parameters in the status bar.

Match Pattern Attributes

The Match Pattern Attributes tool, shown at left, carries the Show Pattern Attributes tool one step further. When you select a previously placed pattern element, its parameters become your active pattern parameters.

Delete Pattern

The Delete Pattern tool, shown at left, removes a previously placed pattern. In the case of grouped holes, however, it does not release the orphan cell. That requires the use of the Drop Complex Status tool.

Summary

With the dimensioning and patterning facilities just described, you can detail just about any drawing imaginable. There is a lot to these two facilities, and it will take time to sort them out. However, you can begin simply and work through these tools and procedures at your own pace.

Even if you have not mastered the dimensioning package, you can begin using portions of it to modify your design. The key here is to keep practicing with the tools. Eventually you will master the various complexities and become adept at their use.

"PUBLISHING" YOUR WORK

Producing Plotted, Printed, and Electronic Deliverables

EVERY DESIGN PROJECT FOR HIRE CULMINATES in a final transmittal to a client. In years gone by, the transmittal consisted entirely of final plots generated on paper (or Mylar) medium. Today, however, other forms of information transmittal are beginning to take hold, including electronic-only deliverables. "Plotting" your final design has been replaced with the encompassing lingo "publishing" your work.

With that said, however, generating old-fashioned paper plots is still the predominant deliverable with most projects. For this reason, you must master MicroStation's plotting system so that you can create the final deliverables required to complete your design project.

MicroStation has always provided a very potent output capability. As new plotters came on the market, Bentley Systems was always the first to provide full plotting support within MicroStation. As a result, MicroStation still provides a wide range of plotting device support, which can sometimes intimidate the first-time MicroStation user. In addition, MicroStation provides full support of the

Windows printing system, which can include everything from a low-cost inkjet printer to high-speed reprographics systems.

Overview of the Plotting Process

Although you could simply select the Print/Plot tool and click on OK to generate a hardcopy to the default printer on your system, the fact is that most hardcopy, commonly called "plots," requires some forethought and planning prior to and during the output generation process. The diagram shown in the following illustration outlines the major steps involved in generating "plot" output. Each step represented is described in the sections that follow.

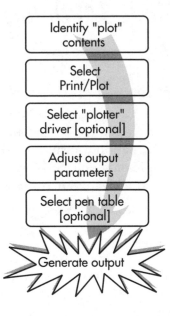

An overview of the steps involved in generating a typical "plot" from a design file.

Plot Versus "Plot"

You may have noticed that the word *plot* bore quotation marks in the caption to the previous illustration. That was done to point out that not all output generated from a design file is necessarily destined for a printer or plotter. In today's Internet economy, it is becoming more and more likely that some or all of your design data will be expressed in some form of web-friendly format.

This is both good and bad. Because plotters and web browsers require dramatically different types of data, there is the possibility that a hardcopy (printed or plotted) of a drawing can appear quite different from its web counterpart. However, that is not the case with MicroStation because both types of output are subject to the same "plotting" operation. Choosing the "plotter" driver is all that is needed to effect the proper data output format, as discussed in material to follow.

Identify "Plot" Content

Before you start the print/plot process you need to identify the portion of the active design file you want plotted. The following are the two distinct methods for doing this in MicroStation.

- By View content
- By Fence content

This means you must either carefully adjust your view shape and zoom scale, which is not always easy to do, or simply place a fence around the portion of the drawing you want plotted. Most users prefer the fence technique.

NOTE: *The view in which you place the fence is very important to the plot process. This view's display attributes are used to determine how to output ("plot") such features as* enter_data *fields, text node symbols, and even solid area fills and patterns.*

A fence has one additional advantage over the view. You can place any sort of fence, including a shape or even a circular one, and the plot will only plot the content of this fence. This gives you very fine control over what appears in your final plot, as indicated in the following illustration.

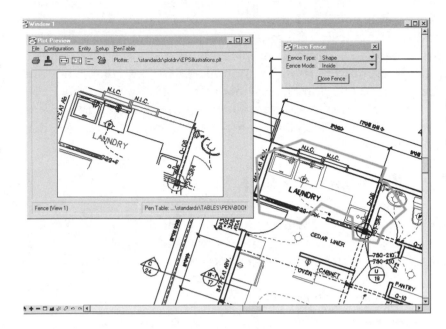

The fence in View 1 (highlighted in gray) dictates the content of the plot as seen in the Plot Preview pane.

Print/Plot Window

Once you have identified the content of your design to be plotted, the next step is to open the Plot window (File menu > Print/Plot). This window, shown in the following illustration, is the primary user interface to MicroStation's powerful plot utility (Batch Plot is the other). As mentioned earlier, the term *plot* is a misnomer, as you are not limited to simply generating pen plotter data. Depending on the driver you select, you can output all sorts of data formats from this single utility (more on this later).

TIP: *You can also open the Plot window by clicking on the Plot icon located in the Standard tool bar or press CTRL-P. You can also enter the command PLOT in MicroStation's Key-in window.*

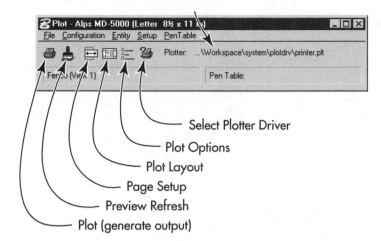

Select Plotter Driver

Plot Options

Plot Layout

Page Setup

Preview Refresh

Plot (generate output)

The entire plotting process is controlled from the Plot window. Note the name of the currently selected Windows printer in the window's title bar. This is a result of printer.plt *as the current plotter driver (also shown in the Plot window's Plotter Driver name field).*

The Plot Driver

As you can see, there are quite a few options associated with this dialog box, starting with the most important parameter, the plotter driver. By default, most users will use *printer.plt*, which directs MicroStation's plot utility to use the current Windows printer device. This provides direct access to the seemingly endless types of printers/plotters directly supported by the Windows operating system.

However, there are times when you may want to redirect MicroStation's plot output to somewhere other than a Windows print device. To select a different output device or file format, you open a special control file called a plot driver file (Plot window tool bar > Plot Driver icon or Setup menu > Driver command). The Select Driver dialog box is shown in the following illustration.

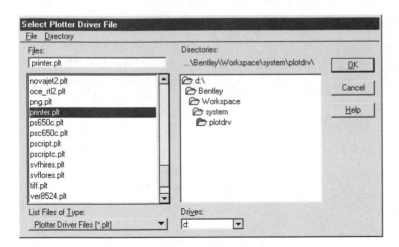

The Select Driver dialog box is used to set the type of plotter data MicroStation will generate. The printer.plt *selection shown is the interface to the Windows printer system.*

Alternate Output Devices and File Formats

MicroStation provides support for a wide variety of alternate output devices and file formats. Table 11-1, which follows, provides a summary of these plot devices and file formats.

Table 11-1: Plotters Supported by MicroStation

Plotter Language	Description
Calcomp 906	Calcomp's pen plotting language used in most of their pen plotter products.
Calcomp 907	Calcomp's second-generation plotting language. Used by many electro-static plotters, including Versatec.
Calcomp 960	Calcomp's original pen plotter language (predecessor of 906). Used with Calcomp's 960 family of belt-driven pen plotters.
DM/PL	Includes the Houston Instruments line of DMP plotters.
ESC/P	Epson dot matrix printer support.
HP-GL	Hewlett Packard's original pen plotter language. Supported HP's entire pen plotting product line.
HP-GL/2	HP's second-generation plotter language. Supported by their later pen plotter products and their DesignJet line of plotters.
PCL	HP's original LaserJet laser printer format.
RTL	HP's third-generation plotter language. Supported by most of their DesignJet line of plotters.

This is not an exhaustive list of plotter types supported by Micro-Station. The best source for this information is the "Printing and Plotting" guide available from the Bentley documentation web site (*http://docs.bentley.com*).

In addition to the plotter data formats just listed, MicroStation's plot utility can also directly generate several other data output formats, most of which are normally associated with publishing to the Web. The sections that follow describe some of the most commonly used formats.

JPEG (Joint Photographic Experts Group)

A common web format used primarily for continuous tone (photographic) imagery. JPEG is also known for its very apparent image degradation effects, especially when image quality is traded for smaller file size. JPEG also does not display well.

TIFF (Tagged Image Format)

Common publishing image format used for raster images. Started life as a fax data format but has expanded to become a de facto standard in the printing business.

PNG (Portable Network Graphics)

Newer web image format designed primarily to replace the GIF raster image format, originally developed by Compuserve. Its use is in slow decline due mostly to the ongoing enforcement of the LZW compression algorithm patent by the Unisys Corporation. LZW compression is at the heart of the GIF image format. Micro-Station does not generate GIF images.

Postscript

Developed and trademarked by Adobe, this common publishing format is used in printing books and documentation. Postscript output is supported by most publishing programs (Adobe Framemaker, Microsoft Word) as well as directly printable on any Postscript-ready printer/plotter. MicroStation also supports Encapsu-lated Postscript output, which includes a TIFF preview image embedded in the output file. This is especially useful in documentation layout to identify the content of a Postscript file while retaining the high quality of the original data.

SVF (Simple Vector File)

A CAD-related vector image format for the Web. Requires a special browser plug-in to view. The use of this format appears to be in decline due to the lack of support by other CAD vendors and the continued availability of only one source for the plug-in.

CGM (Computer Graphics Metafile)

General-purpose vector image graphics format used on the Web and as a data transfer format. This is also the data format used by Bentley's Java-based web viewer, delivered as part of ModelServer Publisher.

NOTE: *If* printer.plt *is not the current plotter driver and you want to direct your plot output to a Windows printer device, click on the Plotter Driver icon and select "printer.plt" from the list provided, as indicated in the following illustration. By default, MicroStation stores its list of supported plotter drivers in* \Bentley\Workspace\System\ Pltdrv.

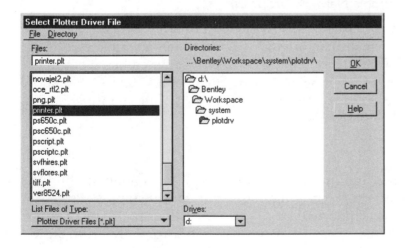

Select the printer.plt *driver to directly access the Windows operating system's printer resources.*

Adjust Plotting Parameters

The second most important step in generating plot output is the setting of various plotting parameters. Plotting parameters are organized into three separate categories, each with its own dialog box. These categories are described in the sections that follow.

Page Setup

If you have selected the *printer.plt* driver, the dialog box you see when you select Page Setup (Plot window > Setup menu > Page) will look very similar to the standard Windows printing dialog box you would use to print from a typical Windows application such as Word or Excel. This dialog box is shown in the following illustration.

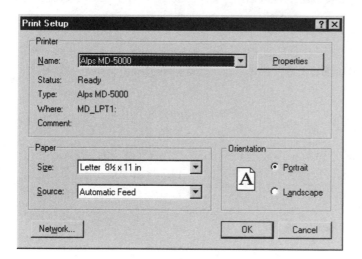

The Page Setup dialog box you see when printer.plt *is your plot driver.*

All of the information you see presented on this page comes from the Windows plot system, including the page sizes and orientation (note the portrait orientation selection in the previous illustration). If you have selected any of the plotter drivers for which MicroStation directly generates output data, the Page Setup dialog box looks quite different, as shown in the following illustration.

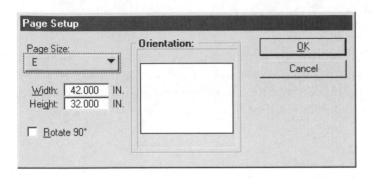

The Page Setup dialog box you see when a direct support plotter driver is selected. The values you see in the various fields will depend on the definitions in the selected plot driver file.

The Page Size field allows you to select your plot output size (each plot driver file can support up to five different page sizes). Selecting one of these page sizes results in the adjustment of the other fields in this dialog box. For instance, selecting the A page name results in a page size of 8.5 x 11 inches. Selecting the OK button changes the page size. Selecting Cancel ignores your selection and reverts to the previously set page size.

Also note the Rotate 90° option on this dialog. This works similarly to the Portrait/Landscape selection in the normal Windows Page setup dialog box.

Plot Layout

When you select the appropriate page name, you are also telling MicroStation the limits of the plotter. This will be important as you set the drawing's output scale. This is the primary purpose of the Plot Layout dialog box (Plot window > Setup > Layout), shown in the following illustration.

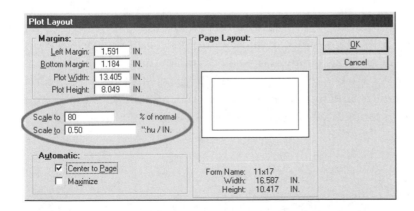

The Plot Layout dialog box is primarily used to set the scale of your output. The two fields highlighted control the scale output of your design.

Pay particular attention to the *Scale to* section of this dialog box. It is here you set the precise scale of your drawing. You can either set it as a percentage of the original "size" (as measured in your design file's master units) or as a "plotter unit to working unit" ratio. The percentage method operates much like any Windows application print setup. However, in the case of MicroStation, whenever you change the percentage field, the ratio field is automatically recomputed.

The ratio field works a bit different. In essence, this field operates according to the following formula, where you define X in the working units of the design file.

```
1 PLOTTER unit = <X> design file working units
```

When setting this ratio, you must take into consideration what your plotting units are (typically inches or centimeters). For instance, if your design file working units are set to feet and inches (a common U.S. architectural measurement), to plot at 1" (plotter paper) = 4' (design file), you simply enter *4* in the ratio field. However, if your design file working units are the same as your plotting units (say, centimeters) and you want to plot something twice normal size, you would enter *0.5* in the design file working units field, *not* 2. This is not a difficult procedure to understand, but it usually takes new users by surprise the first time the drawing comes out at 50%.

To add an extra wrinkle to how MicroStation computes the size of your plot, the current version of the plot utility does *not* let you specify a size larger than the maximum page size selected from the Page Setup dialog box. If you enter a value in either of the *Scale to* fields, MicroStation will simply display the maximum size allowed, period. This also throws first-time (and many experienced) users when it occurs the first time.

TIP: A very common and often confusing problem that occurs is when you try to plot a fixed-size drawing sheet (A-size or A4, for instance) that measures exactly the size of the page. Because most printers and plotters reserve a thin margin around the entire printed page, the true size of the page is some value less than the physical printed page. As a result, MicroStation will always scale your selected drawing to something smaller than what you expect. The solution is to select a smaller area of your drawing centered on your design but that does not exceed the page settings for your particular output device (see following illustration).

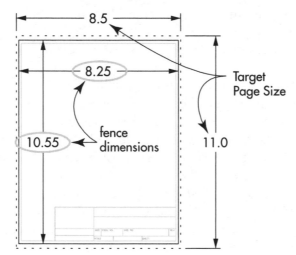

The fence for this A-size drawing is smaller than the actual A-size paper on which it will be printed. Note also the different amount of X and Y margins around the printed border. To plot properly, the Center on Page option must be selected.

TIP: To provide a uniform output every time, place a block shape with its vertices at the location where you want to place the fence. Use the From Element fence type to generate the fence before plotting in the future.

Another technique is to create a "crop line" cell as a point cell (you can only snap to the origin of a point cell, remember?) where a cut line is shown but does not actually occupy the fence corner (see following illustration).

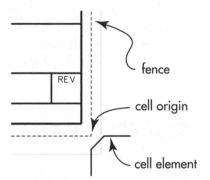

A special crop cell can be used to define a fence snap location by setting it to a point cell during the cell's creation.

Maximizing a Plot

To generate a not-to-scale plot that maximizes the use of the printed page, select the Maximize option. This results in the selected area of your drawing being plotted to the extents of the plotter's page size. This is also the default condition when you first enter the Plot window.

Centering a Plot

If you have not selected the Maximize option, you may opt to use the Center to Page checkbox. This option shifts the entire plotted area to the center of the chosen page size. This is noted by a box drawn on the plot layout frame. In many cases, the plot area you have selected does not use the same height-to-width ratio as the page size. Using the Center to Page option makes the final plot look more like a finished drawing, and may eliminate the need to trim the paper plot after the fact.

TIP: *For nice results on a typical laser printer, use the Center to Page option when plotting letter- or legal-size drawings.*

Plot Options

The final category of plotting parameters deals with additional plot options you may wish to enable or disable. When you open the Plot Options dialog box (Plot window > Setup menu > Options), most of the options shown cannot be changed using this dialog box, shown in the following illustration. Instead, these "options" reflect the current settings of the source view from which you are generating the plot. To change any of the settings displayed, you must exit the Plot Options dialog box and open the View Attributes window instead (Settings menu > View Attributes, or press Ctrl-+B).

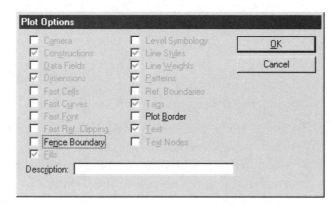

Plot Options dialog box.

There are, however, a few options you control from the Plot Options dialog box. Selecting the Draw Border option results in a box drawn around the extents of the plot (set as part of the page size). This is useful for trimming a drawing to its finished size. The Fence Boundary option plots the fence boundary, which can be smaller than the border.

The Description field allows you to enter a message (50 characters or fewer), which is then plotted just outside the border. This field is very helpful for date stamping or otherwise identifying the resulting plot, especially those "quick plots" everyone generates during the design process. If you decide not to use this feature, just press Enter. The support for this field depends on options set in the plotter driver file, so you may not have access to it.

Previewing Your Drawing Plot

Before actually generating your plot, you have the option of previewing its appearance in the Plot window. This is done via the Preview Refresh icon. When this button is selected, MicroStation actually generates a plot of your drawing, but instead of directing it to the plotter device, it uses the options you have selected along with specific plotter driver data to generate a "thumbnail" view of the plot, an example of which is shown in the following illustration.

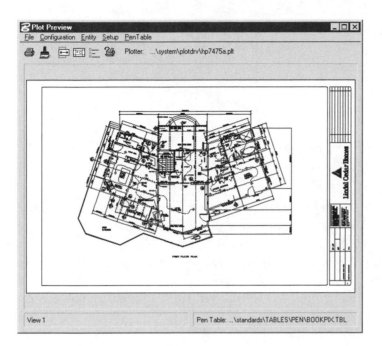

The Preview Refresh button results in a "heads up" preview of your plot prior to actually creating your plot file.

This is very helpful for seeing the effect of line weight definitions within the plot driver file and other key data. For instance, if you are using a single-color output, and your plot driver file has been set up, the result of this should be a black-and-white-only image of your plot.

TIP: *You can close the preview pane by either reselecting the Plot command (via icon or File menu) or by resizing the Plot window to "squeeze out" the preview pane.*

Plotting Your Drawing

With all of your parameters set, you are ready to actually generate some output. Clicking on the Plot icon (or selecting the Plot command from the Plot window's File menu) will initiate the data creation process (see following illustration).

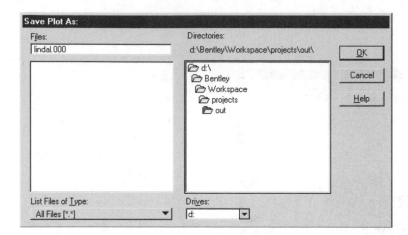

The Save Plot As dialog box appears in most instances when you select the Plot command from the Plot window.

If you have *printer.plt* selected as your printer driver, selecting the command creates and automatically sends the plot output to the Windows printer system. The only indication that anything has occurred is the message *<finished creating plot>*, which appears in MicroStation's status bar. At this point, your plot has been queued up to whatever printer you selected in the Page Setup dialog box.

If, on the other hand, you have a different printer driver selected, you will be prompted to provide a file name for the output data. The extension.000 is a common plot default extension associated with MicroStation's plotter system. It is defined in the plotter driver file, so certain data output types will use an extension more in line with the type of data output being generated (for instance, .jpg for JPEG.PLT).

In many companies, the procedure of first "spooling" the output to a semi-permanent file is still standard practice. This can be helpful for two reasons. First, if something were to go wrong with the plotter (something that happened fairly often with the old pen plotters), you could simply resend the plot data to the plotter.

Second, many companies archive the plot data at critical milestones as a de facto project "snapshot." Any time a question arose about a particular submittal, a quick plot from the archived plot data could provide the answer. With the advent of ProjectBank and other transaction management technologies (not to mention the maintenance overhead of dealing with those incredibly large plot files), this is no longer the only "insurance" a company has. Therefore, this practice is already in decline.

Sending Plot Files to a Plotter

One of the disadvantages to "spooling" your plot data to a discrete plot data file is the extra step required to actually generate the physical plot. How you do this is highly dependent on your computer and network setup. In the "old days," before the almost universal adoption of Windows as the corporate operating system, you could use a DOS utility delivered with MicroStation, called Plotfile. This utility would direct the content of the specified plot data file to the appropriate printer or serial port on your computer. Under Windows, you can direct your plot data to a given plotter, provided it has been properly set up to receive it. The following is a very generic description of how you perform this operation.

1 Open a Command Prompt Windows (sometimes referred to as a "DOS session").

2 Enter the following command.

```
copy /b <drv>:\dirname\plotdata.000\\
  plotservername\plottername
```

The *<drv>:\dirname\plotdata.000* is the plot data file you generated earlier using MicroStation's Plot utility. The *\\plotservername* is the computer to which the plotter is physically attached, and *plottername* is the name of the plotter as shared with its owner.

NOTE: *This plotter must be capable of receiving raw data as part of its configuration. Otherwise, your plotter data may be misinterpreted by the printer software installed on the plot server or your computer.*

For example, assume a company had an HP7585b pen plotter named *hppen1* connected to a computer named *\\plots*. To generate a plot from this pen plotter, you would first make sure to select the appropriate plot device (*HP7585b.plt* would be a good starting point) and generate the plot data (example: *myplot.000*) to a directory named *c:\bentley\workspace\projects\project1\out*. Once created, to send the file to the pen plotter, you would open the Command prompt window (Windows Start menu > Programs > Command Prompt) and issue the following command.

```
copy /b c:\bentley\workspace\projects\project1\out\
    myplot.000 \\plots\hppen1
```

At this point, the plot data is spooled to the specified device and (assuming it has the right paper loaded, pens are in the carousel, and the plotter is online) the drawing will begin to appear.

 NOTE: *Most companies have very specific procedures you must follow to generate plot output. Contact your system administrator for more information.*

Plot Configuration Files

When you create a design file, more likely than not, you will revise it and have a need to recreate plots of the revised design. Quite often, you invest a fair amount of time setting up your design file to generate the desired output. By default, few of your plotting parameters are saved as part of your design file. Instead, MicroStation provides a method for saving and restoring your plot setup in a separate plotter configuration file.

Before creating a plotter configuration file, set up your design file with the desired fence (or view selected), levels displayed, plotter chosen, and pen table identified (see following section on pen tables). Most users create the plotter configuration file *after* they have generated a successful plot. Information contained in a plotter configuration file includes the following.

- Information about the plotted area (fence definition or view)
- Attributes of the view window being plotted
- Location of the fence in the design file
- Display status of levels in the view window being plotted
- Page setup information such as size, margins, and scale
- Pen table information, if used

To save your plotting settings, select Plot window menu > Configuration > New. In the resulting "Save As" dialog box, enter the name of the new plotter configuration file (default extension is *.ini*). Once saved, you can restore your current settings at any time by using the Open Configuration dialog box (Plot window menu > Configuration > Open). A Save command is also available to resave your current settings into an opened configuration file.

 NOTE: *The default directory for the plotter configuration files is deep inside the Bentley software directory structure (\Bentley\Workspace\System\data\). This is not necessarily a good place to place such useful files. A future software update may likely delete the content of this directory without your knowledge. To avoid this, you or your system administrator can redirect the output to a more appropriate directory by modifying the MicroStation configuration variable MS_PLOTINI. This variable contains the directory definition for the default directory, and can easily be modified for placing the files in a safer area of your system.*

 TIP: *You can directly open a plotter configuration file using the following command.*

```
PLOT configuration_filename plot_filename
```

This command can be entered directly in the Key-in window, assigned to a function key or, for the industrious, included in a custom toolbox. This gives you one-button plotting for consistent output every time.

 NOTE: *You can use the same plot configuration file with more than one design file. However, the working units must be the same as the source file used to generate the plotter configuration file, and the target area of all of the design files must be the same.*

Pen Tables

MicroStation provides support for resymbolizing your design file's plotted appearance through its Pen Tables facility. Resymbolization refers to the process of modifying the appearance of plotted drawings without having to actually edit the design file itself. For instance, an architect may want to use a heavier line weight for windows on a building elevation plot, but an engineer may want to keep it a lighter weight. Both needs can be accommodated through the use of Pen Tables without actually editing the design file.

Pen tables are stored as external files and consist of a variety of sections that define resymbolization parameters. When a pen table is used to generate a plot, these sections are compared against each element in the design file and the plot data is resymbolized accordingly. It is important to keep in mind that the design file elements are not edited in any way, only the generated plot data incorporates the changes. With pen tables you can perform the following.

- Alter the plotted appearance of elements

- Control the sequence in which the active design file is plotted in relation to its attached reference files

- Include text substitution and MicroStation BASIC macros as part of pen table sections that define resymbolization

To attach a pen table to your design, use the Select Pen Table dialog box (Plot window menu > PenTable > Load). Pen tables use the default extension *.tbl* and are usually located in their own directory within your current project or the company standards pen table directory (typically *Bentley\Workspace\Standards\Tables\Pen*). An example of a pen table is shown in the following illustration.

Plotting Raster Files

MicroStation provides full support of raster reference files during the plotting process. However, the results you get will depend a lot on your output device or file format. Hardcopy devices fall under two categories: vector and raster. No matter which category your printer or plotter falls under, MicroStation can plot raster reference files to it. Of course, you will get the best results when using supported raster hardcopy devices such as PostScript or HP's DesignJet printers.

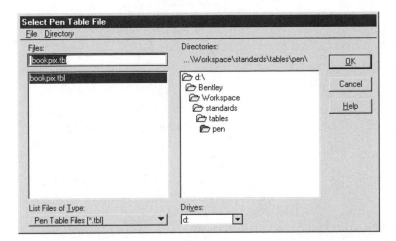

The bookpix.tbl shown was instrumental in generating the high-quality illustrations throughout this book.

NOTE 1: *The* HARDWARE_RASTER, *or the equivalent* SOFTWARE_ RASTER *parameter, found in each plotter driver file identifies whether the hardcopy device is capable of directly handling raster images.*

NOTE 2: *When plotting raster reference files, keep in mind that the raster image is sent first and the vector data is plotted on top of it.*

Bonus: Documenting Level Usage

One method of documenting which levels were plotted is to incorporate a level coupon in the drawing. A simple matrix of lines and numbers, this aide is easy enough to draw. By placing a hash line on each level, the result will be a hardcopy of the displayed levels.

When developing your own coupon, an example of which is shown in the following illustration, it is a good idea to optimize the plotter's motion by drawing the lines and numbers in a serpentine fashion. It is also a good idea to keep this coupon as simple as possible to minimize plotting time spent drawing it. Do not forget to set aside one level (for instance, level 63) for the coupon's skeleton.

TIP: *Using a diagonal line on each level usually results in faster plotting of the coupon. Most pen plotters draw their fastest at a 45-degree angle.*

level coupon

An example of a simple level coupon showing the levels associated with the plotted drawing. This coupon can be found on the INSIDE MicroStation companion web site.

Summary

As you can see, plotting can be very simple or very complex. In this chapter you learned how MicroStation's native plotting capabilities are used to generate a variety of forms of hardcopy and electronic output.

There are other products available from Bentley Systems and other companies that enhance the plotting process. Most of these provide high-powered server-based plotting engines oriented toward generating project and corporate standard output. Examples include Intergraph/Bentley Systems' Interplot, Zeh Corporation's Plot Express, and Equorum's (CADnet) Plot Station. Information about their plotting support products can be found at each company's respective web site:

www.bentley.com
www.intergraph.com
www.zeh.com
www.cadnet.com

This concludes the basics of MicroStation section of this book. In the chapters to follow, the topics will cover more advanced subjects that you will likely encounter once you have started to master MicroStation's basic operations.

PART THREE

3

BEYOND BASIC DESIGN

Advanced concepts for a productive design environment

3D DESIGN

Modeling the Real World

MICROSTATION IS WITHOUT QUESTION AN EXCELLENT 2D design tool. Its construction tools allow you to develop just about any drawing imaginable. But that is not all it can do! With MicroStation you can just as easily create highly accurate 3D models using its wealth of 3D tools and capabilities. In fact, you already know a lot about MicroStation's 3D capabilities because the program integrates all of the 2D tools you have learned so far into the 3D work environment. There are, however, a few things you need to learn about 3D, which is the goal of this chapter.

In this chapter you will get a good overview of MicroStation's 3D design environment. However, because 3D is a subject worthy of an entire book, this discussion is not an in-depth coverage of MicroStation's entire 3D capabilities. However, at the conclusion of this chapter you will have gained a good understanding of how MicroStation can be used to create 3D models.

Getting Started in 3D

Earlier in this book, while creating your first 2D design file, you used the seed file *SEED2D.DGN*. This is also where you start the 3D design process. When you click on the Select button in the Seed File section of the Create Design File dialog box, you are presented with a number of additional seed files. You will note a common theme among the names of the seed files. Almost all of them come in 2D and 3D versions. If you select one of the 3D seed files, the result is a 3D design file.

Why does MicroStation supply two separate versions of seed files, one for 2D and another for 3D? Because 3D is intrinsically more difficult than 2D. That darned Z axis is always getting in the way. If you do not need it, why carry it along?

3D Versus 2D

A fundamental decision you make when creating a design file from scratch is whether to set it up as a 2D or 3D design. Many (some say most) users of MicroStation (and most other mainstream CAD products) do their work exclusively in a 2D environment. As a result, MicroStation segregates 2D and 3D at the file level. This avoids inadvertent "Z" problems, such as skewed lines (a line that starts at one Z value but ends on a different Z value) and circles that do not quite look right (appear as ellipses because the circle's plane is not "normal" to the view). Suffice it to say that segregating pure 2D design work from 3D design work has been with MicroStation a long time, and works.

So, once you have decided to have a go at 3D design, where do you start? Logically enough, you start with the file creation process. Instead of relying on some sort of work environment setting to control the 2D or 3D "switch," MicroStation looks at the design file itself to determine whether it was set up to do 2D or 3D. That way, you cannot accidentally mix 2D and 3D elements in the same file.

Creating a 3D file is simply a matter of selecting the appropriate seed file. Seed file selection is part of the Create New Design File dialog box, invokable within the MicroStation Manager (File menu > New) or within the MicroStation design environment (File menu > New). The Select button for files is shown in the following illustration.

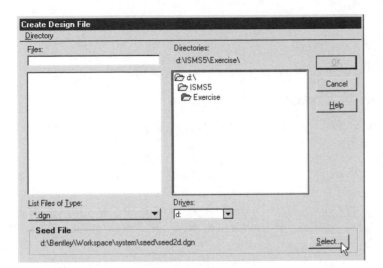

The Select button next to the seed file is used to select an alternative seed file, which includes establishing a 3D design.

To create a 3D design file you must select a seed file with a 3D definition. Several of these 3D seed files are delivered with MicroStation. Look for seed files with *3d* in their names, as indicated in the following illustration.

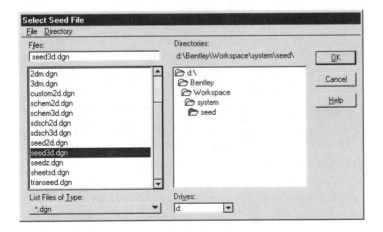

As you can see, it is easy to identify the 3D seed files from the 2D seed files by their names.

TIP: *To quickly set up a 3D design, simply select* seed3d.dgn *as your seed file. This provides the most generic of starting points, one in which you must set up all design settings.*

The following is a brief list of the steps involved in creating a 3D file:

1   Start MicroStation.

2   Select File > New.

3   Enter the new file name (example: *myfirst3d.dgn*).

4   Click on the Select button beside the seed file name.

5   Identify *seed3d.dgn* (or another 3D seed file as appropriate), and click on OK.

6   Click on Create.

ProjectBank provides for 3D files by selecting the *seed3d.dgn* file (or other 3D seed files) in the Create New Design File dialog box, shown in the following illustration.

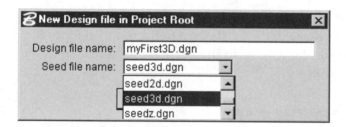

Picking the 3D seed file in ProjectBank DGN's Create New Design File dialog box.

As with all new design files created in a ProjectBank DGN brief-case, you must commit the new file to the server for permanent storage. Whether you are using plain old MicroStation or Project-Bank DGN, the previously outlined process results in a brand new 3D design file ready for work. When you open this newly created 3D file, you will notice a few things that are different about MicroStation, as shown in the following illustration.

The first thing you notice is the number of views. Instead of one planar view, you get several, each normally oriented along a different drawing axis. If you use *seed3d.dgn* as your seed file, you will see four views, labeled Top, Front, Right, and Iso. These views correspond to four of the standard orthographic projection planes. In manual drafting, the orthographic projection drawing technique is used to describe a 3D subject. Introduced in most first-year college, as well as all technical/vocational and high school, drafting courses, orthographic projection operates on the idea

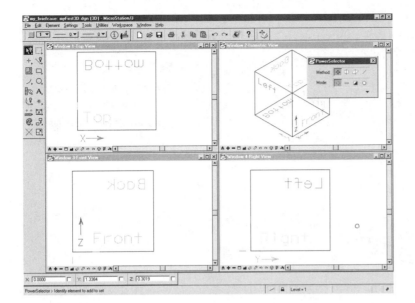

Opening a new design file created with seed3d.dgn presents some new MicroStation features.

that it takes at least two 2D drawings to fully describe a 3D object with each view aligned at a right angle to the other.

The Design Cube

The next thing you will notice about your new 3D design file is the cube object displayed in all four views. This cube is actually just a set of regular MicroStation drawing elements placed in your new design file from the *seed3d.dgn* seed file as part of the file creation process. If you wanted to, you could delete them by using the Select All command (Edit menu > Select All), followed by the Delete Element tool. However, do not do this quite yet.

This "design cube" is a great way of visualizing how the MicroStation environment changes when you go from 2D to 3D. In 2D, you normally work in View 1, which corresponds to the "Top" view in 3D (think of your 2D "drawing" as lying on top of your drafting table). The design file's real 3D design cube is actually much larger than the cube you see on your screen. In fact, it is the same size as MicroStation's normal 2D design plane, except that it is now projected into the third dimension along the Z axis. Another way to conceptualize this is: if the X and Y axes as used in the 2D design process describe a plane, the introduction of a Z axis changes this plane into a cube (see following illustration).

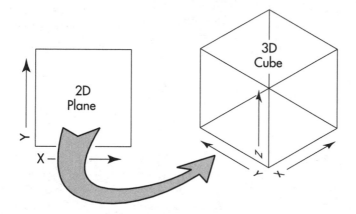

Moving from 2D to 3D changes the design plane into the design cube.

 NOTE: *The terms* design cube *and* in space *are synonymous with* drawing coordinate system. *You will also find that many users also refer to the design cube as the design plane, a holdover from the 2D working environment.*

3D-only Toolboxes

Another enhancement when you open a 3D design file in Micro-Station is the addition of 3D toolboxes. These normally appear grayed out in a 2D design file. The primary tool frame for 3D is 3D Main (Tool menu > 3D Main > 3D Main). This tool frame, shown in the following illustration, is the gateway to four additional toolboxes, which contain a host of 3D-specific tools. Several of these tools are explored later in the chapter.

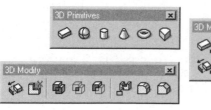

The 3D Main tool frame provides access to toolboxes containing the most-used 3D-specific design tools.

The other major tool set that comes with 3D are the 3D view tools located in the 3D View Control toolbox (Tools menu > View Control > 3D), shown in the following illustration. Several of these tools are related to their 2D cousins. However, there are enhancements specific to the 3D design environment. In addition, there are several tools exclusive to the 3D environment (for instance, Set Active Depth).

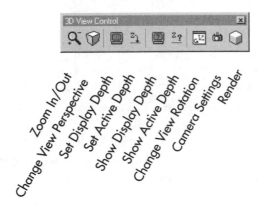

The 3D View Control toolbox contains the view tools unique to the 3D environment.

Before creating your first 3D design file, you need to learn how to navigate in 3D, discussed in the section that follows.

Views in 3D

Working in a 3D model using what is essentially a 2D (or *planar*) device, the video screen presents some challenges. How do you work in 3D using this "flat" device? By using multiple views, of course! Recalling MicroStation's 2D working environment, you can open up to eight individual views, each independent of the others. This means that you can view your design at different scales (zoom in/out) in each view. Equally important in 3D, you can also reorient each view around your design in 3D space. In the following illustration, View 1 shows the 3D cube from a top-down "point of view" (POV).

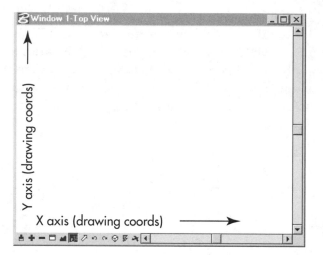

In View 1, Top means that the X and Y axes are aligned, as you would expect in 2D.

In 3D coordinate geometry, the view looks down the design file's Z axis, so that you see the X and Y axes as parallel, respectively, to the top/bottom and left/right edges of the view.

The Front view, on the other hand, shows the X as you would expect, going left to right along the horizontal edge of the view, as shown in the following illustration. The Y axis, however, has changed. Instead of pointing vertically along the edge of the view, the Y axis is now pointing perpendicularly, or away from you. In its place, the Z axis is now seen oriented vertically in the view.

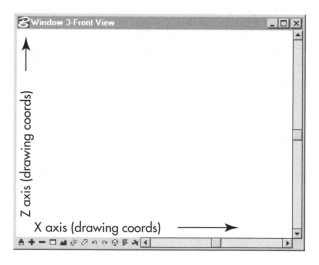

In the Front view, the vertical axis matches the 3D design cube's Z axis.

View 4, the Right view, is oriented at right angles to both the Front and the Top view. This means the horizontal axis is actually Y (right angle to the Top view).

Rotating Views

To help you maintain your orientation within the design cube, MicroStation includes a set of standard views. By name they are:

- Top
- Left
- Front
- Right
- Back
- Bottom
- Isometric
- Right-Isometric

The presentation of the view labels in the table is not accidental. It shows the logical orientation of the various views to one another. You can see how the Left, Front, Right, and Back views provide essentially a 360-degree horizontal rotation around the design cube. Because of the planar nature of views, this ability to "travel" around the three axes of the design cube is critical to your ability to work in the 3D environment. MicroStation provides the following means of invoking these standard views, described in the sections that follow.

- Change View Rotation tool
- Rotate View tool
- *VI=<viewname>* key-in

Change View Rotation Tool

The Change View Rotation settings box, shown in the following illustration, provides an easy-to-follow graphic image of the orthogonal views and their relationship to any of the views you currently have open. After selecting the tool, you identify the view you want to reorient via the View option menu. The cartoon cube will update to show you the current orientation of the selected view (View 1 is the default when you first open the tool).

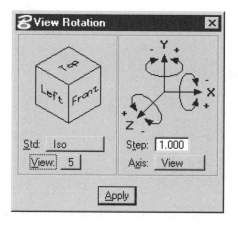

The Change View Rotation settings box represents an effective method of rotating any view.

The Standard View selection menu appears directly beneath the orientation cube. Selecting it presents you with the six standard orthogonal views. Selecting one of these and clicking on the Apply button reorients the view.

In addition to selecting the standard views, the Change View Rotation settings box lets you rotate the view along all three drawing axes. This is done in the right half of the settings box. By clicking on the plus or minus sign along the appropriate axis, you affect the cube's orientation in the left half of the settings box. Selecting the appropriate view and clicking on the Apply button updates that view.

An option associated with this settings box is the Axis option. Normally, you rotate a view with respect to the view's own axes. There are times, however, when you will want to orient the view precisely along the drawing's axes. By selecting Axis Drawing, all rotations are done around the drawing's axes.

For instance, to illustrate the fact that the Z axis pierces the Top view, selecting the Drawing axis followed by a rotation about the Z axis, plus or minus (+ or −), results in the cube rotating about the Top view. Holding down the mouse button makes the cube spin about the Top view.

One of the more unique tools found in any CAD package, the Change View Rotation settings box has been known to provide an entertainment break during intense design sessions. More than any other tool, the view rotation also brings home the action of the Z axis as it relates to the plane of the video monitor. In exercise 12-1, which follows, you will experiment with view rotation along the Z axis.

EXERCISE 12-1: VIEW ROTATION ALONG THE Z AXIS

1 Open *3DCONES.DGN*, shown in the following illustration.

Opened 3DCONES.DGN file.

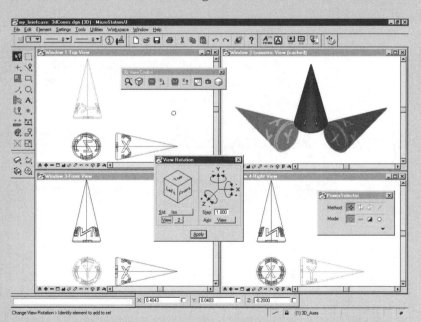

This file contains three SmartSolid elements (cones) that point along the three major axes of the design cube.

2 Open the Change View Rotation settings box (Tools > 3D > 3D View Control > Change View Rotation).

3 Select View 1. Select the Iso view in the Std view field, shown in the following illustration. Make sure the Axis field is set to *View.*

The View box should now be tilted, with the Top, Left, and Front views shown.

4   Click on the positive Z axis arrow and hold down the mouse button. The cube should now be spinning counterclockwise.

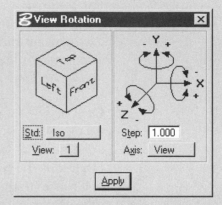

Std view field.

Exercise 12-1 illustrates how the Z axis always faces toward you from the video screen when Axis is set to View. Just like the fingers on your right hand, the direction of rotation sweeps to the left. Now, what happens if you select one of the other axes? In exercise 12-2, which follows, you will explore rotation along the X axis.

EXERCISE 12-2: ROTATING ALONG THE X AXIS

1   Continuing with the View Rotation settings box, select the Top view from the Std view selection box, as shown in the illustration at right.

2   The view cube should now show only the Top view. Now, briefly click on the positive X rotation arrow. Note how the Back view slowly emerges from the top of the cube, as shown in the following illustration.

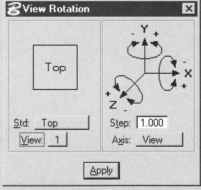

Selecting the Top view.

Result of positive X axis rotation.

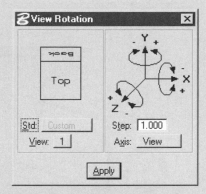

3 Click briefly on the positive Y axis. Another view, the Left one, comes into view, as shown at right.

4 Try rotating the various axes back using the negative direction.

Remember, you must select the rotation in the reverse order in which you originally rotated the view cube. This is because the XYZ axes are oriented with respect to the view itself, not the view cube.

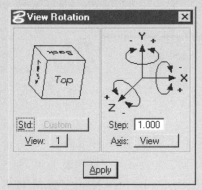

Result of positive Y axis rotation.

Rotate View Tool

Another means of rotating your views is use of the Rotate View tool, located in the view window's view control tool bar. Instead of a dialog box, you directly rotate the view by clicking the data point button in a specific view. The view rotates around the center of the view following the current pointer location. To accept a particular orientation, click a data point again. The following illustration shows the Rotate View tool in action.

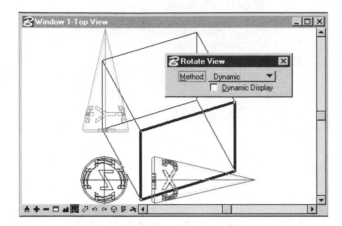

The Rotate View tool is most useful when adjusting your view rotations on the fly. Selecting the Dynamic Display option causes the entire view to update in real time.

TIP: *If you use a combination tentative point/data point instead of simply data pointing with the Rotate View tool, the tentative point location becomes the center of rotation. This is handy when working in close to your design and you need to perform a minute view rotation adjustment.*

When you first try out this tool, you will no doubt spend a few minutes simply rotating your views. This can be downright fun, and time well spent, because you will gain an understanding of how 3D views operate. If at any time you get lost in 3D space, you can simply select the Fit View tool, which will redisplay the entire content of your design file with the current view orientation. You can also directly select the same standard orthographic view rotations, using the Change View Method option menu (located in the tool settings window), shown in the following illustration.

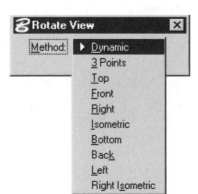

As with the Change View Rotation tool described earlier, you can still reset your view to one of the standard orthographic projection views using the Method option menu.

VI= Key-in Command

The "oldest" method of setting your view to one of the standard orthographic views is to use the key-in *vi=<standard view name>*. With the Key-in browser window open, entering the command

```
vi=top
```

followed by the Enter key will prompt you to select a view. When you do, the orientation of that view will change to the Top view. Typing in *vi=front* and selecting a view will reorient the view to the Front view.

Top-down Design

So, now that you have been introduced to the numerous methods of setting your views, let's explore their orientation a bit further. When you were working in 2D, you probably did not know it but you were actually working in the equivalent of the 3D Top view. The X axis runs horizontal from left to right; the Y axis runs vertical in the Top view. This corresponds to the traditional XY coordinate system. All of the other views are derived from the Top view by rotating about one of the axes by 90 degrees (right angle).

Table 12-1, which follows, summarizes the orientation of each standard view and which axis lies along its horizontal and vertical directions. The table also provides each view's horizontal and vertical axis relationship with the design cube's axes. The axes are labeled from left to right for the horizontal axis, and from bottom to top for the vertical axis. Perpendicular-to-viewplane refers to the axis that sticks straight out of the view toward your nose.

Table 12-1: Orientations/Axes for Standard Views

View Name	Horizontal Axis	Vertical Axis	Perpendicular-to-Viewplane Axis
Top	+X	+Y	+Z
Front	+X	+Z	−Y
Right	+Y	+Z	+X

Table 12-1: Orientations/Axes for Standard Views

View Name	Horizontal Axis	Vertical Axis	Perpendicular-to-Viewplane Axis
Left	−Y	+Z	−X
Bottom	+X	−Y	−Z
Back	-X	+Z	−Y

In addition, there are two non-orthogonal views predefined in MicroStation. They are the isometric views: Isometric and Right-Isometric. These orient the drawing XYZ axes to 30, 90, and 120 degrees, respectively (Right isometric rotates the Isometric view axes by a further 90 degrees around the view's Z axis).

Understanding the relationship of all these views is important. Here is another analogy that might help you understand how they work. In manual drafting, picture looking down on the drawing as it lies on the "top" of the drafting board or desk. The X axis points positively to the right, and the Y axis points positively up.

Now, picture your design as lying on the drafting table instead (for simplicity's sake, imagine a simple box like that shown in the *seed3d* file). Looking down from above, you see the top of your design. The other views are hinged from this top view. For instance, the Front view is hinged from the edge of the box closest to you. The Back view is hinged from the top edge furthest from you, the right view from the right edge, and so forth. The following illustration depicts this hinged approach to the standard view orientations.

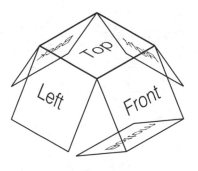

The first six named views and their relationship to one another.

Adjacent Views

Because you must keep your orientation with respect to the design plane straight in your head at all times, a good habit to get into now is to always display at least one additional view that is adjacent to your working view. Adjacent, in this case, means a view that is from a right angle to your working view. With the standard views, this means that if you are working in the Top view, your adjacent views would be Left, Right, Front, and Back. Any of these views can be used to specify a depth with respect to the Top view.

By keeping a view oriented in the general direction you are working, you stand a better chance of maintaining your bearings within the design cube. As you place an element, the element's planar view (in other words, a straight line) should appear. This little visual cue will go a long way toward keeping you sane with 3D.

TIP 1: Start developing standard view combinations you can relate to. For instance, when working in the Top view, you should have the Right view active as well. To make sure they are both looking at the same part of the design, use the Copy View command (Tools > View Control > Copy View icon) to copy one view's orientation into another view. Next, use the view rotation, or VI= key-in, to select the adjacent view. This way, the levels that are echoed on, the zoom scale, and so on are all set the same.

TIP 2: MicroStation provides a visual cue to help you keep your 3D bearings. ACS Triad, when enabled, draws a simple representation of the current 3D axes. These axes are labeled with view-independent text (meaning the text is "normal" to the screen), and therefore you can quickly tell which way is up. The ACS Triad (see following illustration) display on/off is accessible via the View Attributes settings window (Settings menu > View Attributes). By default, the ACS Triad is displayed with its origin set to X0,Y0,Z0. You can move it using AccuDraw's Rotate ACS (RA shortcut) command.

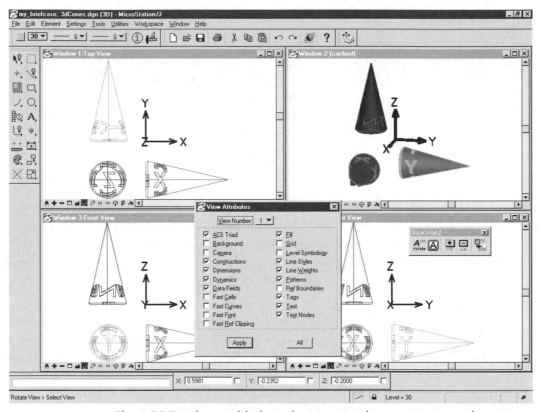

The ACS Triad is enabled via the View Attributes settings window. Note how it appears the same size and text labels are readable in every view.

The Right-hand Rule

Another handy means of keeping your wits about you while dealing with all of these views and their axes is the *right-hand rule*. This rule simply states that all angles sweep in a direction counterclockwise around the positive direction of an axis.

Examine the following illustration. Now, observe your right hand. If you assume your thumb is pointing in the positive direction of an axis, your fingers curl in the positive direction of the angle. In 2D, angle values are normally specified in a counterclockwise direction. Now, assuming your thumb is pointed along the positive direction of the Z axis, which would be directly at your face, and you curl your fingers, guess what? They curl counterclockwise. Looked at this way, using counterclockwise angles no longer seems as arbitrary.

The right-hand rule is derived from the action of your fingers on your right hand. The thumb represents the positive direction of the Z axis.

All of the MicroStation axes exhibit this trait. The key is to remember which direction is positive along any given axis.

TIP: *There is an old CAD designer's trick you may want to try on your own. As a corollary to the right-hand rule, you can visualize the relationship of the XYZ axes using three fingers on your right hand. Holding your right hand vertical, point your thumb to the right. This is the X axis of your design "cube." Point your index finger straight up. This is your Y axis. Point your next finger toward your face. This is your Z axis. The direction each of your three digits is pointing is the positive direction of their axis (see following illustration). If you want, draw a small X on your thumb, a Y on your index finger, and a Z on the third finger (but do not let your colleagues catch you!).*

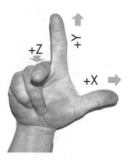

Three fingers on your right hand represent the X, Y, and Z axes pointed in their positive direction.

NOTE: *The terms* drawing coordinates *and* design cube *are interchangeable. The term* drawing *comes from the drafting origins of MicroStation. However, the design cube more closely adheres to design modeling.*

Dealing with Depths

To overcome the limitations of the 2D view plane, MicroStation provides a type of depth control for both establishing 3D coordinates and controlling the view's display volume. There are two distinct types of depth used in MicroStation:

- *Active depth:* A coordinate value set in each view such that, when applied to a data point, provides the third axis coordinate.

- *Display depth:* A pair of values that define the front and back of a view's display volume.

Active Depth

Not to be confused with active level, the active depth sets the third dimension value, along the view's perpendicular axis (the one you cannot see on the screen). A view's active depth essentially provides the third axis value for a data point placed within a given view. That way, you can still use a single data point to establish a complete XYZ coordinate location for your current operation.

Active Depth is often used when you are working in one view at a consistent "depth" within the model. For instance, if you were creating a map with topographic details, you could place the individual contour lines at their proper elevation or depth in a Top view by setting the active depth of that view to the contour elevation height before starting the placement tool. As you data point in this view, the contour would inherit the elevation value from the view's active depth value.

Active Depth Key-in

You can explicitly set the active depth using the *ACTIVE ZDEPTH ABSOLUTE <zvalue>"* *(shortcut: AZ=<zvalue>)* key-in command. You are prompted to identify the view to which you want to apply this new active depth. You can also adjust an existing active depth by using the Active Depth Relative command (shortcut: *DZ= <zvalue>*). However, you should practice with the absolute active depth method before using this variant (it involves adding the new value to the existing one relative to the view's Z axis).

NOTE: *Setting a view's active depth by absolution is always defined with respect to the global origin, X=0, Y=0, Z=0.*

Set Active Depth

The Set Active Depth tool allows you to visually set a view's active depth. By selecting the view, a dynamic plane appears in adjacent views, allowing you to identify with a data point the depth at which you want planar elements to be placed in the chosen view.

For instance, the Top view's perpendicular axis is the Z axis. When you set the active depth for a view oriented to Top to, say, 5 working units, any data point you place in that view will have a Z value of 5.

Display Active Depth

The complement to Set Active Depth is the Display Active Depth tool. Selecting a view with this tool results in the active depth value being displayed in the status bar. Exercise 12-3, which follows, shows you how to set your active depth in a view. Let's try our hand at placing elements at a given active depth.

EXERCISE 12-3: SETTING THE ACTIVE DEPTH

1   Open *CUBE3D.DGN*. Note the standard four views and their orientation, shown in the following illustration.

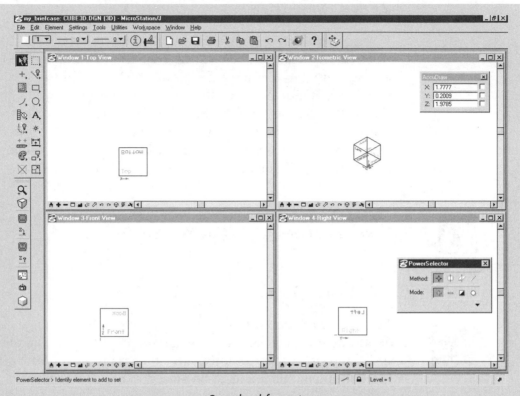

Standard four views.

Next, you will set the active depth in the Front view (View 3).

2 Select the Set Active Depth tool from the 3D View Control toolbox (Tools > 3D > 3D View Control), shown at right. MicroStation prompts you to select a view.

3 Click a data point anywhere in the view window labeled Window 3-Front View.

4 A dynamic dashed box appears in all four views, as shown in the following illustration. This box represents the display cube associated with the Front view. The Iso view really shows the relationship of the Front view to this cube. Compare the relationship of the word *Front* in Window 3-Front View with that of Window 2-Isometric View.

Set Active Depth tool selected on 3D View Control toolbox.

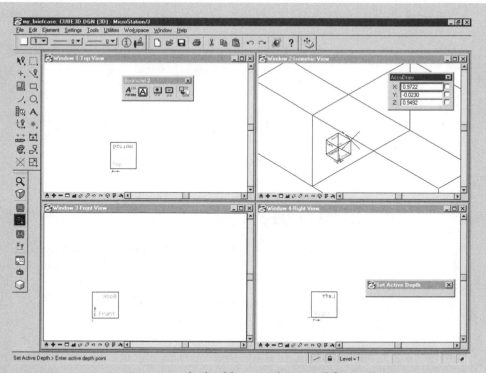

Dynamic dashed box evident in all four views.

MicroStation prompts you to "Enter active depth point."

5   Click a data point in the center of the 3D box in Window 4-Right View.

6   With the active depth set, let's place a circle in the front view.

7   Select the Place Circle by Center tool. MicroStation prompts you to identify the center of the circle.

8   Place a data point in the Window 3-Front View, just to the right of the box.

9   A dynamic circle appears in all four views. Note its orientation in each view. In the Top and Right views it appears as a line. This indicates that both of these views are adjacent to the Front view. Note also how the position of the circle in the Right view is exactly where you clicked when setting the active depth a moment ago.

10 Data point to set the circle's diameter.

11 Of interest is the orientation of the circle in the Isometric view. It appears as an ellipse in this view, which is precisely how you would expect a circle to

appear in a view, rotated 60 degrees from the Front view. Now place a circle in the Isometric view.

12 Data point once in the Isometric view at the corner of the box.

Note how the resulting circle of this exercise appears exactly the same in the three opposing views. This is, of course, because of the Isometric view's 60-degree orientation to the Front, Top, and Right views.

Display Depth

The other view-related depth attribute is *display depth*. Because you are viewing a volume of space in each view, you need a method for controlling how "deep" into the cube you want to peer. If you did not have this capability, you would have all elements that cross your line of site appearing in your view.

Unlike the real world, in which detail is less distinct with distance, the design plane's elements remain sharp and in focus at all times. Thus, when you are trying to work inside a complex 3D construct, such as that shown in the following illustration, you can get interference from the elements placed earlier. Because these elements are all over your design file, you cannot avoid selecting one if it crosses your view, no matter how far down the perpendicular axis it is placed. This can become very frustrating and counterproductive.

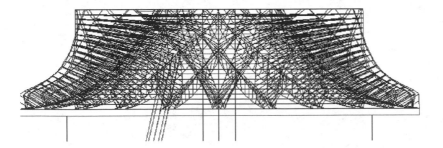

Without control of your display depth, you would not be able to find your way around a complex design.

Recognizing this problem, MicroStation provides a mechanism for clipping out a section of the perpendicular axis so that you can display only a portion of your design file lying within the bounds of the display depth. The Set Display Depth tool provides an interactive method of setting your view's front and back display clipping planes (another way of saying depth). You set a view's display depth by providing two data points in a second view oriented at a right angle to the target view (the "adjacent view" concept).

Display Depth Key-ins

As with Active Depth, you can explicitly set a view's display depth using the *set ddepth absolute <frontplanevalue>, <backplanevalue>* key-in (shortcut: *DP=<frontplanevalue>, <backplanevalue>*). You can also set the display depth relative to the existing values using the *set ddepth relative...* key-in (shortcut: *DD=*).

Set Display Depth

You can also set a view's display depth interactively by defining the required depth in an adjacent view using the Set Display Depth tool (Tools menu > View Control > 3D > Set Display Depth). By selecting two points in an adjacent view, you tell MicroStation which portion of the design cube you wish to see. In exercise 12-4, which follows, you will manipulate a view's display depth using both the interactive Set Display Depth tool and the Display Depth key-in.

EXERCISE 12-4: SETTING THE DISPLAY DEPTH

Using the design file from exercise 12-3, let's set the display depth in the Front view.

1 Select the Set Display Depth tool from the 3D View Control toolbox, shown at right.

2 MicroStation prompts you with "Select view for display depth." Select Window 3-Front View.

The dashed box delineating the display depth for the Front view will appear in all views. MicroStation prompts you to define the front clipping plane. You must identify the front and back clipping planes in proper order.

3 Data point just to the left of the "Right" text string in Window 4-Right View. MicroStation displays a plane identifying the front clipping plane and prompts you to define the back clipping plane.

Selecting the Set Display Depth tool.

4　Data point just to the right of the "Left" text string in Window 4-Right View.

The words *Front* and *Back* disappear from the Window 3-Front View, along with the squares representing the front and back planes. These two elements are now outside the display depth you selected for the front view.

5　Select the Set Display Depth tool again, and select Window 2-Isometric View.

6　Placing data points inside the cube and then outside, define the depth in the Top view around the top right corner of the cube.

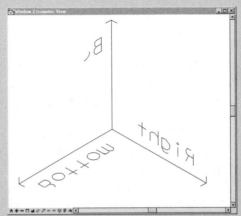

What happened to the view? Instead of a slice of the box being shown, you get what appears to be a corner of the box, as shown in the illustration at right. Because the display depth is calculated with respect to the view, and not the drawing, the "slice" you defined with your data points cuts through the heart of the box. You see only a portion of the box.

Result of defining the depth in the Top view at the top right corner of the cube.

Display depth can also be manipulated using the *Set DDepth Relative* or *DD=* key-in. This command will add or subtract the value entered to the front and back planes of the selected view. This command is cumulative and is easier to remember as "delta depth," referring to the difference between what you see and what you want to see.

7　Open the Key-in window (Utilities menu > Key-in). Type in:

```
DD=60,-60
```

8　MicroStation prompts you to select a view. Data point in Window 2-Isometric View.

The Isometric View should now display the entire cube again. The distances provided with the *DD=* command were added to the display depth you defined earlier. The *–60* was "added" (negative number pushes the plane away from the view) to the backside of the display depth, and the *+60* was added to the front. The *Set DDepth Absolute* (*DP=*) key-in command is used to set the absolute front and back clipping planes for a given view.

 V8: MicroStation version 8 will introduce a new view-independent volume display feature called "view-volume." By using an existing element to identify how much of the design volume you want to view, this features eliminates the view clip settings.

Working in 3D

The sections that follow discuss placing elements, drawing coordinates, using tentative points, the 3D tentative point versus the 3D data point, commonly used 3D tools, 3D-specific tools, 3D construction tools, 3D element modification tools, and 3D utility tools. Exercises are included for practice in using these tools.

Placing Elements

Now that you have a basic understanding of orientation in the 3D design cube, the next item to tackle is how you actually create your design. When working in 2D, you directly specified the coordinates for your design using the data point and explicit data entry (in other words, AccuDraw). When working in 3D, you do the same thing but with a twist. Because each view is only 2D (or planar) in nature, it is impossible to explicitly enter the X, Y, and Z values of an element using only a single data point. A view can provide only two of the three values needed to flesh out a 3D coordinate location (which two depends on your view's orientation, discussed further in material to follow).

MicroStation provides a number of means of specifying 3D coordinates, depending on what you are trying to accomplish. The following describe these various methods.

Active depth: Set each view's perpendicular axis to a predetermined value, which is automatically applied to a data point's X and Y values. Each view has its own active depth value (discussed in the previous section).

3D data point: Two data points are used to identify a single point in space. The first data point identifies two of the coordinates (which two depends on the view's orientation where the data point is placed); the second data point defines the third coordinate in a view at a right angle to the first view.

AccuDraw: Manipulate AccuDraw's compass to establish the X, Y, and Z values of a data point by using a combination of compass axis rotation and various shortcut keys. This is by far the most popular method of specifying coordinates in 3D.

Auxiliary Coordinate System (ACS): A powerful method by which an independent coordinate system can be defined and used to define both absolute and relative locations within the design cube. The ACS has the added benefit of defining the coordinates as rectangular, spherical, or cylindrical. AccuDraw draws upon the ACS system to perform many of its operations.

Key-in: Use the *XY=<xval>,<yval>,<zval>* key-in to explicitly define a coordinate location within the design cube. Coordinates can also be placed relative to any view using the *DX=* key-in. This method has been largely replaced by AccuDraw.

Each method has its strengths and weaknesses. Which one you use will depend a lot on your particular design situation. Because this is an introduction to 3D and not an exhaustive treatise on the subject (that would easily be a book on its own!), the Active Depth and AccuDraw techniques of 3D coordinate input are discussed in detail, with only an overview provided for the other techniques.

Drawing Coordinates

Even when twisting and turning views, the 3D coordinates you generate as part of the design process are always stored in the design file as drawing coordinates. View and other relative coordinates are transformed to their drawing coordinate values prior to storage in the design file.

Relax! If, during the course of your design session, you get turned around and create your design along the wrong axis, you can easily rotate the results to the proper axis using PowerSelector and the Rotate Element tool. In most cases, all you have to do is rotate your design 90 degrees and you will be back on the right track.

Using Tentative Points

As mentioned before, all MicroStation element placement and modification tools function properly in the 3D environment. As you adjust your Active Depth and place elements, you begin to feel confident working in 3D; that is, until you try using the tentative point. Using the tentative point does require some forethought and understanding of how it treats your design cube.

When you click a tentative point near an object, the resulting data point takes on not only the X and Y values of the "snapped-to" object but its Z depth. Thus, if you were to place a circle by diameter by snapping to two lines at different depths, the resulting circle will be skewed between the two lines. (See the following illustration.)

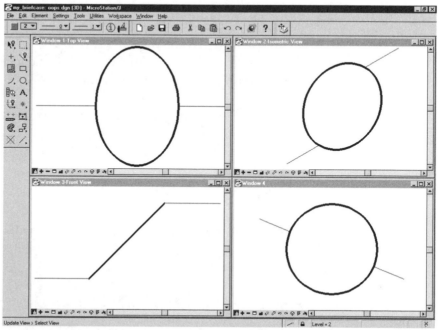

This circle is perfectly round. (Note View 4.) Because it was placed using a tentative point between two lines at different depths, it looks distorted in the other three views.

V8: MicroStation version 8 will introduce a new feature called "sticky-Z," which will eliminate skewed elements (such as shown in the previous illustration) when working in 3D.

The 3D Tentative Point and 3D Data Point

No matter how often you rotate views and select active depths, there are times when you just want to select a point in space. MicroStation provides this capability in the form of the 3D tentative point and the 3D data point. These are accessed with a combination of the Alt key and the appropriate mouse button. Table 12-2, which follows, specifies the 3D mouse button assignments.

Table 12-2: 3D Mouse Button Assignments

Function	Default Assignment
3D Data Point	Alt - Mouse Button 1
3D Tentative Point	Alt - Mouse Button 2

If you hold down the Alt key on the keyboard while clicking either a data point or tentative point, a dashed bore line will appear in the views adjacent to the one in which you clicked a data/tentative point, as shown in the following illustration.

Clicking a 3D data point or 3D tentative point again, but this time in an adjacent view, results in a data point or tentative point at the intersection of your first data/tentative point and the second one.

"Everyday" Tools in 3D

The sections that follow describe tools you will work with routinely in the 3D environment. Exercises provide you with practice in using these tools.

Placing Elements in 3D

Now that you have been introduced to the way you navigate in 3D, working with the regular MicroStation 3D drawing tools is not that much different than in 2D. Because most drawing tools are

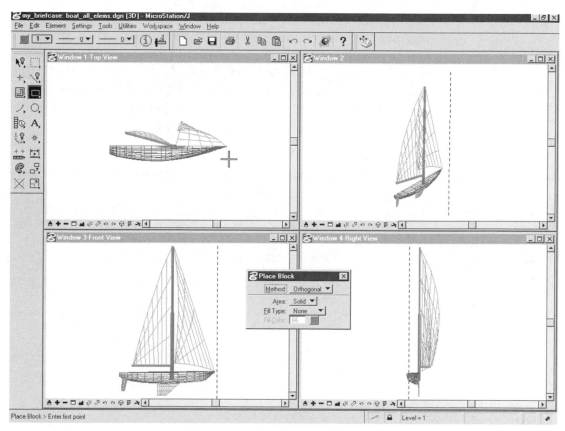

The 3D data point can be used to select a point in free space. Note how the cursor in the Front view is lined up with the bore line on the Top and Right views.

planar by nature, they work well within the constraints of the view. The only real consideration is the active depth where the selected tool will place its output. It gets trickier when you want to place elements "off the view plane."

Getting Started with AccuDraw 3D

By now you are probably beginning to understand that the key to working in 3D is defining the "plane" along which you want to place your element. Doing this using the active depth and multiple views is effective, but there are times you will wish there were an easier way. There is! If you thought AccuDraw made drawing in 2D easier to do, wait until you see what it does in 3D!

As you will recall, one of the powerful features of AccuDraw is its compass and the various ways you can manipulate it. In 3D, multiply this by another dimension. In addition to rotating the compass, you are able to reorient it along any plane within the design cube (see the following illustration). This is accomplished using several shortcut keys. Table 12-3, which follows, outlines shortcut keys for 3D compass orientation in the design cube.

Table 12-3: Shortcut Keys for 3D Compass Orientation in the Design Cube

Key/Plane Association	Orientation
T (top plane)	AccuDraw X axis along the design cube X axis; AccuDraw Y axis along the design cube Y axis
F (front plane)	AccuDraw X axis along the design cube X axis; AccuDraw Y axis along the design cube Z axis
S (side plane)	AccuDraw X axis along the design cube Z axis; AccuDraw Y axis along the design cube Y axis
V (view plane)	Aligns AccuDraw axes normal to the view

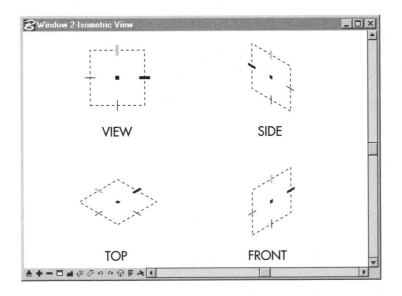

The AccuDraw compass reorients itself in response to the Top, Front, Side, or View (shown in Isometric view).

Using AccuDraw lessens the need to work in two views during the design process. Most users find that once they get the hang of using the T, S, F, and V shortcut keys, they use AccuDraw for 90% of their 3D input.

NOTE: *Users that have been around a long time (including the author of this book) remember a time when there was no AccuDraw. Back then, working in 3D was tedious and very error prone. The introduction of AccuDraw in MicroStation 95 is truly one of the major milestones in the history of MicroStation.*

Another important shortcut you use a lot in 3D is the O (origin) key. As in 2D, this key repositions the AccuDraw compass to the current location of the cursor. This comes in handy when used in 3D, especially when you are performing an element placement relative to something else in the design. Reorienting and relocating the AccuDraw compass from a tentative point is a very common operation. It provides you with the ability to precisely locate the start point of an element at any point in space relative to any other point in space.

There is a lot more to AccuDraw than can be adequately explained in a few paragraphs. As you explore the 3D world, you will quickly learn how to control AccuDraw's behavior via several additional shortcuts and settings. More importantly, you will also learn how various tools respond to AccuDraw's actions, each according to its requirements. These various AccuDraw behaviors are explained as each tool is described in the sections that follow.

Using Familiar Tools in 3D

Before moving into the 3D-specific tools and their operation, it is important to understand how the 2D tools introduced earlier in this book work in 3D. For the most part, 2D tools work as they did in 2D. By default, 2D (or "planar") elements are placed parallel or normal to the view at the current active depth associated with the view in which you place them.

For instance, when you place a circle in a 3D file, it appears like an ordinary circle. However, by definition a circle is planar ("flat" to a given plane). If you place a circle in the Right view, you will see a circle in that view. However, if you look at that circle in an adjacent view (Front, for instance), the circle all but appears invisible or as a line. Why? Because the right view plane is perpendicular to the front plane. The reason you may see two dots, two line segments, or one line segment in the front view depends on the current settings of the display depth, as described previously (you see some or all of the circle intersecting the Front view).

AccuDraw does not affect the planar nature of 2D elements. However, it does allow you to place those elements on a plane other than the one parallel to your views. The tool works with AccuDraw to place the element on the plane dictated by AccuDraw. This is also known as *in-context* behavior. AccuDraw has an option to turn off such contextual behavior, but most users find it too useful to eliminate. One of the major benefits of this behavior is the ability to create these elements in a pictorial view, such as that shown in the following illustration.

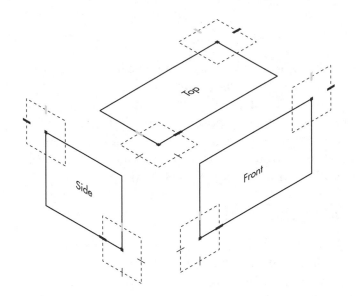

Here, you can see how AccuDraw's compass orientation can have a profound effect on how the Place Block tool creates its elements (shown in Isometric, pictorial view).

Pictorial view refers to any view that is not a recognized MicroStation standard view. It also includes the Isometric and Right-Isometric views and any standard view in which you have turned on a camera. The reason it is called pictorial comes from the pre-AccuDraw days. Back then, you almost always worked in one of the standard views, but you might open the Isometric or rotate an existing view to give you a better 3D perspective of the design. However, because users could not place elements in these views with confidence (active depths do weird things in arbitrarily rotated views), they tended to view these rotated views as pretty pictures, thus the term *pictorial.*

In exercise 12-5, which follows, you will design simple table in 3D using the Place Block tool. First, however, you will familiarize yourself with navigating a bit in the "seed" file for this design. After that you will have the opportunity to create the top and legs of the table with simple 2D shapes. Let's get started.

EXERCISE 12-5: ORIENTATION IN THE DESIGN CUBE

1 Open *Table3D.dgn.* Three objects have already been placed in the exercise file to help you get oriented within the design cube, as shown at right.

2 Use the view control tools to look around the design. Try the Rotate View tool to view the axis markers from different angles.

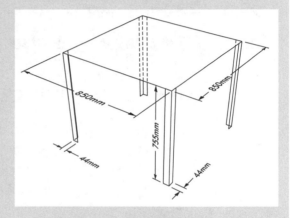

Table to be designed.

3 Before continuing with this exercise, reset the view content using the Saved Views utility. Open the Saved Views utility window (Utilities menu > Saved Views).

4 In the lower list box of this window, select *Start.* Make sure the Dest View field is set to 2. Click on Attach.

The settings in the saved view *Start* replace View 2's settings, as shown in the following illustration. In exercise 12-6, which follows, you will start designing the table.

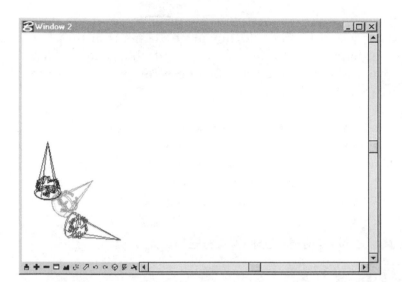

View settings established for the design process.

EXERCISE 12-6: DRAWING THE TABLETOP USING ACCUDRAW

At this point, you are ready to begin the design process. This is a very simple design consisting of a planar top and four legs (two shapes to each leg).

1   Select the Place Block tool.

2   In View 1, place a data point near the center of the view. Note how the dynamic block is oriented "normal" (planar) to your view.

3   With your input focus in the AccuDraw window (press Esc key until it is), press the T key. The orientation of the shape changes to the Top plane.

4   Using AccuDraw, complete the placement of the top of the table using the dimensions 850 mm by 850 mm (master units are meters). The result is shown at right.

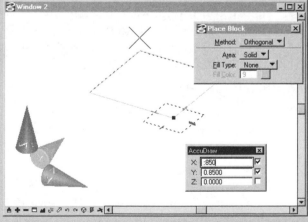

Tabletop with dimensions 850 mm by 850 mm.

EXERCISE 12-7: CREATING TABLE LEGS USING ACCUDRAW

The next item of business is the first leg. You will use the same Place Block tool to create one side of one leg, shown at right.

1   Snap to and data point at the corner of the top you just placed that is closest to you. Do not data point yet! Note how the orientation of the dynamic block is still set to the Top plane.

2   With the input focus on AccuDraw, press the F key. The block changes orientation to the Front plane.

Table leg to be created.

3   Complete placing the first edge of the leg with the dimensions 44 mm wide by 755 mm long. You need to complete the leg (there are only two sides to it for now).

4   Snap/data point to the same corner of the top. The dynamic block is still oriented along the front plane.

5   With input focus on AccuDraw, press the S key. The dynamic block is now aligned along the Side plane.

6   Snap to the bottom of the leg but do not data point! Press the O key instead. This resets the AccuDraw compass origin to the bottom of the leg.

7   Staying indexed to the compass' horizontal axis (the "floor" of the table, as it were) enter *:44* (remember, the master units are meters) in the X field and data point.

You now have one leg of your table! Using the technique just described, go ahead and place the other three legs with shapes along the Front and Side of your table. The finished table is shown at right.

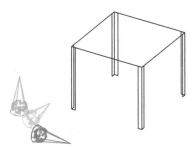

Completed table.

See how easy it was to move around in 3D using AccuDraw? You could have placed those same legs without AccuDraw by manipulating the active depth in two adjacent views. However, it would have been much more tedious and seriously error prone.

NOTE: *When not using AccuDraw, most first-time users end up with geometry at different "depths," and when the design is rotated about, components of the design seem to float in space.*

As you can see, even a simple 2D tool such as Place Block responds well to AccuDraw's actions. Now, imagine if there were tools that took full advantage of the 3D environment and integrated well with AccuDraw's ability to change directions on the fly. That is the subject of the next section.

3D-specific Tools

There are several additional tools that become available only when you are working in a 3D design file, most of which give you the ability to create elements with a new fundamental characteristic: volume. Element types created by these tools include slabs (3D boxes), cones (also known as cylinders), spheres, and toroids. In addition, there are tools to take any 2D graphic element and extrude or revolve it into a 3D volumetric element.

Surface Modeling

For those needing to define truly complex free-form surfaces in 3D space, MicroStation comes equipped with a complete suite of 3D surface tools, shown in the following illustration. For the first-time MicroStation user, these tools can be very intimidating. They require you to have a good grasp of how you want to put your design "skin" together (a surface can also be thought of as the skin of a 3D volume). Surface modeling tools define volumetric objects by defining scores of individual key points on the surface of the design.

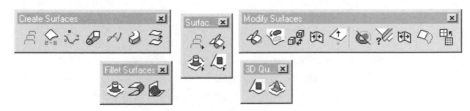

MicroStation provides a wealth of surface modeling tools.

As a first-time user of MicroStation, chances are good you have not encountered the need to create a 3D free-form surface yet. However, there are entire industries that rely on the designer's ability to explicitly define the surface of an object, point by point. One such industry are the folks that create all those fancy bottles used to hold the various consumer products, from food products such as mustard or ketchup to soaps to practically anything that involves holding a liquid. These sorts of products are nearly the exclusive domain of the surface modeling environment. Surface modeling tools are also used for such forms as the hull of a boat, shown in the following illustration.

Boat hulls are often modeled using surface modeling tools.

NOTE: *Working with 3D surfaces is a subject unto itself, and is well beyond the scope of this introductory text.*

Solid Modeling

Recognizing that working with 3D volumetric objects can be tedious and error-prone to create and modify (especially modify), Bentley introduced solid modeling tools with MicroStation/J. Called Smart Solids, these special element types and tools provide a much more user-friendly 3D modeling environment when compared with surface modeling. Instead of explicitly defining a 3D element by points on its surfaces, solid modeling utilizes a more building-block approach.

Solid modeling uses a series of primitive building-block objects you combine or subtract from one another to arrive at your final design. An example of this is the fan blade assembly shown in the following illustration. The advantage of solid modeling is its more intuitive approach (most users were exposed to building blocks as infants). In MicroStation, this solid modeling borders on fun!

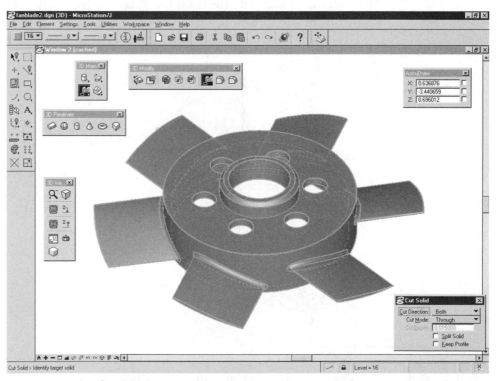

This fan blade assembly was created primarily from primitive solids, mostly cylinders.

MicroStation's Smart Solids

MicroStation does not make a hard and fast distinction between what is a solid and what is a surface. Yes, there are tools specifically designed for creating and manipulating surfaces. However, as you go through the process of creating your 3D design, MicroStation decides what is the best way to represent your design (solid or surface).

This is normally done behind the scenes and is the cornerstone of the "Smart" in Bentley's Smart Solids and Smart Surfaces technology. In reality, most 3D designs consist of a combination of moderately complex surfaces and easily identifiable solids. The fact that MicroStation can relate these two representations so seamlessly is one of the reasons 3D design is almost fun!

The following illustration shows a Smart Solid that is constructed from a solid and a surface. In this instance, a free-form surface that is somewhat representative of a terrain model has been "subtracted" from a 3D cube solid. The result is very impressive!

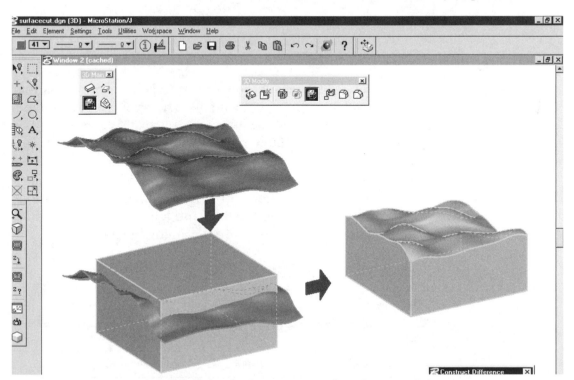

An example of subtracting a surface from a solid.

NOTE: *Bentley Systems licensed the Parasolids software library from XXX, which is at the heart of the Smart Solid system. This provides an interesting level of compatibility with other solid modeling packages. In addition, Bentley has maintained its license for the ACIS solid model library so that it can import other ACIS (SAT file format) designs.*

The 3D Toolboxes

You access most of the 3D design tools from the 3D Main tool frame (Tools > 3D Main > 3D Main), shown in the following illustration. From this tool frame, you have four separate toolboxes organized by type of 3D design operation.

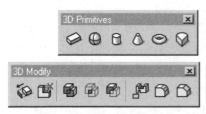

The 3D Main tool frame is the nexus for the most often used 3D design tools in MicroStation.

3D Primitive Element Tools

The first set of 3D-specific tools you will likely encounter is that used to create 3D "primitive" elements. The term *primitive* is somewhat of a misnomer because these tools actually generate rather complex data structures that resemble the simple basic 3D shapes you are already familiar with. MicroStation uses the term *primitive* to differentiate these elements from the more complex Smart-Solid solids that result when you perform further manipulations of these primitives (such as filleting, and joining solids). Table 12-4, which follows, summarizes the 3D primitive element tools.

Table 12-4: 3D Primitive Element Tools

Tool Name	Description
Place Slab	Places a 3D rectangular shape
Place Sphere	Places a circular volume of revolution

Table 12-4: 3D Primitive Element Tools

Tool Name	Description
Place Cylinder	Places a cylindrical surface or shape
Place Cone	Places a cone-shaped object
Place Torus	Places a donut or toroidal shape
Place Wedge	Places a pie-shaped projection

The sections that follow describe several of these 3D primitive element tools.

Place Slab

Recognizing that many element projections performed in a design session result in orthogonal shapes (i.e., cubes and boxes), MicroStation provides the Place Slab tool. By defining the diagonal corners of one side and a depth using either an adjacent view or AccuDraw, as shown in the following illustration, you can create 3D boxes in a hurry.

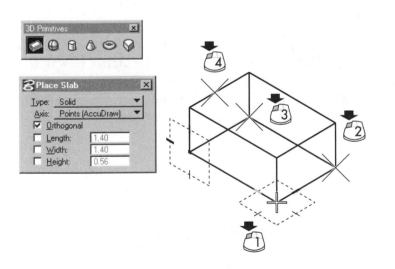

Place Slab tool shown in Isometric view with AccuDraw set to Top plane.

The Axis option allows you to pre-orient the slab to one of the major axes, either the drawing X, Y, or Z, or the view X, Y, or Z. As with many of the Place Element tools, the size of the resulting object can be predetermined using Length, Width, and Height options.

Place Sphere

Used to place a sphere in your design, the Place Sphere tool offers the same axes options as the Place Slab tool. In addition, you can select Radius to predefine the size of the sphere, as indicated in the following illustration.

Place Sphere is used to create spheres in MicroStation 3D. You can use AccuDraw as well as the Radius tool setting to precisely set the sphere's radius.

Place Cylinder

A cylinder is a recognized primitive solid. In 3D, you create cylinders using Place Cylinder, shown in use in an Isometric view in the following illustration. It normally appears as a circle projected along a right angle to the circle's plane. You can skew the cylinder's Z axis by turning off the Orthogonal tool setting.

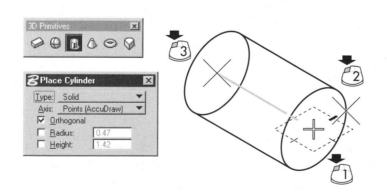

Place Cylinder shown in Isometric view with AccuDraw set to Top plane.

The Type option takes a little explaining. Before SmartSolids were introduced in MicroStation, all 3D elements were represented exclusively as surfaces (also known as wireframe solids). Because it is a surface, you have the option to "cap" off the ends of your cylinder. With true solid models, this is not an option. For all intents and purposes, capped cylinders are solids, and non-capped cylinders are surfaces.

Place Cone

Another primitive 3D element type is the cone, an example of which is shown in the following illustration. Although most people would think of a wizard's hat as the prototypical cone, in reality, it is better described as a solid (or surface) projected between two circles of different radii. The classic wizard's hat results from one of the two circles having essentially a zero radius.

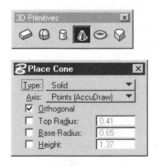

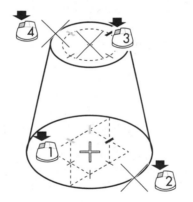

Cone created with Place Cone.

As with the Place Cylinder tool, you can use AccuDraw to precisely set the diameters and direction of the cone. Pressing Reset for the diameter of the second radius will result in a pointed cone.

Place Torus

The torus is an interesting 3D shape. Reminding most users of a donut (a morning pastry popular in the United States), the Place Torus tool can create both continuous and partial toroidal elements, an example of which is shown in the following illustration.

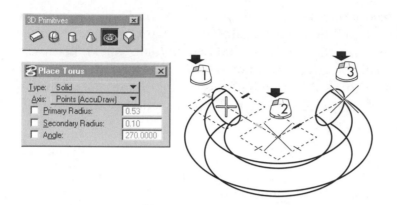

Torus created with Place Torus.

The first data point defines the start of the primary radius, the second data point defines the center of the torus, and the third data point defines both the sweep (angle) of the partial torus and the cross section (secondary) radius.

To create a closed torus, enable the Angle option and set it to 360. To create a partial torus of a given sweep, set the number of degrees counterclockwise from your first data point in the same Angle tool setting.

Place Wedge

The Place Wedge tool is used to create 3D elements that resemble a section of another "pastry" type, the pie. Alternatively, you can think of the wedge as a "wedge of cheese." As indicated in the following illustration, you specify the outer edge of the wedge "wheel," followed by its center point, in a clockwise direction. Finally, you set its height with a third data point. As with other 3D tools, you can preset all of these dimensions in the Tool Settings window.

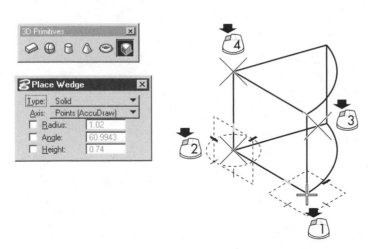

The Place Wedge tool is intuitive in its operation.

3D Construction Tools

In addition to the placement of 3D primitive elements, you need a means of taking existing 2D geometry and turning it into a 3D volumetric element. MicroStation includes several very powerful tools for doing just that. These tools are described in the following sections.

Extrude (Construct Surface of Projection)

In its simplest form, the Extrude tool creates a complex 3D shape of your target 2D element by extruding this shape along a third axis. This is not unlike the effect of squeezing a tube of toothpaste, where the cross section of the resulting "squeeze" is the same as the orifice in the tube (the children's toy Playdoh is another example).

The first decision you must make when setting up this tool is whether the outcome should be a solid or a surface. Most users will opt for a solid, which automatically generates a SmartSolid definition.

This tool's operation was revolutionized with AccuDraw. Instead of having to specify axes in adjacent views (top/front or front/ right), AccuDraw provides a more intuitive input of the direction and depth of the extruded element. An example of this tool in action is shown in the following illustration.

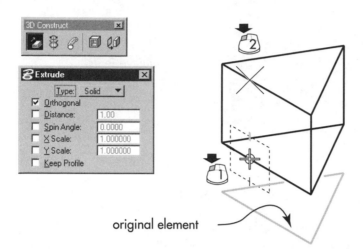

In its simplest form, the Extrude tool creates simple extruded 3D elements from your source 2D element.

original element

Remembering that data points always extrude your selected 2D element from its current "depth" to the view's active depth, an element you extrude that is already at the active depth will create a flattened projection. First-time users who place all of their data points in the same view generally end up with one or more of these types of extruded elements. This can be very confusing. It is highly recommended that you practice with this tool in a pictorial view (for instance, the Isometric view) with AccuDraw.

The Keep Profile option in this tool's tool settings is very important, especially when you are first learning how this tool works. When selected, it retains your source element so that you can have another try at it if the extruded results were not quite what you had in mind. If you leave this option off, the Extrude tool will automatically delete the source 2D element you selected to be extruded.

The X and Y Scale and Spin Angle options can dramatically affect the outcome of the extrusion process, as indicated in the following illustration. The Scale option scales the shape of the extruded end of a 3D element, whereas the Spin Angle option applies the angular "twist" to the object. You can specify more than 360 degrees of "spin," which can result in some very interesting output. Keep in mind that the twist is linear (straight line), and therefore large spin angles over short distances can result in very strange output.

X & Y scale = .75

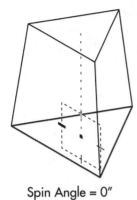
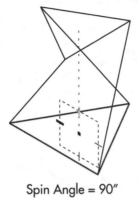

The Spin Angle and X/Y Scale effect on the Extrude Element operation.

Spin Angle = 0″   Spin Angle = 90″

TIP: *The point at which you select the element to be extruded will affect the result, especially with Spin Angle and X/Y Scale enabled. Use the Center Snap mode to produce a symmetrical outcome.*

Construct Revolution

Another major type of 3D element construction is the (surface or solid of) revolution. Whether it is a solid or a surface, this 3D element type starts with a 2D element (the profile) and spins or revolves it around a defined center point, as indicated in the following illustration. The cross section of this toroidal-like element matches the source 2D element. If your 2D element is a closed shape, the output will be toroidal. If it is an open shape, you get more of a lathe-like output, with the selected "profile" turned about the axis you define.

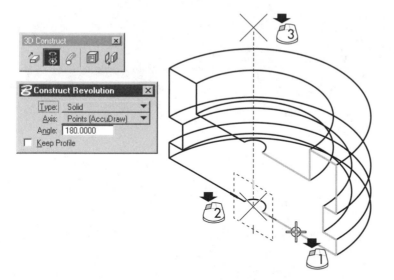

The complexity of the outcome of the Construct Revolution tool depends on the 2D element you select as its profile.

Your second and third data point define the axis around which the revolution operation will take place. The Angle tool setting defines the "sweep" of the revolution. This is not the same setting as Active Angle used in the Rotate Element tool.

When you rotate a profile through 360 degrees, you get a closed shape or solid. When viewed in wireframe display mode (the default), this results in a very simple outline of the object, which can be difficult to interpret. If you want to see more of the revolved solid/surface, perform a series of revolve operations with an angle of 45 or 90 degrees. Multiple data points after the first revolution will result in multiple profiles being displayed around the revolved element. In the following illustration, the revolved element on the left shows the result of a single 360-degree angle of rotation, whereas the element on the right shows one that was stepped through 45 degrees.

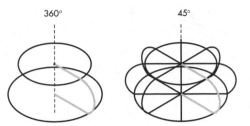

360°   45°

Rotated shape showing 45-degree rotation versus 360-degree rotation.

As with the Extrude tool, you have the option of keeping the original profile (Keep Profile). Again, it is recommended that you enable this option when first trying out this tool, so that you can get a feel for how it works. Furthermore, you should use AccuDraw to define the axis of rotation when first trying out this tool, as there is nothing that says you have to keep the axis of the rotation perpendicular to the selected profile.

Extrude Along Path

A powerful but confusing tool to initially understand is the Extrude Along Path tool. This tool generates a complex solid or surface (again, depending on the output you select) that starts with a path and a profile. The path can be any 2D planar element or 3D nonplanar, non-surface element, such as the 3D linestring or curve string. Note that there is no restriction put on whether the path is open or closed. This flexibility of selecting the path element is what makes the Extrude Along Path tool so powerful.

By Circular

You have two definitions for the profile used for the extrusion. The simplest is the circular or tube-like definition, an example of which is shown in the following illustration. When you set the Defined By tool setting to Circular, two additional parameters are needed, both of which can be input using the data point. However, when first trying out this particular option, use the tool settings fields. Trying to input the inner and outer radii using the mouse can be somewhat confusing.

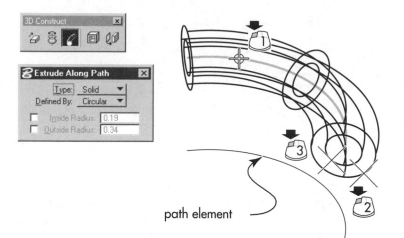

The Circular method results in a tube or pipe-like structure.

path element

If you select only an outside radius and turn off the inside radius option, the resulting extrusion will be solid throughout. Turning on and setting a reasonable value (less than the outside radius) will result in a tube-like extrusion. This is very handy for creating pipes and other design elements that have a wall thickness as part of their geometry.

By Profile

The other major way to generate an extrusion along a path is using the By Profile definition. Where Extrude Along Path gets really interesting is when you decide to define your own profile to be extruded. For this method, you must prepare a profile element to be used as part of the extrusion process. The orientation of this element is not crucial to the outcome, as indicated in the following illustration. However, most experienced users try to orient it at a right angle to one end of the path element. This way, you will get a rough idea of how the extrusion should be situated at its conclusion.

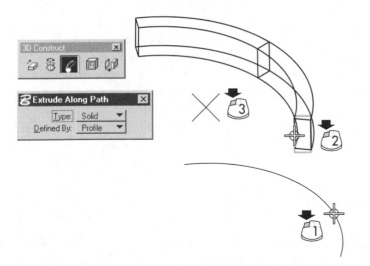

Note how the profile does not have to be adjacent to the path.

You have one other major decision to make before starting the Extrude Along Path operation. How you select the profile will greatly influence the outcome of this operation. If you simply data point on the profile, the resulting element will originate from the location of the profile but will extrude along the path element.

If, on the other hand, you snap to the data point on the profile, the snap point will become the delta distance from the path for the extrusion to be generated. This results in an offset extrusion that follows the path element at the given distance. This is a subtle but powerful feature of this tool.

Imagine having a curve that represents the center line of, say, a bridge. This path element is used to generate many different physical structures in 3D (the bridge deck, the supporting super-structure, and the rails and curbs), shown in the following illustration. By allowing you to define any number of profiles and then extrude these along one common path, the result is a clean design in the least number of design operations. This also allows you to use the selection set to identify all of your profiles at one time and "cast" them along the common path.

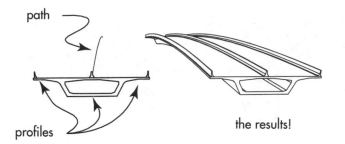

path

profiles

the results!

In this example, a single common path was used to extrude several profiles to create a bridge deck and superstructure. The tunnel was then subtracted from the bridge deck using the Construct Difference tool.

Thicken to Solid

Similar to the Extrude tool, Thicken to Solid creates a SmartSolid element from a 2D planar element profile, as indicated in the following illustration. In this case, the thickness is applied toward one "side" of the profile or centered on the profile itself. This is a handy tool to take essentially flat geometry and give it true thickness or depth.

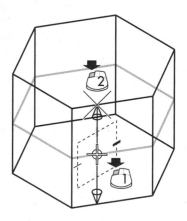

Thicken to Solid is used to add substance to existing 2D geometry.

NOTE: *Prior to the V8 release of MicroStation, this tool would not work on non-rational (b-spline) surfaces.*

Shell Solid

The Shell Solid tool works with certain 3D primitive elements such as the cylinder, slab, and most extruded elements. This tool "thickens" the targeted solids by applying a uniform thickness to the existing geometry. In addition, you are given the opportunity to selectively remove faces of the solid so that you can gain access to the thickened interior of the new SmartSolid element. This is accomplished by data pointing on the highlighted faces after the initial element selection data point, as indicated in the following illustration.

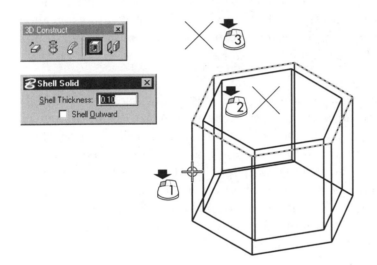

The Shell Solid tool is used to "hollow out" the inside a 3D solid. You identify faces to be removed as part of the shelling operation with your second and subsequent data points.

NOTE: *Shell Solid's dynamic highlighting of faces to be eliminated is often your first introduction to this new type of dynamic behavior. As you move the pointer over the already selected element, faces of that element are automatically highlighted. Data-point select the highlighted face for further action (in this case, deletion). You can expect this dynamic behavior to find its way into more and more MicroStation tools as MicroStation continues to evolve.*

3D Element Modification Tools

To this point, the 3D tools you have seen have all resulted in the creation of new solids or surfaces. In this section, you will be introduced to 3D's equivalents of the Modify Element tools used in 2D design.

NOTE: *The standard element manipulation tools, such as Move and Copy, also work on 3D elements. It is only when modifying the individual faces of these 3D elements that you have to work with special-purpose tools. In the future, you can expect these functions to be merged within the standard 2D modification tools.*

Modify Solid

The Modify Solid tool allows you to reposition any "face" or facet of a solid in a manner similar to the way Modify Element moves individual segments of a line string. It does not, however, allow you to move a single vertex within a solid. Your first data point identifies the target solid.

As with the Shell Solid tool, the faces of the selected solid highlight as you move the pointer over them. A data point on a given face selects it for further modification. Depending on the face, you will be prompted to reposition it along a specific axis from the body of the solid. AccuDraw helps with this repositioning by placing the compass at the point identifying the face and orienting it per the axis along which you can move the face, as indicated in the following illustration. You can enter exact values via AccuDraw or data point for the change.

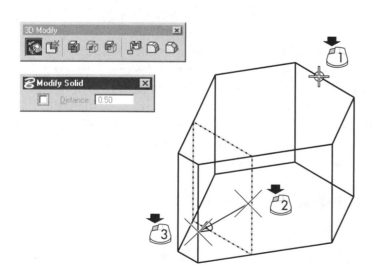

Modify Solid "moves" individual faces of the solid.

NOTE: *The Modify Solid tool is very particular about what sort of changes you can make to a solid or whether they can be performed at all. For instance, the Extrude Along Path tool often results in solid definitions so complex that Modify Solid cannot establish a face's direction of modification.*

Construct Union

Construct Union is one of three Boolean tools you can use to take two or more 3D elements and combine them into one solid, as indicated in the following illustration. In the case of Construct Union, the two or more 3D elements are combined, the result being the sum of all the parts, assuming they touch or overlap.

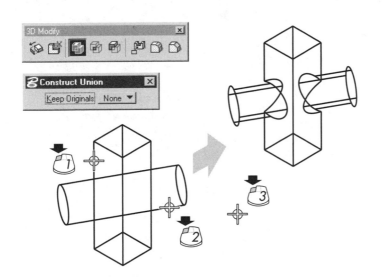

You can data point on each 3D solid or use a selection set to identify the target elements ahead of time.

You will note that when you combine solids in this way, the Construct Union tool removes all redundant faces and generally provides the most efficient single definition for the given 3D geometry.

The Keep Originals Setting

The Keep Originals tool setting is a very important element of the Construct Union tool and the other Boolean construction tools, described in the following material. It provides control over the disposition of the elements you select as part of the tool's operation. The None option deletes all of the source elements as part of the tool's operation. All retains all of the shapes as they are, in addition to generating the new unioned solid. First retains the first element you identify, and Last retains the last element.

Construct Intersection

When executed, the Construct Intersection tool will affect only that portion of any and all solids you select that overlap or intersect with the other 3D elements. The Keep Originals/All option is often used with this tool, as you are likely to use this tool to ascertain the volume of overlap between adjacent elements, as shown in the following illustration.

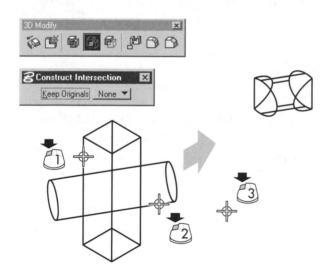

The Construct Intersection tool can be used to accept two or more 3D elements to calculate their common volume of intersection.

Construct Difference

The Construct Difference tool is one of the author's favorite 3D "sculpting" tools. By subtracting one shape from another, you essentially "sculpt" one solid using a second solid as your "knife," as indicated in the following illustration. With this tool, you identify the target element, and then the element to be subtracted from the target (presuming it already overlaps, of course).

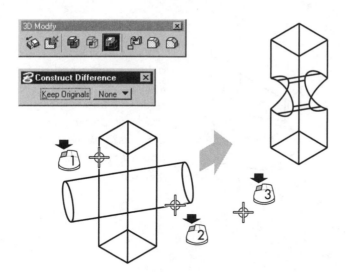

Construct Difference is used to cut away the first solid identified with the volumes of subsequent identified solids.

The Keep Originals/Last option might be your most often used option with this tool because you may need to clear out the volume in the first element to make room for the fit of the second solid.

Cut Solid

Another favorite tool, the Cut Solid tool, acts more like a hot wire when applied to a block of wax. Similar to the previous Construct Difference tool, Cut Solid prompts you to first identify the solid you want to "cut." Next, you identify the 2D cutting elements you want to apply to the solid. The direction the cut is applied depends on the plane of the 2D element. It will always cut perpendicular to the plane of the cutting element, as indicated in the following illustration.

The results you get from this tool depend a lot on the orientation and relationship of the cutting element(s) with the target solid element. For instance, if the cutting element lies entirely within the solid and does not intersect any of the faces, the result is no cut. If you think about it, that would be the same as cutting into the body of the model but not through its sides. The cut would heal as soon as you withdrew the cutting "knife" (or wire).

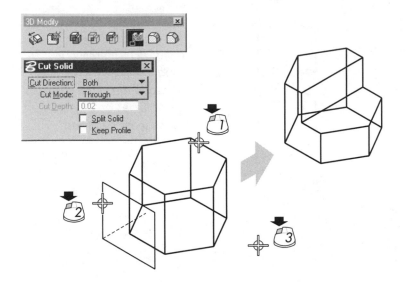

The Cut Solid tool in action. The dashed line is shown as a reference for the lower left corner of the cutting element against the target element.

If, on the other hand, the cutting element does intersect the sides of the solid, you essentially end up with two (or possibly more) pieces of the solid. Part of the Cut Solid operation is to dispose of the unwanted portions of the cut. This is done by identifying the solid on the portion of the element you want to keep. This way, the portion sliced by the cutting element away from your data point is automatically deleted; that is, unless you select the Split Solid option, you will create new solids for each piece of the original solid sliced by the cutting element.

Cut Direction

This option controls the direction in which the cut takes place. The Through option cuts in both directions from the plane of the cutting element, Forward cuts toward the solid from your viewpoint, and Back cuts toward you. The secret to the Forward and Back options is the position of the cutting element with respect to the target solid. If you place the cutting element inside the solid and select either the Forward or Back cut direction, the result will be a pocket cut out of the target solid.

Cut Mode

By default, the Code Mode is set to Through. This means the cut goes all the way through the target solid. When set to Define Depth, you enter a value in the Cut Depth field. When the cutting element is entirely outside the solid, the cut depth is "punched" into the face of the solid to the depth desired (the depth is measured from the first point of contact with the target solid).

Fillet and Chamfer Edges

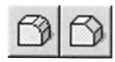

These two tools perform the same task as their 2D brethren, Fillet Element and Chamfer Element. However, instead of modifying a 2D element, the Fillet Edges and Chamfer Edges tools modify the faces of a solid along their common edge. The result is the addition of a new face (curved for fillet, flat for chamfer) and the modification of the two adjacent faces.

Both tools work in a similar manner. You select the target solid, and then the individual edges you want filleted/chamfered. As with several of the previous tools, the edges are interactively selected as you move the pointer over the solid. A data point in space initiates the fillet/chamfer operation, as indicated in the following illustration.

The Fillet tool prompts you for the radius value, whereas the Chamfer tool requires one or two distances back from the edge to chamfer the faces. Select Tangent Edges is a handy option that provides a fast way of selecting all of the common edges around a

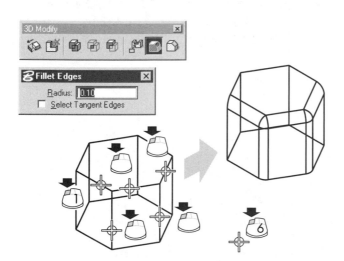

Data points identify the edges to be filleted. The final data point invokes the fillet process.

particular face. With this option selected and the target element identified, the Fillet/Chamfer Edges tool will "follow" the edge currently under the pointer so that you can fillet/chamfer a common edge around a face.

Remove Face and Heal

Remove Face and Heal is a rather interesting tool. It provides you with the ability to selectively remove an entire face of a solid. This can have profound results on the appearance of the targeted 3D element. It is a very powerful tool that, when used judiciously, can perform functions not easily duplicated elsewhere in MicroStation.

For instance, the Fillet Edges option introduces additional rounded faces to a solid as part of its operation. Sometimes you may want to only round over select sides of a face but not others, or reverse the entire fillet operation (assuming it was done in a past design session). The Remove Face and Heal tool allows you to reverse selected rounded faces while retaining others from the same Fillet Edges tool.

The Method tool setting controls how the faces are selected. The Faces option allows you to pick the individual faces you want removed, whereas the Logical method removes all faces that were generated from the same operation (such as removing the same fillet applied to all sides of a face or solid). The Logical Groups option in action is shown in the following illustration.

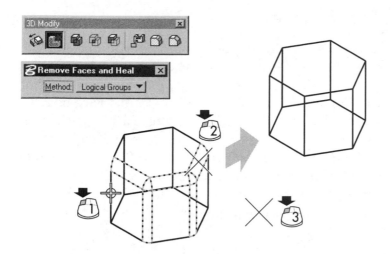

The Logical Groups option enables you to remove all faces created with the same operation (in this case, Fillet Edges).

3D Utility Tools

The last category of 3D tools collects those miscellaneous tools that do not neatly fit in any of the other three categories (Primitive, Construct, and Modify). Most of these are also low-usage tools that benefit the long-term user rather than the person just learning to work in 3D. However, for the sake of completeness, a short description of each follows.

Align Faces

Align Faces performs the same function as Align Edges for 2D elements. It aligns the selected face of the second element with the selected face of the first element. This allows you to develop several separate solid elements and then pull them together along common faces.

Change SmartSolid Display

Change SmartSolid Display allows you to set the number of wireframe rule lines used to represent the curved portion of a solid. This can help you better visualize the wireframe representation of most 3D solids but not all (Revolve Element, for instance).

Extract Face or Edge

Sometimes in the course of the design process you may need to extract one of the faces of a solid for use in other operations. For instance, you may want to use the face of one solid to "cut" another solid using the Cut Solid tool. The Extract Face or Edge tool does just that. When you select Extract Face, each face is highlighted under the pointer after you select the target solid. Selecting a highlighted face will result in the 2D elements that represent this face being generated as a complex shape, observing the element attributes you set in the tool settings box. The Extract Edge option extracts the single edge elements one at a time.

Other 3D Work Environment Considerations

The following are additional aspects of the MicroStation 3D design environment.

Fence Operations

There will be times when you will need to use a fence operation in 3D. When you do, there is a simple rule. What you see is what you affect. In other words, if your display depth is set to display only a portion of the design, your fence manipulation command will only operate on those elements present.

You still need to be aware of the fence lock condition, as this is still active. Fence clip lock will cut those displayed partial elements that fall within the fence. This includes solids and surfaces. The fence is projected perpendicular to your view. If you place a fence in an Isometric view, and then use the Delete Fence Contents with clip mode, the fence will act as a cookie cutter right through whatever elements it crosses in the Isometric view. This can have some interesting (and sometimes entertaining) effects on your design.

Boresite Lock

When using tools such as Change Element Attributes or Delete Element, you may want to grab only those elements at the active depth of your view. This way, other elements that may be present on the view, but not at your active depth, will not be highlighted when you click near them.

This is in marked contrast to the fence commands that will affect any element visible and within the fence. To turn Boresite off (thus restricting element selection to your active depth), select Boresite Lock from the Locks option menu (on the status bar) or the Locks settings box.

Using the Fit View Command

One last note about view control. When you have gotten really lost in your design cube, as a last resort you can invoke the Fit View view control with the Expand Clipping Planes option selected. In 3D, this control looks for the extents of your elements in all three directions and sets the chosen view to display them. It also resets your active depth for that view to the center of the display depth. In this way, you have a starting point from which to navigate to the area of interest. In addition, the View Previous view control comes in handy when you get "all twisted around."

Summary

As mentioned in the opening of this chapter, 3D is a subject in and of itself, more than worthy of an entire book. In fact, you could write several books on the subject of applying 3D modeling techniques to all sorts of design situations. Hopefully, this short introduction to the subject will pique your interest in 3D and will help establish a "framework" around which you can build your 3D modeling skills.

INTRODUCTION TO IMAGE RENDERING
Getting Those Photorealistic Images

MOST READERS OF THIS BOOK will have undoubtedly seen some of the 3D illustrations that grace MicroStation's sales literature, or the cover of *MicroStation Manager* magazine, examples of which are shown in the following illustration. All of those images were generated using MicroStation-native 3D tools and photorendering capabilities. In this section, you will be introduced to MicroStation's basic rendering tools and techniques.

What Is Rendering?

Rendering is the process of creating a photograph of your 3D design data. Normally during the design process, most people use the wireframe display, in which the edges and some of the key features of each design object are displayed. This display method has the advantage of speed; that is, it does not take long to update your screen. However, in real life, most of these objects would be opaque. Depending on your point of view, objects closer to you should obscure those they are in front of along your line of sight.

*Examples of recent
MicroStation
Manager covers
featuring pictures
rendered with
MicroStation.*

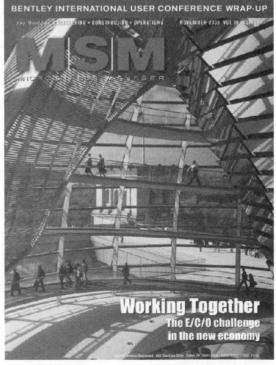

For MicroStation, displaying your 3D objects in this real-world manner requires a lot of extra computational "horsepower." Depending on the realism you need, this cost can be trivial or very expensive (read, time consuming). To provide the best balance of computational requirements versus final appearance, MicroStation comes equipped with several rendering algorithms, or modes, examples of which are shown in the following illustrations.

Starting with the now-familiar wireframe mode, you can opt to render your views in varying degrees of visual realism, from simple cartoon-like hidden-line output to extremely realistic ray-traced imagery that is nearly indistinguishable from photographic work. The output you choose depends on your imagery needs and the available computer resources.

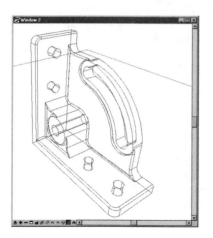 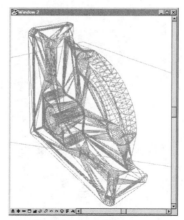

Left: Wireframe
Right: Wiremesh

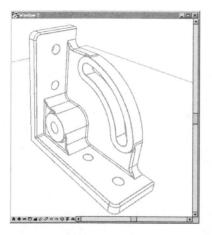 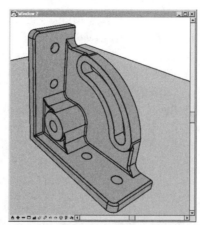

Left: Hidden Line
Right: Filled Hidden Line

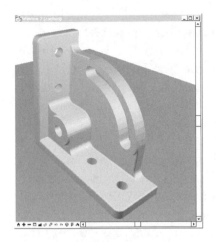

Left: Constant
Right: Smooth

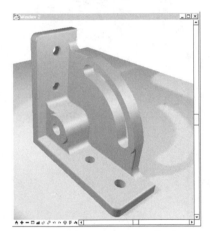
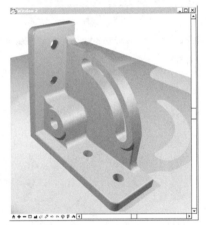

Left: Phong
Right: Raytrace

Rendering as a Design Tool

Generating rendered images is useful for more than just creating interesting pictures. Often, when you are dealing with extremely complicated models, blatant spatial errors are not always apparent in wireframe mode. However, viewing that same model using one of the more advanced rendering options can uncover potential design flaws. By generating a rendering of this area of the design, shown in the following illustration, you can see that there is an unintentional intersection between a pipe and a wall.

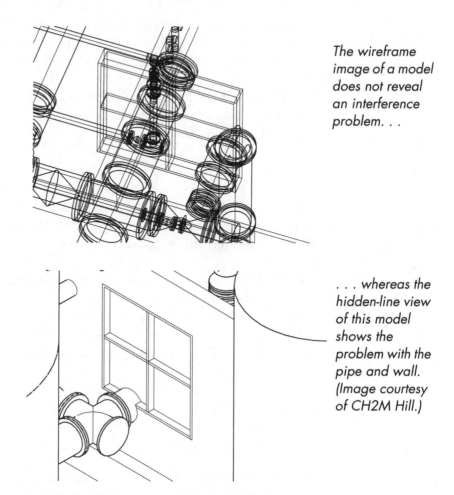

The wireframe image of a model does not reveal an interference problem. . .

. . . whereas the hidden-line view of this model shows the problem with the pipe and wall. (Image courtesy of CH2M Hill.)

MicroStation's Rendering Methods

Technically, what you see in the MicroStation's view windows are rendered images of your design data (model). By default, MicroStation generates a wireframe rendering of your design; however, as mentioned previously, there several other rendering options or methods available. These include:

Wireframe: The default display mode in which only edges of the 3D design elements are displayed. All elements appear transparent.

Wiremesh: Similar to wireframe, with the addition of curved elements displayed with a mesh over their surface. Helps establish subtle undulations in some surfaces.

Hidden Line: First "solid" rendering mode, faces of objects are rendered opaque in the same color as the background color. Edges of elements are shown in the element color.

Filled Hidden Line: Faces of objects are rendered opaque, but in the same color as the element. The edges are highlighted in a color 0 (normally, white).

Constant: Curved elements are broken up into facets and rendered with varying degrees of color, depending on the lighting conditions and pattern map assignments. Sometimes called faceted rendering because curved elements such as spheres take on a distinctly faceted appearance.

Smooth: Also known as Gouraud shading, this rendering mode shows curved surfaces in fine detail, with the color value calculated at the edges of the element surface or face and "blended" toward the other edges of the element face. This is the highest-quality rendering mode directly supported by QuickVisionGL (OpenGL).

Phong: High-resolution rendering mode in which the color of each pixel of the image is calculated on its own, based on lighting and material assignments. Also supports shadows and true variable material transparency.

Ray Trace: High-fidelity rendering nearly indistinguishable from real life. Computationally intensive ray-trace rendering takes into consideration the reflectivity of each element surface, light dissipation, refractivity, and other factors.

Radiosity: A refinement of the ray-trace mode, radiosity adds the characteristic of color properties of each surface as light is reflected. This results in very realistic artificially lit scenes (for instance, brown objects result in an amber light reflection).

The higher the quality of imagery, the more "expensive" it is to produce. This cost is not linear. With Ray Tracing, Radiosity, and Particle Tracing, it can take several hours to render what in Smooth takes just a few seconds!

MicroStation's QuickVisionGL Rendering Technology

For a few of the rendering modes, there is an exception to the rule that better rendering means longer computing time. MicroStation's QuickVisionGL technology provides a bridge between high-performance, OpenGL-accelerated graphics hardware and MicroStation's rendering methods.

You have probably seen those "shoot-em-up" games, such a Quake or Duke Nukem, and no doubt marveled at how quickly the 3D, textured images respond to the player's input. Games such as these often take advantage of OpenGL-accelerated graphics hardware to dramatically improve the quality of play.

OpenGL is essentially a 3D software "library" (or protocol) that lets software such as MicroStation harness the power of specialized graphics hardware for the purpose of quickly rendering complex models. OpenGL supports only certain rendering methods, and is therefore not a solution for every rendering need.

When enabled, QuickvisionGL can accelerate the wiremesh, hidden-line, constant, and smooth rendering output by using OpenGL-enabled video cards, of which there are literally dozens to choose. In addition, Microsoft's latest round of Windows operating system software provides software and hardware OpenGL support so that even simple operations, such as screen savers, operate much more smoothly. To see if MicroStation takes advantage of any OpenGL technology in your computer, perform exercise 13-1, which follows.

EXERCISE 13-1: DETERMINING OPENGL SUPPORT

1   Open the Rendering View Attributes dialog box (Settings menu > Rendering > View Attributes), shown at right.

2   Select Smooth Shading from the Display option menu.

3   Make sure Graphics Acceleration is enabled.

4   Select the Settings button (adjacent to the Graphics Acceleration option).

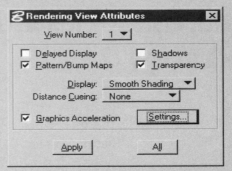

Rendering View Attributes dialog box.

The QuickVisionGL Settings dialog box appears. If there is no OpenGL graphics card in your computer, or if the video driver associated with it has not been properly installed, you will likely see the GL Context line shown in the following illustration.

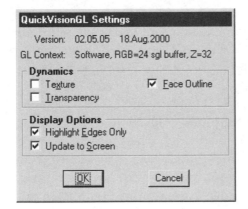

Note the GL Context line: Software.

This means that the rendered image will be generated by your computer's main processor alone. This is not all bad because of the vast improvement in computer performance in the last few years. However, if you know that you have an OpenGL-enabled video card in your system and you do not see settings similar to those shown in the following illustration, chances are you have a video driver problem within Windows 9X(ME), NT, or 2000, all of which support OpenGL in their core operation.

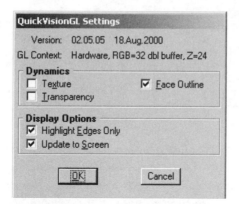

Note the GL Context line: Hardware.

QuickVisionGL accelerates the display of several rendering modes, from Wiremesh to Smooth Shading (but not Phong or Ray Trace). It can also render any pattern maps you assign to objects in your design, as well as transparent objects (type of pattern map). QuickVisionGL will also display element edges and even hidden lines.

Change View Display

You can quickly change your rendering mode by using the Change View Display tool located in each window's view controls, shown in the following illustration. To change a view's rendering mode, simply select the Display mode option in the tool settings window. If a mode supports edges and hidden edges display, these two options will also be available.

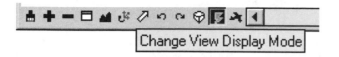

Change View Display tool.

The fastest way to take advantage of QuickvisionGL is to change one or more of your view window's default rendering modes. Although you can do this in the Rendering View Attributes settings box, this is more easily accomplished by using the Set View Display Mode view control from the view's view controls, which is shown in the following illustration.

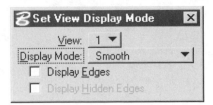

Set View Display Mode view control.

TIP: *At the time of publication, no OpenGL video card vendors provide hardware acceleration of the Phong render mode. If selected as the view's display mode, Phong will slow down the display of this view in most window-related operations. Select Smooth Shading for the best balance of appearance versus display performance.*

If the particular rendering mode you choose supports it, you can also turn on the Display Edges and Display Hidden Edges options, which can help further define the rendered image. These options are shown in the following illustrations.

Smooth Shading provides support for displaying element edges and hidden edges.

The Render Image Tool

There are times when you do not necessarily want to set a view to a high-cost rendering mode but you want to do a quick rendering of the view. Located in the 3D View Control toolbox, the Render View tool provides you with the ability to generate a one-shot rendering of any view, or of a fence or single element within a view, as shown in the following illustration.

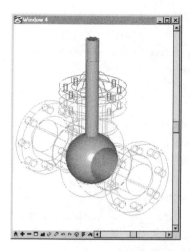 

Example of rendering a single element and rendering fence content.

When you update the view using any of the view tools, the rendered image is replaced by the current rendering mode set in the view's rendering attributes settings box.

Introduction to MicroStation's Advanced Rendering Features

MicroStation has had rendering capabilities from its inception. Version 3 implemented rendering by spawning a separate application that processed a 3D design file, whereas in version 4 rendering was tightly integrated into the normal MicroStation graphics environment. Version 5 saw the introduction of several major rendering enhancements, including the first availability of true photographic rendering add-on products (e.g., MicroStation Masterpiece). With each generation, MicroStation offers more incredible rendering capabilities.

Using Cameras in MicroStation

There is one other aspect to a realistic rendering, and that is applying a visual perspective to your model. This refers to the phenomenon of the vanishing point or horizon. When viewing "the real world," distant objects tend to merge or vanish toward a horizon. This is due to the properties of the eye and principles of optics. Suffice it to say that if this perspective point of view is not taken into consideration, no manner of rendering mode tuning would result in a truly realistic image.

In MicroStation, the perspective aspect of the rendering process is controlled by the placement and configuration of one camera for each view within the 3D model. A camera is an excellent metaphor for this process. In essence, whenever you are preparing to render a given view, you are essentially setting up a "photo shoot" not unlike that which occurs in nearly every photographer's studio on a daily basis. As such, you must pick out the appropriate place to set your camera with respect to the subject of the photo "shoot," set up the lighting, and choose the appropriate lens (the degree of foreshortening to be applied to the image)—and away you go!

Setting Up a Camera in a Hurry

The quickest method of setting a view's camera is to use the Change View Perspective tool (see following illustration) located in each view's view controls tool bar. By visually adjusting the view's perspective angle, you are essentially changing your camera "lens" on the fly. You can either adjust this perspective using an animated cube (dynamic option turned off), which is quite often the preferable method, or dynamically adjust the actual view "scene" (dynamic option enabled). The latter is best used in simple models, where MicroStation does not have a large burden to redisplay the dynamic view; otherwise, you will quickly lose your bearings and end up with a highly distorted view.

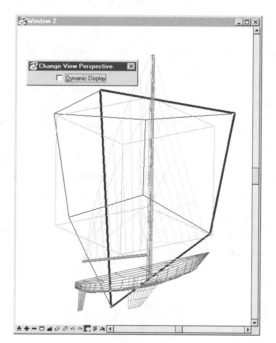

Change View Perspective tool in action.

Precise Camera Setup

The much more precise method of adjusting a view's camera settings is to use the Camera Setup option (Settings menu > Camera > Setup). From this dialog box, you set the target (what the camera is pointing toward) and the "eyepoint" of the camera (where the camera is set up, as on a tripod), using any of the methods of specifying a location in your design file (e.g., AccuDraw, 3D data point, or active depth). Finally, you choose the "lens" of the camera just as you would a 35-mm SLR camera. You can set your own focal length and angle of view cone, as well as select one of a number of preset lenses, from "fisheye" to telephoto/telescopic.

There are several tools associated with adjusting each camera's parameters. You can move the camera itself, change the target (where the camera is pointing, similar to the Pan View view control), change its lens, or turn it off completely.

Pattern Mapping

MicroStation provides extensive support for the use of pattern maps. In essence, pattern mapping takes a raster bitmap image and drapes it over the surface of the elements being rendered. Pattern mapping, also known as material or texture mapping, brings a level of realism to MicroStation that must be seen to be believed.

As an example, to render a brick facade on a building, you assign a brick pattern map to the elements that constitute that facade, such as that shown in the following illustration. Then, when the image is rendered, the normal color attribute associated with those elements is replaced by the assigned brick pattern. The result is a photorealistic image. Pattern and bump maps are part of MicroStation's material definition and assignment facility.

Facade with brick pattern.

Bump Maps

In conjunction with pattern mapping, MicroStation supports an additional texture enhancement called bump mapping. You may have noticed that in the previous example the brick facade appeared a little "flat" (more 2D than 3D in appearance). Because brick has a rough texture that is almost impossible to describe with a single raster image, MicroStation utilizes a technique in which the primary bitmap (the brick) is highlighted with a secondary highlight (or map) of the texture's "bumps" (thus the name). Depending on the lighting and other factors, the bump map can greatly enhance the final appearance of the textured element in the image, an example of which is shown in the following illustration.

Bump map rendering of the brick facade.

Casting Shadows

Shadows also provide additional realism to a rendered image. Phong and Ray Trace modes support the generation of shadows for local light sources you insert into the design file itself and for simulated solar lighting. Incorporating shadows into a rendered image improves the appearance of architectural models by providing a good feel for what the structure will look like on a given day and time. This is also useful when evaluating the effect of the sun on building overhangs (also known as solar studies).

Transparency

Elements can also be assigned a variable degree of transparency. Objects behind such transparent elements become visible according to the amount of transparency assigned to the obscuring element. This is controlled by the material definition and assignment settings.

Anti-aliasing

The Anti-aliasing feature reduces the jagged edges ("jaggies") that are particularly noticeable on low-resolution displays. This is accomplished by blending the sharp edges of an object into the background color.

Rendering Setup

Although you can simply render any view at any time, good rendering results require a fair amount of setup and "tweaking." There are several settings dialog boxes associated with setting up the rendering process. After a fair amount of setup, you render a view with one of several rendering methods. To improve the quality of the rendering process, MicroStation provides numerous rendering settings in the settings dialog boxes. These settings, described in the sections that follow and shown at right, are accessed via the Settings menu (Settings > Rendering).

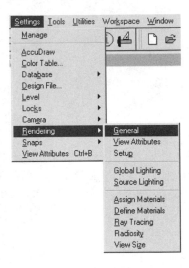

General Rendering Settings

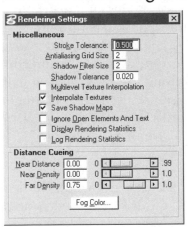

This settings box is the first place you start when fine-tuning your rendered image output.

The first of the rendering settings dialog boxes, shown in the previous illustration, is where you set the more advanced features of the rendering process. The most notable option found here is the set of Distance Cueing values. These are used to simulate the effect of atmosphere on your rendered image. You can adjust the distance cueing from none, such as for a clear day with zero humidity, to a foggy San Francisco morning. In addition, you can select the color of the fog itself (maybe a brownish color to simulate smog?).

Rendering View Attributes

Used to set specific view attributes related to rendering, view attributes options have a profound effect on your rendered image's appearance. Three options found here are of the most interest, which are shown in the following illustration. The Pattern/Bump Maps option activates the use of MicroStation's advanced rendering capability, the ability to use pattern maps to add texture to your rendered objects.

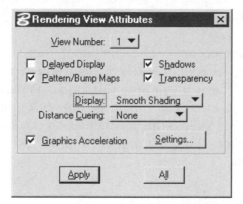

This settings box is important for turning on fundamental rendering attributes for each view. Note, especially, the Pattern/Bump Maps, Shadows, and Transparency options.

Shadows turns on MicroStation's shadow-casting simulator used with the high-end rendering methods. You can get reasonably good shadows, which are useful in adding an extra level of realism to any rendered image. Transparency allows you to render "see-through" objects, such as windows and glass vases.

The Display field controls whether you want to enable real-time rendering on a specific view. Normally set to Wireframe, you can elect to have a view always shown as a rendered image. This is great when you have a simple image with some confusing 3D features. However, the trade-off is the increased time it takes for MicroStation to keep this view updated. Distance Cueing enables the "atmosphere" effect previously described.

Lighting

The sections that follow describe lighting parameters associated with rendering. These include options that address global lighting, controlling lighting (including ambient, flashbulb, and solar light), and source lighting.

Global Lighting

Many of the rendering methods perform not only a hidden-line function but calculate the brightness of the various surfaces within your design. How bright an object appears depends on a number of variables. The most important aspect is the object's angle to the light source.

When you select any of the shading options with the default lighting source, MicroStation sets the light source as that of the viewer. This would be as if you were shining a flashlight at the view. And just as with a flashlight, the image does not quite look real. Think of how things look in the headlights of a car. To better control the look of a shaded image, MicroStation provides a number of lighting options.

Controlling Lighting

When you select the Rendering settings dialog box from the Settings pull-down menu, you are presented with a number of lighting options, shown in the following illustration of the Global Lighting dialog box.

You have three choices: Ambient, Flashbulb, and Solar. Although you can turn all three on, this diminishes the effect of the individual lighting options.

Ambient Lighting

An artificial light source, this option controls the uniform illumination surrounding your rendering. Similar to general office lighting, Ambient Lighting is good for highlighting model details that would otherwise be too dark to see. The default value of 0.10

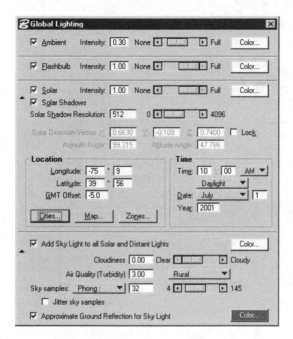

The Global Lighting settings box is used to set all of the parameters related to lighting in your design.

(i.e., 10%) is a good value. Anything over 0.4 will wash out the colors of your rendering.

Flashbulb

The default lighting source, this is the light shone on the elements from your viewing perspective. This can be adjusted from 0.0 to 1.0 (100%). Lower values darken the colors used to shade surfaces.

Solar Light

A very interesting option, the Solar Light source simulates the direction and intensity of light as it would appear coming from the sun. By setting the latitude and longitude, date and time, you can approximate how your project would appear on a given day. This is most appropriate for architectural renderings, but can be applied to any project. You could enter this information via the fields provided. However, a fun alternative is to call up the map using the Map option and directly select your location, as indicated in the following illustration.

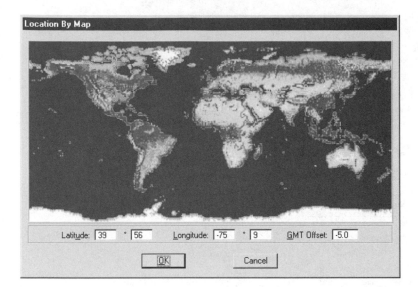

The Map option allows you to select your project's location from a world map.

You can also choose your location from a list of cities of which MicroStation knows the whereabouts. However, this is not as much fun as the "You Are Here" selection from the world map.

You can also specify whether you want solar shadows. This option is applicable to the Phong rendering method only, and requires that the Shadows option be enabled in the View Attributes settings box.

Source Lighting

In addition to the global lighting just described, MicroStation supports fully customizable source lighting. This refers to the ability to place lighting sources within your design that can, in turn, cast light upon, and around, your model. The source lighting settings box is shown in the following illustration.

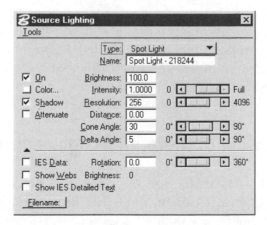

This settings box is crucial in placing light sources within your drawing. It includes both settings for light sources and commands for placing the lights themselves.

The Source Lighting settings box assists you in the placement and configuration of these sources, providing you with quite a few options. MicroStation uses specialized cells to simulate these light sources. The Tools pull-down gives you a number of commands for the placement and editing of these light sources, as indicated in the following illustration.

Source lighting placement and editing options.

Defining and Assigning Materials to Your Design

Rather than limiting your rendering capabilities to the use of color and some minor texture control, MicroStation incorporates truly sophisticated tools that give you fascinating control over the appearance of the final rendered image. Although you can still render images using the basic colors of the elements themselves, with the use of material assignments and pattern/bump maps, you can create images that are almost impossible to distinguish from normal photographs. Called Photorendering, the use of scanned-in samples of real-world textures means that you can apply stucco, shingle, and other effects to your model, just as you would to a physical model of a project.

Overview of Material Assignment and Definition

On the surface (pardon the pun), this process may look rather magical, and it can be somewhat intimidating. However, if you look at how the system works, it becomes obvious that it is really rather simple. The following is a summary of the steps involved:

1   Define material palettes.

2   Assign pattern maps to specific material definitions.

3   Assign materials to specific parts of your design.

4   Select rendering options.

5   Render your view.

Material Definition Files

To start this process, you need to have material definitions ready. MicroStation comes with a series of predefined materials for use in your design. However, there is nothing stopping you from creating your own or modifying existing definitions. The hierarchy of the various files associated with materials is shown in the following illustration.

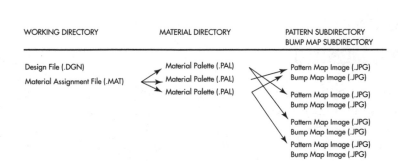

A "road map" of the relationship among the various file types associated with the material definitions and their assignment to your active design file.

As you can see, material assignments are made in a separate file using the .MAT extension. Essentially a text file, the material assignment file contains a list of which material palettes to use, and what specific materials to map to which object.

Materials are mapped to a combination of levels and colors. In other words, each level can have up to 256 materials assigned to it (one for each color). This works out to 16,128 distinct materials, which is more than you will ever use.

Material Palettes

Sort of like an artist's palette, a material palette file contains definitions for individual materials, including their various parameters and, most importantly, the names of the pattern maps associated with them. As you can see in the preceding diagram, a palette can point to many different pattern map files, but each material found in the palette can have only one pattern map.

MicroStation is delivered with a number of predefined material palettes, summarized in table 13-1, which follows. The palettes (found under the *USTATION* directory in the *MATERIALS* subdirectory) are organized along the type of material definitions they contain.

Table 13-1: Material Palettes Delivered with MicroStation

File Name	Description
BACKDROP.PAL	Various backdrops useful as active backgrounds
BACKYARD.PAL	Fences, mulch
BRAKE.PAL	Used by the brake design file in the default workspace
CARPET.PAL	Samples of carpeting material
DOOR_WIN.PAL	Samples of doors and windows for mapping to rectangular objects
FABRIC.PAL	Fabric samples
FINISH.PAL	Variety of material finishes, such as plastic, glossy, and metal
FLORA.PAL	Trees, bushes, and other plant life

588 **Chapter 13: Introduction to Image Rendering**

File Name	Description
GLASS.PAL	Samples of glass, evidencing transparency
GRANITE.PAL	Samples of granite patterns
HOMEOFIC.PAL	Book covers, computer fronts, clocks
KITCHEN.PAL	Typical kitchen appliances
MARBLE.PAL	Samples of various marble patterns
MASONRY.PAL	Various masonry patterns
METAL.PAL	Definitions of metallic materials
PEOPLE.PAL	Examples of people for use with shape elements (this is also where the author can be found)
RUG.PAL	Samples of rugs
SURFACE.PAL	Miscellaneous samples of surface treatments
TILE.PAL	Examples of various tiles
TOWER.PAL	Used by the "tower" design file in the learning workspace
VEHICLE.PAL	Examples of automobiles, including a Jeep
WATER.PAL	Examples of water (no, really!)
WOOD.PAL	Wood varieties

As you can see, these files contain a wide variety of material definitions. In most cases, a pattern and even a bump map file are associated with each material sample.

Define Materials Settings Box

Material definitions are created via the Define Materials settings box (Settings > Rendering > Define Materials), shown in the following illustration. You can, of course, create or modify each material definition if you desire. However, before you go out and wholesale change the world, it is highly recommended that you back up the palettes delivered with MicroStation, or create your own material palettes.

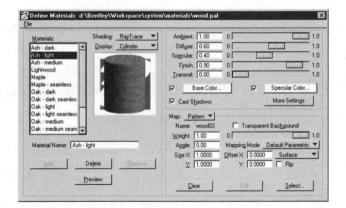

The Define Materials settings box.

Using the Define Materials settings box to create and edit individual material definitions, a bitmap image in any of the supported formats (such as TIF, JPG, and so on) can be assigned to a material. Note the various settings associated with each material definition.

As you can see, there are numerous settings associated with each material; more than can be covered in an overview. However, you should note the use of the Pattern Map section, which is used to designate the .TIF image file to be used with the material name.

Assign Materials Settings Box

After you have decided which materials you want to use, you must assign the chosen materials to your design file. This is done using the Assign Materials settings box (Settings > Rendering > Assign Materials), shown in the following illustration.

The Assign Materials settings box is the main tool for assigning individual levels/colors to specific materials.

A distinctly visual settings box, Assign Materials provides you with a preview of the material you are currently working with. To assign a specific material, perform exercise 13-2, which follows.

EXERCISE 13-2: WORKING WITH THE ASSIGN MATERIALS SETTINGS BOX

1 Open the desired material palette (Assign Materials > File > Open Palette).

2 Select the desired material from the Palette listing (middle column).

3 Assign the material to a specific level/color (Assign Materials > Tools > Assign or Assign by Selection).

The Assign command requires you to enter the level and color for the assignment. An alternative is the Assign by Selection command. This command allows you to click on an element to which you wish to explicitly assign a material definition. This overrides the normal level/color assignment and is good for exceptions. However, as a general rule you should use the level/color assignment, as this allows you to assign a material to related elements (a brick facade for the front of a building, for instance).

Saving Rendered Images

You probably render individual views using MicroStation's rich set of rendering tools. What do you do, however, when it is time to create hardcopy of such renderings? You could use MicroStation's built-in screen capture command (Utilities > Image > Capture) to save a rendered view to disk. This, however, limits the final resolution of the image to that of your video screen.

Most color-capable output devices print at a higher resolution than most video monitors. Your video display is usually limited to between 72 and 90 dots per inch (dpi). Color hardcopy devices (laser printers, dye sublimation printers, and thermal printers) can print anywhere from 200 to 400 dpi. When you print your screen-captured image at these resolutions, the result is either an

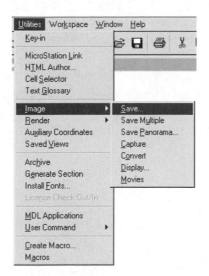

Use Save Image to direct rendered image output to a file.

image that is a fraction of the page size or a very blocky-looking hardcopy.

To get around this, MicroStation provides a mechanism for rendering images at any size. Called Save Image, this feature is found on the Utility pull-down menu. Invoking this command results in the Save Image dialog box, shown in the following illustration. Here, you select the various options associated with this rendering process.

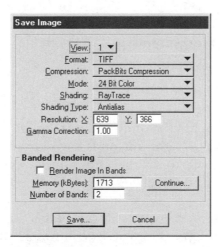

The Save Image dialog box provides the mechanism for creating larger-than-life (view) images from your design file model.

Although the Shading option may appear to be your most important consideration, other key decisions are also made here. For the process to work at all, you need to prepare a view with all of

the parameters necessary to orient your view. This includes lighting, camera parameters, levels, and so on.

Next most important is the format of the resulting image file.
Because different printers and graphics programs require different data formats (Tell me about it. This book required the use of
no less than three different formats!), you must select the appropriate file format.

- Img
- Img (24-bit)
- Ingr. COT
- Ingr RGB
- Ingr RLE
- JPEG (JFIF)
- PCX
- PICT
- PostScript
- Sun Raster
- Targa
- TIFF (Compressed)
- TIFF (Uncompressed)
- Windows BMP
- WordPerfect (WPG)

As you can see, MicroStation supports quite a variety of formats.
Choosing the one right for you is beyond the scope of this book.
However, it is safe to say that with a little research and trial-and-
error, you will find the image type most suited to your needs.

The Compression field is used with only a couple of the formats,
most notably the JPEG format. By controlling the amount of signal
loss, you can balance the space the image occupies on your hard
drive against the quality of the image. This can be a real consider-

ation when generating large, 24-bit (16.7 million possible colors) images, where a single file can run into the multimegabyte range!

The Mode field allows you to adjust the color depth generated in the final image. As just mentioned, MicroStation supports millions of colors. However, if your final output is a black-and-white document (such as this book), there is no sense in generating a 16.7-million color image just to crunch it back into a grayscale image. It also reduces the time it takes to generate the image. The Shading field offers you the same basic rendering types as the View Rendering functions.

The Resolution fields are the real reason you selected the Save Image dialog box in the first place. This is where you can set the exact size of the final image. By keying in a number in the X or the Y field, you set that aspect of the final image. One point to note is that the ratio of X to Y is set by the view you have selected in the View field, so you will want to make sure it is adjusted to the correct X-to-Y ratio.

Gamma Correction is a little-understood adjustment field. When printed, most screen-shot images appear very dark. This is due to the difference in how colors are created. Without going into a long dissertation on transmissive versus reflective colors, this difference requires you to do some "tuning" of the output. This is the purpose of the Gamma field. It allows you to bring up the middle tones of your image without lightening true black or darkening true white regions. A higher number results in a lighter and better-printed image.

TIP: *Selecting the Save button starts the rendering process. You should be warned that selecting a large image size, Phong, and 24-bit color will result in a long computation session. It could be hours or even a day before MicroStation returns from this process. You should plan such a session at the conclusion of your workday, or prepare yourself for a boring couple of hours.*

Exporting Visible Edges

Before concluding this brief look at the rendering capabilities in MicroStation, there is one more often overlooked but very nice "rendering" technology in MicroStation. Called Export Visible Edges, this tool creates, without a doubt, the highest-quality illustrative output possible from MicroStation.

Instead of rendering a particular view as a raster image, Export Visible Edges generates a vector 2D or 3D that consists of those element edges visible from a given point of view (including camera perspective), an example of which is shown in the following illustration. The result is very clean illustrative output useful for any variety of purposes, including product design, conceptual design, and technical illustration. In fact, all of the line-art illustrations in this book were generated using Export Visible Edges and creative use of MicroStation's plotting output capabilities!

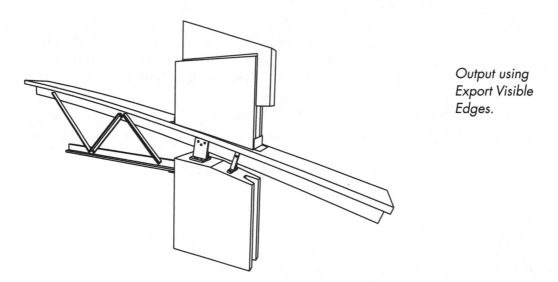

Output using Export Visible Edges.

Furthermore, once you have the 2D visible edge output (preferred by the author over 3D), you can selectively apply additional element attributes to the results, including simulated hand-sketched line styles, as well as further annotation (text, leader lines, callouts, and so on). An example of annotation is shown in the following illustration.

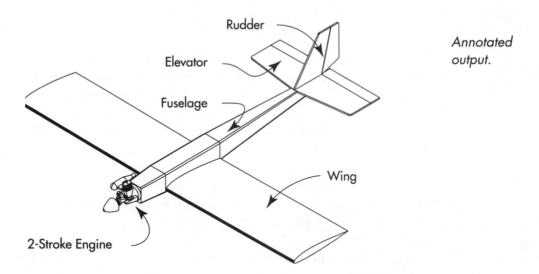

Annotated output.

The Export Visible Edges process is started by setting up a given view, as you would for any rendering. Primarily this means setting up your camera properly. Next, you open the Export Visible Edges dialog box (Files menu > Export > Visible Edges). There are a lot of parameters associated with this tool, but only a few of them need to be adjusted from their default settings, as indicated in the following illustration.

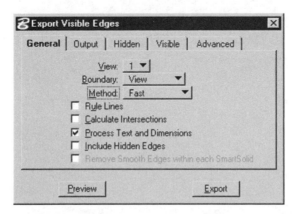

Export Visible Edges dialog box settings.

Once you click on Export, MicroStation processes the output from your view into a new design file, with the same settings as the active file. After processing the output, all that remains to be done is to open the "hidden line" file (default extension is *.hln*) and evaluate the outcome.

Summary

This chapter has explored the scope of image rendering in MicroStation. Image rendering as a "science" and as a topic of special focus in MicroStation is worthy of a volume in itself. This chapter has presented an overview of the nature of rendering as a practice, as well as the significant practical areas to be considered in using MicroStation's rendering capabilities. Among these are MicroStation's rendering modes, OpenGL and MicroStation's QuickVisionGL rendering technology, advanced rendering features, defining and assigning materials to a design, saving rendered images, and exporting visible edges.

In applying MicroStation's rendering features, you will want to consider such things as view settings; the use of "cameras"; pattern mapping, bump maps, shadows, transparency, and anti-aliasing; and rendering setup and lighting parameters. In the area of defining and assigning materials, the chapter provided an overview of material definition files, material palettes, material definition settings, and material assignment settings. The chapter concluded with an explanation of the Export Visible Edges tool, used to achieve the highest-quality MicroStation image output. Now, go forth and create those dazzling, photorealistic pictures!

CUSTOMIZING MICROSTATION

Becoming a Power User

A KNOWLEDGE OF MICROSTATION FUNDAMENTALS is essential to creating any design drawing with the software. To get any useful work done, you must know how to create a new design file, place primitive graphic elements, manipulate your views, and snap to existing elements. This chapter assumes you are already familiar with these concepts. But even a good working knowledge of the drawing and editing commands does not a power user make.

There is much more to MicroStation than just the flexibility of its basic drafting tools. Understanding why the tools take on the default settings they do, where to change defaults so that Micro-Station works the way you prefer, and how several commands can be grouped to act like a new single command is the first step in taking control of MicroStation's design environment and command set.

There is a belief that customizing MicroStation requires programming skills. On the contrary, there is set of interactive, dialog-box driven tools that enable you to mold MicroStation to your company, discipline, project, or user needs without such skills. Customizing MicroStation can be thought of as the task of changing its default behavior. To be sure, MicroStation supports a sophisticated programming environment, but an enormous amount of customization can be done without even venturing there. The focus of this chapter is the exploration of end-user customization tools. The chapter also briefly discusses the MicroStation programming interface.

TIP: *Although the demanding nature of a production CAD environment in a design office leaves little time for exploring MicroStation customization, it offers an ideal opportunity to identify processes that do not seem to work the way you want. Always keep a "wish list" notebook handy, and when you find yourself repeating the same sequence of commands or inadvertently making the same mistake over and over, jot it down. Going through your notes later, when you have time to customize MicroStation, you will find a treasure chest of ideas to implement as custom tools.*

The first section of this chapter is titled "Sorry, But I Don't Program," which introduces you to the easy-to-use, interactive customization tools in MicroStation. Here you will learn about:

- User preferences
- Workspaces
- Settings Manager
- Function keys
- Custom line styles
- Glossary

The second major section of this chapter is titled "Programming MicroStation," which highlights the more involved methods of customizing MicroStation that require some background in programming. These aspects are not covered in detail here. However, after reading this section, you will at least understand how MicroStation can be programmed, and the level of effort needed to tackle the task. In this section you learn about:

- User commands
- The MicroStation development environment

- MicroCSL
- Dynamic link modules (DLM)

Sorry, But I Don't Program

All customizable options described in this section require no programming experience. You will be able to establish CAD standards at your office with the Settings Manager, modify configuration variables so that different projects have easy access to cell libraries specially made for them, or create custom line styles, all from within MicroStation itself. For some tasks, however, you may want to start by copying existing configuration files and modifying them with a text editor (more on this later).

User Preferences

The Preferences dialog box (Workspace menu > Preferences) is the main control center for managing nearly all user preferences. The dialog box is split into two panes; the left pane organizes the preferences by category, and the right pane provides access to the individual settings associated within each category, as shown in the following illustration.

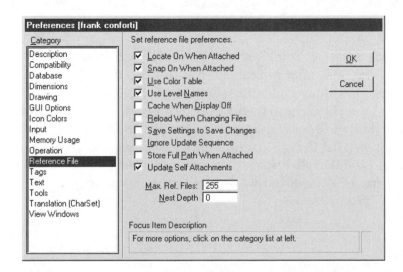

Preference categories and settings within the Preferences dialog box.

Some common preference items you may wish to change are:

- The look of MicroStation's dialog boxes, whether Motif or Windows

- Reference file settings

- Text display options

- Appearance of the user interface

- Line weight display settings for the screen

To change a user preference, simply highlight the appropriate category in the left pane of the Preferences dialog box and set the desired preference field in the right pane. MicroStation provides a Help pane along the bottom of the Preferences dialog box that shows a short description of each set of preferences.

NOTE: *Workspace preferences are automatically saved to* <username>.upf *in your current MicroStation Home directory (normally* \Bentley\Home\). *If you delete this file, MicroStation resets the preferences to their default values.*

In addition to individual user-type preferences, MicroStation supports an entire array of resources and settings that can be customized for specific projects or engineering disciplines. In addition, the very look and feel of MicroStation can be adjusted. All of this customization is collectively referred to as the MicroStation *Workspace*.

The Workspace is a combination of special variables you set before, during, and after MicroStation is initialized that, in turn, point to various resource files. Resource files include cell libraries, customized interface components, font resources, line style definitions, and other files too numerous to list. Workspace configuration is so powerful that all of the Bentley Engineering products (Triforma, Modeler, Plantspace, Geographics, and so on) make extensive use of customized workspace components to enhance their users' productivity.

Workspaces

Depending on what you use MicroStation for, the collection of files you need—such as cell libraries, seed files, and symbology resources (font and line styles)—are different from the set of files another user may need. Thus, if you are working on a mapping project, the working units in your seed files would be different than if you were working on a commercial building project. Similarly, if your project calls for a frequent interface with an AutoCAD user, the default set of fonts and line styles you use is likely to be different from someone using MicroStation exclusively.

The collection of data files (cell libraries, fonts, seed files, dimension styles, multiline definitions, named levels, and the like) is referred to as a named workspace. MicroStation can be installed with any number of sample named workspaces ready for you to use as is or to customize. In keeping with the modular nature of a workspace, MicroStation uses a structured directory hierarchy to hold the various components that constitute a workspace.

A MicroStation workspace is a custom environment designed for easy access to interface components and workspace modules for a particular task or user. MicroStation uses a variety of special configuration variables to define a workspace. A configuration variable is much like an algebraic equivalent. It is essentially a name a program such as MicroStation can look up to find out a specific piece of information.

For instance, the configuration variable *MSDIR* points to the directory where MicroStation was installed. From this "value," MicroStation can find all of its constituent components by appending subdirectories to this value. There are literally dozens of configuration variables associated with MicroStation. In addition, MicroStation honors environment variables (Microsoft's name for the same thing) set in the operating system. In other words, if you set an environment variable in Windows NT's System Environment Variables dialog box that has the same name as a configuration variable in MicroStation, MicroStation will use that value.

Configuration variable definitions are stored in a variety of configuration files located in several locations under the *Workspace* directory. For a default installation, the various subdirectories (where project and user configuration files are stored) are located in the directory *\Bentley\Workspace*, as indicated in the following illustration.

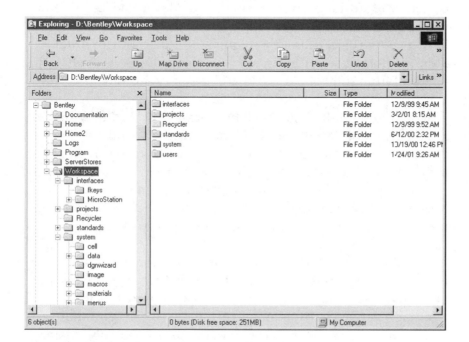

MicroStation normally organizes workspaces under the \Bentley\ Workspace directory.

NOTE: *MicroStation supports five layers of configuration files: system, application, site, project, and user. There is an order of precedence in how these configuration files are read and interpreted.*

TIP: *The project and user configuration files have the PCF and UCF file name extensions, respectively. These are plain text files, and you can learn a great deal about creating your own workspaces by examining them closely.*

Generally, when you first start up MicroStation from a fresh installation, you are starting from a "generic" project as user "examples" (this may vary a bit from site to site though). Note that the design files shown in MicroStation Manager are located in *\Bent-*

ley\workspace\projects\examples\generic. To use the sample civil workspace (if installed), you simply select it from the Project option menu in MicroStation Manager. Alternatively, in a command or DOS window, key in:

```
C:\USTATION -wuCIVIL
```

Note also that the design files listed in MicroStation Manager at startup of the Civil workspace are located in *\Bentley\Workspace\Projects\Examples\Civil\Dgn\.* The command line switch *-wu* in the previous key-in stands for *workspace user,* and it must precede the name of the user configuration file you wish to use. A configuration file defines the search path for various files that constitute the workspace.

Another method of selecting a new user configuration file is through the User menu. Selecting Command Window > User > Workspace > Select Default Workspace brings up the Workspace dialog box, which lists all user configuration files installed on your system. The Workspace dialog is shown in the following illustration.

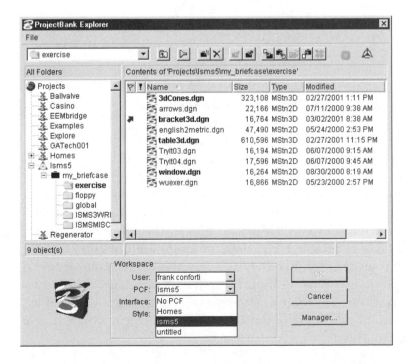

The Workspace dialog box makes it easy to pick the user configuration file for your future sessions.

As an example of a user configuration file, the following is the content of the sample *ARCH.PCF* file supplied with MicroStation.

```
#------------------------------------------------------------------
_USTN_PROJECTDESCR = Architecture Example
#------------------------------------------------------------------
# Set architecture search paths.
#------------------------------------------------------------------
MS_DEF                    < $(_USTN_PROJECTDATA)dgn/
MS_CELL                   < $(_USTN_PROJECTDATA)cell/
MS_CELLOUT                = $(_USTN_PROJECTDATA)cell/
MS_CELLLIST               < $(_USTN_PROJECTDATA)cell/*.cel
MS_CELLSELECTORDIR=         $(_USTN_PROJECTDATA)cell/
MS_GLOSSARY               < $(_USTN_PROJECTDATA)data/*.gls
MS_SEEDFILES<               $(_USTN_PROJECTDATA)seed/
MS_SYMBRSRC               > $(_USTN_PROJECTDATA)symb/*.rsc
MS_CELLSEED               = (_USTN_SYSTEMROOT)seed/seed2d.cel
MS_DESIGNSEED             = archseed.dgn
MS_LEVELNAMES=              $(_USTN_PROJECTDATA)data/
MS_SETTINGSDIR<            $(_USTN_PROJECTDATA)data/
MS_SETTINGS               = $(_USTN_PROJECTDATA)data/archset.stg
MS_SETTINGSOUTDIR         = $(_USTN_PROJECTDATA)data/
```

The lines in the previous listing that have the # character in the first column are comments. The variables in parentheses and preceded by the $ symbol are configuration variables that expand to their directory path value when used in an assignment statement. Look at the definition for the *MS_DEF* variable in the previous listing. (MicroStation looks for design files in the directory defined by the *MS_DEF* variable.) The default design file directory for the ARCH project under *Examples* is *<drv>:\Bentley\Workspace\projects\examples\arch\dgn*. This location is arrived at via a fairly circuitous route. Working in reverse (lower-level configuration variables first), the *MS_DEF* assignment for the Arch project workspace is decoded this way:

```
_USTN_WORKSPACEROOT = <drv>:\Bentley\Workspace\
_USTN_PROJECTSROOT = $(_USTN_WORKSPACEROOT)PROJECTS\
_USTN_PROJECT = $(_USTN_PROJECTSROOT)EXAMPLES\
_USTN_PROJECTNAME = arch
_USTN_PROJECTDATA = $(_USTN_PROJECT)$(_USTN_PROJECTNAME)\
    MS_DEF = $(_USTN_PROJECTDATA)dgn
```

As you can see, *MS_DEF* is not nearly as simple to decode as it would appear in the *Arch.pcf* file. This is due to the "layer cake" arrangement MicroStation uses for determining the location of the different workspace components, depending on your local, system, project, and user settings.

NOTE: *Though an experienced user may find it more convenient to use a text editor to directly modify and create configuration files, there is a more convenient method of dealing with them. The Workspace submenu (see following illustration) under the User menu offers items that invoke a dialog box to let you edit, create, and select workspace users, and to modify user and project configuration files.*

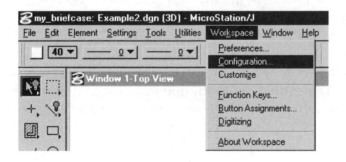

The Workspace submenu offers items for editing the workspace user configuration file, working with the Preferences dialog box, and customizing the MicroStation interface.

In addition to the configuration files, a workspace also includes user preferences and a user interface resource. It is worth noting that each workspace user configuration file has a corresponding user preference file, so that each workspace user can maintain his or her own preference environment.

The user interface of a workspace consists of icons and menus the user interacts with to access the various tools available within MicroStation. All of these interface elements can be modified with the help of the interactive Customize tool via its own dialog box (Workspace menu > Customize). You can modify the following interface elements with this tool:

Toolboxes: You can modify any of the native MicroStation toolboxes with Customize, or you can create your own toolboxes. Custom toolboxes can borrow native MicroStation icons, or you can create your own icons and assign them a command (key-in).

Tool frames: Tool frames look like toolboxes, but they are of a fixed size and are a repository for toolboxes. You can modify any of MicroStation's native tool frames, or you can create you own.

Menu bar: MicroStation's pull-down menu bar is also customizable. Because a Windows application can have only one such menu bar, Customize lets you edit this existing menu bar, but not create your own.

View border: You can modify the view toolbox located on each view window. However, you can place view control tools on this tool bar only.

Any of the items you choose to modify with Customize, whether a menu option or a command icon, can be assigned a command key-in or a sequence of commands strung together (separated by semicolons). The upper left option button in the Customize dialog box lets you choose any of the four previously listed item types for further work. Depending on which option you select, the list box on the left displays all available tools or menus. The list box to the right displays available tools in a specific toolbox or menu you select from the option button just above it.

The Copy button between the two list boxes lets you copy an available tool or menu from the list of available items on the left to the items list you are editing on the right. Double clicking on an item in the right-hand list opens that item's properties for you to edit, as indicated in the following illustration.

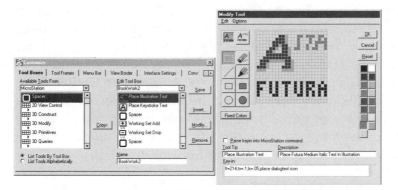

Double click on an item in the list on the right of the Customize dialog box to edit its properties.

Several sample customized interfaces are provided with MicroStation that are invoked when you log on in one of the example workspaces. For instance, the Arch workspace includes a new Architecture item under the Tools menu that, when selected, opens a custom tool frame.

TIP: *To display the user, project, interface, and preferences files your current session is using, select the About Workspace option from the Workspace menu.*

Settings Manager

The Settings Manager should perhaps be called the "Symbology Manager," as most users are likely to use it in that capacity. Then again, perhaps the term "Settings Manager" is the more appropriate name, because you can control other settings with it in addition to active symbology, such as dimension styles, multi-line element definitions, text settings, and drawing scales.

Of what use is the Settings Manager, given that the active color (*CO*=), weight (*WT*=), line style (*LC*=), and level (*LV*=) can be easily changed at any time? The answer lies in examining how design offices have standardized their use of symbology. This is done for creating drawings that are consistent and that are easier to manage from project to project.

Most engineering drawings are still produced on a monochrome plotter, and so the element color and level play no role insofar as the hardcopy is concerned. Thus, most beginners tend to draw virtually everything on the same level with the same color, only taking care to set element weights and line styles to correspond to project standards. Of course, paying attention to the colors and the levels on which elements are placed helps you create more flexible drawings. Just think of the Select By tool for element selection by color, and of the need to turn off the levels containing text and dimensions prior to plotting, as examples of the flexibility offered by standardizing all symbology components for your drawings.

Considering the differences in drafting requirements for different departments within a company, most CAD standards are discipline specific, and share some elements among them. Table 14-1, which follows, illustrates in concept what most standards offer as guidelines. The table has deliberately been kept small for simplicity.

Table 14-1: Drafting Standards for Civil Drawings

Item	Level	Color	Weight	Line Style
Border	50	4	2	0
Roads	6	3	1	0
Storm Sewer	21	5	2	0
Tree Line	6	2	1	Tree Line

Let's see how such a standard can be implemented with the help of the Settings Manager. You use the Edit Settings tool (see following illustration) to create settings groups, and you use the Select Settings tool to activate them. To invoke the Edit Settings tool, select Command Window Settings > Groups > Edit.

The Edit Settings tool lets you create a group and associated settings entries.

Exercise 14-1,which follows, outlines the steps for creating a settings group named Civil in an existing or new settings file.

EXERCISE 14-1: CREATING A SETTINGS GROUP

1 Select File > Create > Group from the menu in the Edit Settings window. This creates an unnamed group, whose name can be edited to read Civil in the edit field under the group list window.

2 With the Civil group highlighted, select Edit > Create > Linear to create an unnamed item associated with the Civil group. Repeat this command three more times to create a total of four items for this exercise.

3 One by one, highlight each of the four unnamed items in the lower list box and edit their names in the edit field below to read Border, Roads, Storm Sewer, and Tree Line, respectively.

4 All that remains now is the task of assigning settings to each of these items. Double click on a settings item to invoke the Modify window (see following illustration), which lets you assign various settings, such as color, weight, line style, and level. Use table 14-1 to assign settings for each of the four items.

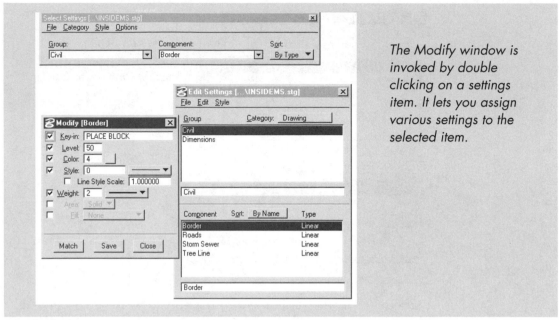

The Modify window is invoked by double clicking on a settings item. It lets you assign various settings to the selected item.

As you can see, creating a settings group is tedious but not at all difficult. Such settings groups, once created, are put to use by the Select Settings tool. You invoke the Select Settings window by selecting Command Window > Settings > Groups > Select. The Select Settings tool probably opened up with your recently created settings group already loaded. If not, you can use the File > Open command from the Select Settings window to open your settings group. The following illustration indicates the use of this command.

NOTE: *Settings group files have the .STG file name extension, and, for the default workspace are stored in the* C:\USTATION\WSMOD\ DEFAULT\DATA *directory.*

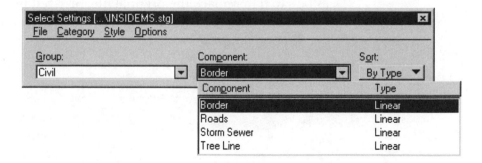

Highlight a group in the top panel to display the settings items associated with it in the bottom pane. Notice how clicking on the settings items in the bottom pane changes the symbology and issues a command making the implementation of company standards a simple task.

Whereas the traditional method of creating sidebar menus for implementing company standards was a little tedious to write, and only larger companies with development resources usually ventured into creating them, the Settings Manager lets you manage settings (at whatever office scale) by making the task so simple. Even novice users will have no trouble conforming to project standards.

Function Keys

Of all of the types of menus MicroStation supports (pull-down, digitizer, sidebar, tutorial, and function key), the function key menu is the easiest to customize. The function key menu is really just a text file that contains command key-ins for all of the ways a function key can be used. Thus, on a personal computer with an enhanced keyboard, this means a total of 96 keystroke combinations: F1 through F12 when pressed alone, or when pressed in conjunction with Shift, Control, or Alt, or any combination of these keys, including Shift+Control+Alt+F1. By default, the Function Key definitions are stored in *\Bentley\workspace\interfaces\fkeys*. The default function key file name is *funckey.mnu*.

MicroStation provides an interactive tool for creating and modifying function key menus (Workspace menu > Function Keys). There is a File menu to let you open existing function key menu files for modification, as well as a Save and Save As. Modified files can be saved to replace the original file, or saved in another file to preserve the original. An example of this functionality is shown in the following illustration.

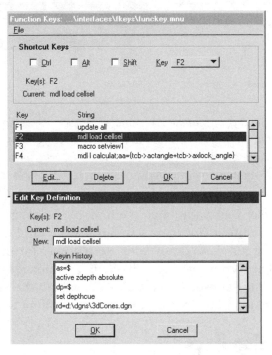

Function keys are easy to program. Here, F5 is being reprogrammed as Update View, followed by a data point in View 1. Notice how commands can be concatenated with a semicolon separating them.

Programming your new function key menu is just a matter of selecting the appropriate function key and clicking on Edit to invoke an edit field for supplying the command text. The command text can be any valid MicroStation key-in. You can also combine several commands on a single line by separating them with a semicolon.

Custom Line Styles

MicroStation provides eight built-in line styles, as well as support for an unlimited number of custom line styles stored in an external resource file. The term *external* refers to the fact that the dash-dot pattern that defines custom line styles is maintained in a file external to MicroStation. If others are to use line styles you create, the resource file must be made available to them. For a default MicroStation installation, the external line style resource file is located in

:\Bentley\Workspace\system\symb\LSTYLE.RSC

However, because MicroStation uses a rather complicated config-uration variable, *MS_SYMBRSRC*, line style definitions can also be loaded from several other locations, depending on whether you have created any additional line styles. The "default" definition for *MS_SYMBRSRC* is

```
$(_USTN_SYSTEMROOT)symb\*.rsc;$(_USTN_SITE)symb\*.rsc;$(_USTN_PROJECT
DATA)symb\*.rsc
```

From a quick scan of this definition you can see that line style def-initions (and other symbol-related resources) can be found in sev-eral site- and project-specific locales. Custom line styles are used like the built-in line styles. Virtually all graphic elements that can be drawn with a built-in line style can also be drawn with custom line styles. In fact, the shortcut method of invoking both types of line styles is the same. You key in *LC=name*, where *name* is a num-ber from 0 to 7 for the built-in line styles, or the alphanumeric name of a custom line style. Another way of activating custom line styles is by selecting Element > Line Style > Custom from the menu to bring up the Line Styles settings box, as indicated in the following illustration.

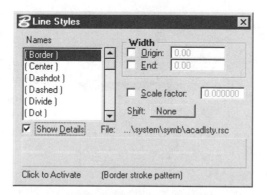

Double clicking on a custom line style name in this settings box activates it.

Custom line styles offer an immense amount of flexibility. They can consist of a stroke pattern, and they can incorporate symbols in their definitions. Additionally, the same line style definition can be placed in your design file with different scale, shift, or width settings.

NOTE: *Clicking on the Show Details checkbox in the Line Styles settings box expands the window to provide access to the three custom line style placement parameters: scale, width, and shift.*

Being able to scale a custom line style before placing it in a design file means that you can scale up or down the stroke, or dash-dot pattern, for a single instance without modifying the line style definition. Similarly, being able to shift it or change its width means that the defined origin of the style can be shifted or its width changed. Custom line styles can also have varying widths.

Using custom line styles is a trivial matter. However, deciding on the line styles to be created, and actually creating them, takes some advance preparation. Several sample custom line styles are provided with MicroStation, which include tree line, railroad track, earth line, and telephone line.

A custom line style can be defined in terms of three components: a stroke pattern, a point symbol, or a compound definition that combines stroke patterns and point symbols. You define line styles using the Line Style Editor (Command Window > Element > Line Style > Edit), shown in the following illustration.

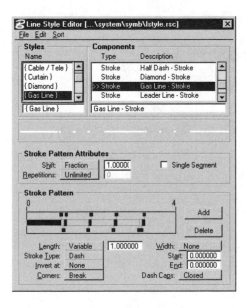

The Line Style Editor invoked from the Element menu lets you create new, or edit existing, custom line styles.

The process of creating custom line styles is like that described for the Settings Manager. You must first open an existing line style resource file, or create a new one using the File menu in the Line Style Editor window. Exercise 14-2, which follows, outlines the process of creating a line style with a stroke pattern component.

EXERCISE 14-2: CREATING A LINE STYLE WITH A STROKE PATTERN COMPONENT

1 Create an unnamed line style using the Edit > Create Name menu option, and edit its name in the edit field under the Styles list box.

2 Create an unnamed line style stroke component using the Edit > Create Stroke Pattern menu option. Again, you will want to edit the unnamed component's name to be more descriptive.

3 Link the newly defined style name with the stroke component by highlighting both and selecting Edit > Link from the menu.

4 Add the necessary number of stroke patterns (dash and gap sequence) to the stroke component created in step 2. To do this, highlight the stroke component and click on the Add button in the Line Style Editor's lower half, titled Stroke Pattern.

5 Edit each of the stroke patterns created by first highlighting the section and then selecting appropriate values for its length, type, and width. This completes the line style definition, and you can now save the line style resource file.

The procedure for creating line styles based on a point symbol component or a compound component is similar except for the specific attributes that define it.

NOTE: *To create your own point symbol for use in line styles, draw it using primitive elements such as lines and arcs, place a fence around it, define an origin for it (just as you would for a cell definition), and key in* **CREATE SYMBOL** <symbol_name>. *This adds the symbol to the active line style resource file, making it available for inclusion in line style definitions.*

Glossary

Glossary is a helpful text placement utility used to place standard text expressions in the design file by clicking on their abbreviation from a customizable list displayed in a dialog box interface. You can use this tool to maintain a list of commonly used text expressions so that you do not have to key them in when needed.

You invoke Glossary by either selecting Command Window > User > Utilities > Glossary or by keying in *MDL LOAD GLOSSARY* at the Command Window prompt. The following illustration shows the Glossary utility.

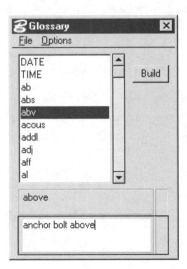

Below the list of abbreviations in the Glossary utility are two panes. The top pane displays the full text expression corresponding to the selected abbreviation, and the bottom pane displays the text expression built so far, prior to placement in the design file.

To place text in your design file from the glossary file, highlight the abbreviation for the expression desired, click on the Build button, and place a data point at the desired location. The sample glossary file supplied with MicroStation is named *EXAMPLE.GLS*, and is located in the *<drv>:\Bentley\Workspace\system\data* directory. However, there are several additional sample files available in the various example workspaces (e.g., *Arch.GLS* in the Arch project).

TIP: *The glossary file is workspace specific and is pointed to by the configuration variable* MS_GLOSSARY. *To edit configuration variables, select Command Window > User > Workspace > Modify User Configuration.*

A Glossary file is a plain text file with a simple structure. A # character in the first column of a line starts a comment. *$date* and *$time* are special variables that represent system date and time. A set of two lines constitutes a glossary entry. The first line contains an abbreviation for the glossary entry, and the second line contains the expanded text expression corresponding to the abbreviation. The following is a sample Glossary file:

```
STEEL
All structural steel to conform to ASTM A-36 specifications.
CONC
All concrete to have 4,000 PSI 28-day compressive strength.
TIMBER
All timber to be Douglas Fir or equal.
DATE
Today's date is: $date.
TIME
The current time is: $time.
```

Programming

As you have seen thus far, a great deal of customization of the design environment is possible. Nevertheless, there are times when you may want to roll up your sleeves and delve into the world of programming in MicroStation.

The most powerful way of taking control of command input and of integrating drawings to design calculations is through the use of MicroStation's programming interfaces. Each of the three ways of programming MicroStation is introduced here. This section serves merely to introduce you to the tools. No attempt is made to teach the terminology, language, or its syntax. However, if you wish to find out what it takes to program in MicroStation, and which interface will be the more appropriate one for you, read on.

MicroStation BASIC

MicroStation BASIC is a powerful but easy-to-use programming language suited to the needs of both the casual user and CAD support personnel. It replaced the archaic User Command programming environment that dated back to the Intergraph VAX-based days of the early 1980s. Using a simplified English-like syntax, MicroStation BASIC provides a macro recording capability that captures a user's activities for playback at a later time. This has the advantage of allowing you to perform a proposed "dry run" through a design procedure you want to automate, and then fine tuning the recorded results.

Although you edit BASIC macros using either the built-in macro editor or a standard text editor such as Windows' Notepad, MicroStation actually compiles the program into a machine-readable format. This is performed automatically whenever you run the program. A BASIC macro is stored as a text file with the *.bas* file extension, whereas the compiled version of the program can be found as the same file name but with the *.ba* extension.

By default, MicroStation stores its BASIC macros in subdirectories pointed to by the *MS_MACROS* configuration variable. Normally, these are *<drv>:\Bentley\Workspace\system\macros* and *<drv>:\Bentley\ Workspace\standards\macros* (the system subdirectory is reserved for macros delivered with MicroStation, and the standards subdirectory is for user-developed macros). The following is the "system" subdirectory as delivered with MicroStation/J.

Subdirectory *D:\Bentley\Workspace\system\macros*

06/18/98	03:03p	3,766	cellmod.bas
06/18/98	03:03p	1,542	cmd.bas
06/18/98	03:03p	3,474	currtran.bas
06/18/98	03:03p	17,252	dbform.bas
06/18/98	03:07p	327,680	dbform.mdb
06/18/98	03:08p	5,097	dbform.txt
06/18/98	03:03p	3,243	dbprofil.bas

06/18/98	03:03p	3,503	dbquery.bas
06/18/98	03:03p	7,822	ddedb.bas
06/18/98	03:07p	65,536	ddedb.mdb
06/18/98	03:03p	4,788	demo.bas
06/18/98	03:03p	10,711	elemshow.bas
06/18/98	03:03p	1,574	fopen.bas
06/18/98	03:07p	327,680	gis.mdb
06/18/98	03:03p	2,609	hilite.bas
06/18/98	03:03p	1,114	inputbox.bas
06/18/98	03:03p	10,446	locate.bas
06/18/98	03:04p	9,197	mdlbasic.ba
06/18/98	03:03p	5,721	mdlbasic.bas
06/18/98	03:03p	1,971	msgbox.bas
06/18/98	03:04p	10,986	refinfo.ba
06/18/98	03:03p	7,819	refinfo.bas
06/18/98	03:03p	15,506	report.bas
06/18/98	03:03p	2,825	scaletxt.bas
06/18/98	03:03p	1,493	select.bas
06/18/98	03:04p	17,538	table.ba
06/18/98	03:03p	16,523	table.bas
06/18/98	03:08p	511	table.txt
06/18/98	03:04p	3,171	tower.ba
06/18/98	03:03p	9,253	tower.bas
06/18/98	03:04p	14,336	towerdat.xls

NOTE: *Although MicroStation does recompile the .ba version of a file whenever a change has been detected in the source code, do not delete the .ba version of the file. If you do, certain non source code resources, such as dialog box definitions, that are stored within the .ba version will be lost.*

With that said, you can, however, run a BASIC macro using only the compiled (*.ba*) version of a file. This is an important consideration when deploying finished macros to other members of a workgroup or company site. Providing only the compiled version of the macro ensures the consistency of operation, as no edits can be made to the macro without the text file (*.bas*) file.

You run a macro in MicroStation in one of two ways. The first method involves opening the Macros window (Utilities menu > Macros), selecting the appropriate macro and clicking on the Run button, as indicated in the following illustration.

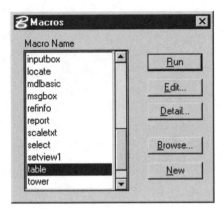

Running a Macro using the Macros option and the Run button.

You can also run a macro directly from the Key-in browser using the following command line:

```
macro <macroname>
```

If the macro is found in the MS_MACROS search path, the macro is executed. You can use this method in function key definitions, even as an icon in a custom toolbox. If this is the first time the macro has been run (no *.ba* found), you will get an Alert message box, shown in the following illustration.

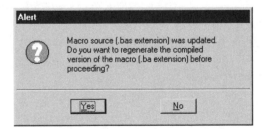

Alert for first-time run of a macro.

Clicking on Yes will cause MicroStation to compile the macro and run it. This is the normal response. At this point, the macro takes over control of MicroStation. It can bring up dialog boxes, prompt users for input, perform data point operations, and even invoke other commands and macros. In other words, it automates normal MicroStation operations. Upon conclusion of the macro's execution, control is returned to the user.

NOTE: Because MicroStation BASIC invokes tools just like a user would, when the macro terminates, the program can still leave a tool active. For this reason, it is recommended that when the macro concludes its operation it should invoke an innocuous tool such as PowerSelector or Select Element.

MicroStation's Record Macro Feature

Usually, the first step in writing a macro is working out the order of tools and commands you want to include in the new macro. The fastest way to capture this "workflow" is to record it as a macro. The Create Macro feature in MicroStation essentially records the user activities into a new macro file that you can then polish into the final form.

The Create Macro > Utilities menu > Create Macro option brings up the Create Macro dialog box, in which you enter the name and a short description of the macro you are about to record. The description is very helpful because it gives you an opportunity to describe the intent of the new macro, as well as the recorded actions. Later on, when you are editing the results, this intent can be invaluable. The Create Macro dialog box is shown in the following illustration.

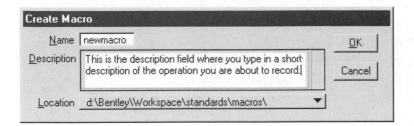

Create Macro dialog box.

Once you have entered the name of the macro and have pressed the OK button, the Create Macro process begins. The small VCR-like button macro window, shown in the following illustration, lets you know that the recording process in underway.

Macro button window.

You can stop (terminate macro recording, close macro file), pause (pause recording mode, keep macro file open), or resume the macro recording process via this simple window.

The macro record function does not actually record the position of your pointer. Instead, it monitors the event queue within MicroStation so that when you invoke a tool, open a dialog window, or generate a mouse button press (data point, reset, or snap) this event is recorded to the macro. When you select a tool, the macro records the actual command MicroStation uses to "fetch" that tool. Later, you can (and will) edit these recorded steps to tune it to your specific needs.

TIP: *One benefit from MicroStation's Create Macro function is the ease with which you can find out how MicroStation performs specific actions. By recording a given action, you can review the subsequent macro and identify the specific commands MicroStation executes for that action. This includes opening dialog boxes, and using a variety of design environment settings, including those that are design file specific.*

Editing Your Macros

Once you have recorded a macro, the next step is to review its content. This is accomplished by editing it. Although you can use a standard text editor such as the Window Notepad application, most first-time BASIC users will modify their macros using Micro-Station's built-in Macro Editor. This is invoked from the Macros window (Utilities menu > Macros). After identifying the macro to be edited (if you do this right after a Create Macro session, the resulting macro will automatically be highlighted), you click on the Edit button, shown in the following illustration.

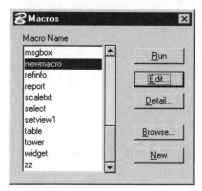

Macro showing Edit button.

The BASIC editor appears with the source code for the selected macro in the text editor window. This editor, shown in the following illustration, provides several tools for the budding programmer. These are located in the menus and tool bar integrated into the editor window. The editor contains all of the usual text editor tools, such as copy, cut and paste, undo, and a find string function. In addition, it has several tools specifically designed for the macro development process. These include a dialog box editor, in which you create custom dialog boxes invoked by your macro, and a set of debugging tools critical to finding code errors during execution of the macro.

BASIC editor.

MicroStation's BASIC Extensions

In addition to the normal BASIC programming language components, MicroStation BASIC also implements several features unique to MicroStation. Called MicroStation BASIC Extensions (MBE, for short), these functions are easily identified in your code by the *MBE* prefix. These extensions are broken down into several categories:

- Constants (*MBE_SUCCESS*, *MBE_FileReadOnly*)
- Data types (*MbePoint*)
- Functions (*MbeGetInput*, *MbeWritePrompt*)
- Objects (*MbeSettings*, *MbeElement*)
- Collections (*MbeView*, *MbeRefFile*)

Without getting into a complete programming textbook here, suffice it to say that these extensions to the BASIC language are at the heart of the macro system and its ability to control MicroStation.

Using MicroStation BASIC's Help

One of the best references on MicroStation BASIC is the online help provided with MicroStation itself. To access this extensive reference resource, go to the MicroStation program group (Microsoft Windows Start menu > Programs > MicroStation) and select the MicroStation BASIC Help icon. Clicking on this icon brings up the BASIC Help window, shown in the following illustration.

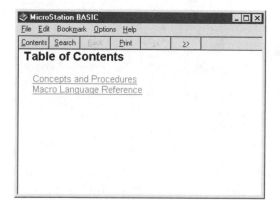

Initial BASIC Help window.

The help file is organized into two major categories:

- Concepts and Procedures
- Macro Language Reference

Concepts and Procedures is a good place to start when learning the basics of macro creation. The Macro Language section is indispensable when you need to find a function to perform a specific action. Examples of Macro language help entries are shown in the following illustration.

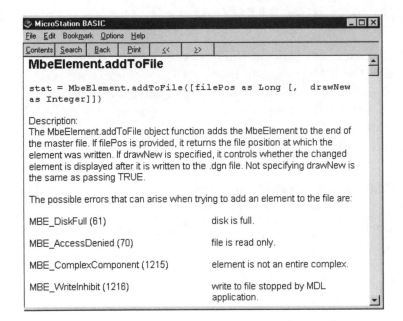

Macro language help entries.

The search capability is another important aid. Most of the Macro Language Reference entries include short code examples you can directly copy/paste into your macro for further refinement.

MicroStation Development

MicroStation Development Language (MDL) is the ultimate developer's tool for creating commercial software applications that run "within" the MicroStation environment. As such, MDL is an implementation of the C programming language, the de facto professional developer's programming language of choice. When running, applications written in MDL are almost indistinguishable from MicroStation. In fact, most of MicroStation's graphic tools are written in MDL.

Writing an MDL application is much like developing a C language program for any GUI. Interface components of an application (such as dialog boxes and icons) are referred to as "resources." MDL provides easy access to the four resource editors: Command Table Editor, Dialog Box Builder, Icon Editor, and String List Editor. MDE also includes a debugger and a "make" utility. If your application needs these resources, and you have built them with the appropriate editor, you can invoke the Resource Source Generator within MDE to create source files for your application's resources.

In addition to the resource files, you must also create MDL source files to impart functionality to your application. Several standard source files (including files that define constants, variables, and structures) are also part of MDL.

Once all resource and source files have been created, you invoke the Resource Compiler, *RCOMP.EXE*, to process the resource files and create compiled resources. At this stage, you also process the MDL source files through the Source Compiler, *MCOMP.EXE*, to generate an object module, followed by the linker, *MLINK.EXE*, to generate files ready to be combined with your program's resources. The resource librarian, *RLIB.EXE*, accepts compiled resources and source files to generate the final MDL application, which has the *.MA* extension.

Even though MDL applications are compiled into a binary format, they are not directly executable files. These binaries are interpreted by MicroStation as they execute. The MDL functions that call upon built-in MicroStation graphical commands execute at full speed, but computational functions execute relatively slowly because they are interpreted.

MDL applications you create are easily portable to other platforms. The MDL source from one platform usually requires a simple recompile to get it running in the host environment. MDL applications you create are invoked within MicroStation by issuing the command

```
MDL LOAD filename
```

where *filename* is the name of your *.MA* application.

As you have seen in this brief discussion of MDL programming, creating MDL programs requires an understanding of programming concepts in general, and the C language in particular. If you are serious about creating powerful MDL applications, the investment in learning C will pay itself back in the professional quality of your add-on applications.

J/MDL: MicroStation's Java Implementation

The J in MicroStation/J refers to the introduction of Java into MicroStation. More accurately, a Java Virtual Machine (JVM) was incorporated into the core of MicroStation, thus providing the ability for Java developers to run pure Java applications entirely within MicroStation. However, if that was all you could do, there would be no real advantage to using Java (after all, you get a Java Virtual Machine in the major web browsers).

The real advantage comes in the enhancements or the DGN package, which provides an interface between the Java Virtual Machine (where the Java code actually executes) and the Micro-Station internal engine. Dubbed JMDL, this new programming environment led directly to the development of products such as ProjectBank and Viecon.

What Is Java, Anyway?

Java is an object-oriented programming environment developed and licensed by Sun Microsystems. Since its introduction in 1995, it has become one of the most widely adopted programming languages to date. Much of the "dot com" business environment was built on the Java model of "Write one; run anywhere," meaning that you could invest in development of a particular software resource knowing it would run on current and future computer platforms without modification.

There are two types of Java programs. The first is the Java application, which runs much like any standalone application. Its only dependency is the Java Virtual Machine. It generates its own user interface, and manipulates data much like any other native program running on your system.

The other type of Java program is the Java applet. This is a program that is "hosted" by another program, in essence removing it one step further from your core computer operations. This is intentional. By isolating a Java applet into its own "sandbox," you prevent potential harm that could be caused to your system or data files. Most Java programs the average user encounters are of the Java applet sort. MicroStation/J supports both Java applications and Java Applets.

Java Applications

Java applications are compiled into "class" files (extension *.class*). In MicroStation, you run such applications using the following key-in command:

```
java <appname>
```

Note that you do not specify the *.class* extension. Because Java is a true object-oriented environment, the one major requirement for a Java program to run is that it be locatable by the Java Virtual Machine at the time of its initialization. This is normally done when you first start up MicroStation/J. If you store your Java applications in a subdirectory outside those normally associated with MicroStation, you must either create or append that directory name to the configuration variable *CLASSPATH*.

NOTE: CLASSPATH *definitions get more complicated than simply defining known Java class repositories. Because Java supports direct execution of Java applications stored in Java Archive (.JAR) files, you must also append the name of each .jar file in which you may have class files stored. Fortunately, most of the details of maintaining the* CLASSPATH *are handled during the installation of the individual software components written in JMDL.*

Java applets are run within MicroStation using the following key-in:

```
appletviewer <filename>.html
```

Note that you cannot run an applet directly but must instead create a minimal *html* web page that provides critical meta-information for the applet. At the very minimum, the HTML web page must contain the following line:

```
<applet code="appletclassfilename"></applet>
```

where *appletclassfilename* is the name of the applet class as found either in the *classpath* variable or in an identified archive (identified by the *codebase=* keyword). Applets do not have to reside on the local machine. You can run an applet from the Internet if you know the URL (universal resource locator) name.

Summary

This was the briefest of introductions to the many ways you can customize MicroStation. There is a lot more to this subject, including more on how to write your own JMDL, MicroStation BASIC and, real soon, Visual BASIC applications (part of the MicroStation V8 release). For more information on customizing MicroStation, consult the web address cited in the Introduction. A list of additional resources on this and other MicroStation related subjects is continually updated.

That concludes this exploration into the fundamentals of MicroStation. At this point, you simply need to spend some time with this most powerful of all design products on the market today. As you become familiar with its operation, you will come to appreciate the logical and consistent manner in which MicroStation functions, as well as how easy it is to create and manipulate even the most difficult designs. Welcome to the family!

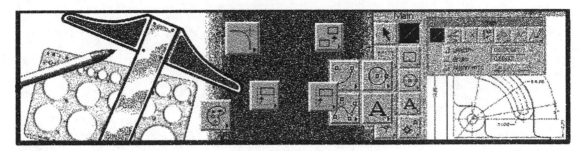

INDEX

Symbols

! on status bar 42
.000 extension 490
.ba file extension 619
.bas file extension 620
.CEL extension 333
.ini extension 493
.MAT file extension 587
.PCF file extension 602
.UCF file extension 602

Numerics

3D design 499–563
 AccuDraw, drawing exercises 534, 535
 AccuDraw, getting started 529–531
 converting 2D elements to 3D 545–553
 creating 3D file 501–503
 creating 3D file, with ProjectBank 502–503
 depth controls 518–525
 depth controls, exercises 519–522, 523–524
 design coordinates 526–527
 design cube 503–504, 518
 design cube, orientation exercise 533
 fence operations 562
 modifying elements 553–561
 mouse button assignments 528
 placing elements 525–526
 primitive element tools 540–545
 seed file selection 500–501
 smart solids 539
 solid modeling tools 538
 specifying 3D coordinates 525–526
 surface modeling tools 536–537
 toolboxes 504–505, 540
 utility tools 561–562
 video cards and 14
 views 505–518
 views, rotation exercises 509–511
 vs. 2D design 109, 500
3D Main tool frame 504
3D simulation, isometric tools 152
3D View Control toolbox 505
3DCONES.DGN, view rotation along axes exercise 509

A

AA= key-in 267
Above Element method, text placement 266
absolute coordinates, placing elements at 184

AccuDraw
 3D compass orientation shortcuts 530–531
 3D design, getting started 529–531
 activating 64
 axis indexing 122–123, 178–179, 182
 axis indexing feature 73
 compass 122, 123–124, 125, 180
 context sensitivity 123–124, 131
 drawing basics 175–185
 enabled in v8 121
 in-context behavior 532
 introduction of 8
 introduction to 120–129
 Keypoint Snap Divisor 182
 Origin shortcut 180–181
 Previous Distance lock 78
 setting Keypoint divisor 182–183
 shortcuts 183–185
 shortcuts for 124–126
 Shortcuts window 178
 Smart Lock 179–180
 snap shortcuts 183
 snap window 159
 specifying 3D coordinates 526
 starting 121–123
 table legs drawing exercise 535